GOD AND GAIA

God and Gaia explores the overlap between traditional religious cosmologies and the scientific Gaia theory of James Lovelock. It argues that a Gaian approach to the ecological crisis involves rebalancing human and more-than-human influences on Earth by reviving the ecological agency of local and indigenous human communities, and of nonhuman beings.

Present-day human ecological influences on Earth have been growing at pace since the Scientific and Industrial Revolutions, when modern humans adopted a machine cosmology in which humans are the sole intelligent agency. The resultant imbalance between human and Earthly agencies is degrading the species diversity of ecosystems, causing local climate changes, and threatens to destabilise the Earth as a System. Across eight chapters this ambitious text engages with traditional cosmologies from the Indian Vedas and classical Greece to Medieval Christianity, with case material from Southeast Asia, Southern Africa, and Great Britain. It discusses concepts such as deep time and ancestral time, the ethics of genetic engineering of foods and viruses, and holistic ecological management.

Northcott argues that an ontological turn that honours the differential agency of indigenous humans and other kind, and that draws on sacred traditions, will make it possible to repair the destabilising impacts of contemporary human activities on the Earth System and its constituent ecosystems. This book will be of considerable interest to students and scholars of the environmental humanities, history, and cultural and religious studies.

Michael S. Northcott is Professor Emeritus of Ethics at the University of Edinburgh, Scotland; and Guest Professor at the Indonesian Consortium of Religious Studies, Gadjah Mada University, Yogyakarta, Indonesia, and the Evangelische Theologische Faculteit, Leuven, Belgium.

Routledge Environmental Humanities

The *Routledge Environmental Humanities* series is an original and inspiring venture recognising that today's world agricultural and water crises, ocean pollution and resource depletion, global warming from greenhouse gases, urban sprawl, overpopulation, food insecurity and environmental justice are all *crises of culture.*

The reality of understanding and finding adaptive solutions to our present and future environmental challenges has shifted the epicenter of environmental studies away from an exclusively scientific and technological framework to one that depends on the human-focused disciplines and ideas of the humanities and allied social sciences.

We thus welcome book proposals from all humanities and social sciences disciplines for an inclusive and interdisciplinary series. We favour manuscripts aimed at an international readership and written in a lively and accessible style. The readership comprises scholars and students from the humanities and social sciences and thoughtful readers concerned about the human dimensions of environmental change.

For more information about this series, please visit: www.routledge.com/Routledge-Environmental-Humanities/book-series/REH

GOD AND GAIA

Science, Religion and Ethics on a Living Planet

Michael S. Northcott

Routledge
Taylor & Francis Group
LONDON AND NEW YORK

earthscan
from Routledge

Designed cover image: Michael S. Northcott

First published 2023

by Routledge
4 Park Square, Milton Park, Abingdon, Oxon OX14 4RN

and by Routledge
605 Third Avenue, New York, NY 10158

Routledge is an imprint of the Taylor & Francis Group, an informa business

British Library Cataloguing-in-Publication Data
A catalogue record for this book is available from the British Library

ISBN: 978-0-367-62775-1 (hbk)
ISBN: 978-0-367-62774-4 (pbk)
ISBN: 978-1-003-11075-0 (ebk)

DOI: 10.4324/9781003110750

Typeset in Bembo
by Deanta Global Publishing Services, Chennai, India

CONTENTS

FIGURES

ACKNOWLEDGEMENTS

I planned the outline of this book in the autumn of 2019 during a visit to the Indonesian island of Sumbawa, whose Tambora volcanic eruption in 1815 led to three years without summer in Europe, and global crop failures. The deep winter of 1815 led to the loss of nine-tenths of Napoleon's army as it attempted a retreat from Moscow. The cold summer of 1815 was also the time in which Mary Shelley dreamt up the spectre of *Frankenstein*, which was a darkly prophetic novel about the coming impacts of modern science and technology on humanity and the earth. In 1815, no one knew that particles in the atmosphere from the huge explosion had reduced global sunlight for three years and cooled the planet. Only later did Earth System scientists realise that this is what had happened and so Sumbawa is a fitting place to have begun a book on the topic of the Gaia theory, also known as Earth System science, and its religio-cultural implications.[1]

Having obtained a contract from Routledge, I wrote the book during 2020–2022 when I remained in Indonesia, and mainly in Yogyakarta in Java and in Canggu and Ubud, Bali, due to international travel restrictions and border closures. My interactions with colleagues and students at the Graduate School of the Universitas Gadjah Mada were both offline and online throughout that time as I was lucky to be living in communities less affected by restrictions on social interaction than my former abode of Edinburgh, Scotland. I therefore had many informal face-to-face conversations on the topics in the book, including with Willem and Muriel Loots in Kampung Seseh, Bali, with Victor 'Nightjar' in Ubud, and with Iskandar Warowuntu, founder of the permaculture Bumi Langit Institute in Imogiri near Yogyakarta.

Finding readers for my draft manuscript was more of a challenge far from colleagues in Europe and the United States. Besides the editors of the series in which it appears for Routledge, who offered overall encouragement, my deepest thanks for reading through parts or wholes of what follows in a penultimate draft are to Richard H. Roberts of Stirling in Scotland, Alastair McIntosh of the Center for Human Ecology, energy analyst and author Wilf Wild, and Paul Tyson, Research Fellow of Queensland University, Australia. Paul gave me detailed feedback on the whole draft, pointed out mistakes, and encouraged me to pursue and undergird further the broad outlines of the connective thread – the argument – of the book. I am also thankful to John Milbank with whom I have had stimulating conversations over many years around the subjects of Gaia and God, and who gave me insightful leads as I researched the historic background to the modern Gaia theory in Indian, Greek, and Christian philosophy. I am also thankful to colleagues at the Indonesian Consortium of Religious Studies at Universitas Gadjah Mada with whom I discussed some of the ideas presented here and especially Zainal Bagir and Dicky Sofian.

This book includes excerpts from previously published papers which were in the main the products of conference papers I was invited to give, and I am grateful to the hosts of those conferences for their hospitality in Trondheim, Norway; Munich, Germany; Lancaster, Cambridge and London, England; Paris, France; and to the publishers of the book chapters for their permissions to reprint them here in modified form.

They are as follows:

Chapter 1 includes revised excerpts from 'Eschatology in the Anthropocene: From the Chronos of Deep Time to the Kairos of the Age of Humans.' In Clive Hamilton, François Gemenne and Christophe Bonneuil eds., *The Anthropocene and the Global Environmental Crisis: Rethinking Modernity in a New Epoch*, Abingdon, Routledge, 2015, 87–99; and from 'On Going Gently into the Anthropocene.' In Celia Deane-Drummond, Sigurd Bergmann and Markus Vogt eds., *Religion and the Anthropocene*, Eugene, OR, Cascade Books, 2017, 19–34.

Chapter 2 includes substantial excerpts from 'Biofuel Energy, Ancestral Time, and the Destruction of Borneo: An Ethical Perspective.' In Jonathan Chaplin and Marc Ozawa (eds.), *Good Energy: The Economics and Ethics of Energy Sustainability*, Cambridge, Cambridge University Press, 2019, 247–57; and excerpts from 'Reading Genesis in Borneo: Work, Guardianship and Companion Animals in Genesis 2.' In Nathan MacDonald, Mark W. Elliot and Grant Macaskill eds., *Genesis and Christian Theology*, Grand Rapids, MI, Eerdmans, 2011, 190–203.

Chapter 4 includes excerpts from '"Behold I Have Set the Land Before You" (Deut 1.8): Christian Ethics, GM Foods, and the Culture of Modern Farming.' In Celia Deane-Drummond, Bronislaw Szerszynski with Robin Grove-White eds., *Reordering Nature: Theology, Society and the New Genetics*, London, T&T Clark, 2003, 85–106; and 'Crop Science, the Heisenberg Principle and Resistance to Genetically Modified Organisms.' In Phil McNaughten and Suzanna Carellero eds., *Governing Agricultural Sustainability: Global Lessons from GM Crops*, London, Routledge, 2015, ch. 13.

Chapter 5 is a revised and expanded version of 'COVID-19, Human Ecology, and the Ontological Turn to Gaia.' In Alexander Hampton (ed.), *Pandemic, Ecology and Theology*, London, Routledge, 2020, 90–102.

Chapter 7 includes excerpts from 'The Ontological Turn, Religious Tradition, and Human Cosmological Fittingness.' In Michael S. Northcott and Steven van den Heuvel eds., *Fittingness and Environmental Ethics*, London, Routledge, 2023.

Note

1 See further Gillen D'Arcy-Wood, *Tambora: The Eruption That Changed the World*. Princeton, NJ, Princeton University Press, 2014.

INTRODUCTION

I first encountered James Lovelock in person at a multi-disciplinary conference at the University of Exeter in 2019 organised to include a celebration of his one hundredth birthday. I was at that point unfunded for academic conferences and as it was summer, I opted to camp close to the River Exe in the rather beautiful valley it forms around the hilly heights of Exeter City. During the conference Lovelock engaged in an animated 90-minute dialogue with the host of the conference – Earth Systems scientist Professor Timothy Lenton – and with a large public audience (Figure 0.1). His extraordinary energy, his passion for truth, and his sense of humour were all on display, apparently undiminished by his years. I have long suspected that Lovelock's Gaia theory is the most significant advance in scientific thinking about the environment in the twentieth century but until I met him, I had never sat down and read his work at length or considered its analogies with pre-scientific perspectives in philosophy and religion. I subsequently did so and decided the time was opportune to investigate the Gaia theory from a religious ethics perspective, the results of which investigation I present here.

There are a few books by theologians which claim to treat of the Gaia theory. The best known are by feminist theologians Rosemary Ruether and Anna Primavesi. Ruether's book while it has the title *Gaia and God* does not offer a theological engagement with the science of Gaia.[1] Instead, Lovelock's naming of his theory after the feminine Greek goddess of the Earth is taken as a jumping-off point for consideration of cultural representations of the divine feminine, her implication in the origins of the physical Earth and in the lives of nonhuman and human creatures, all the way from Egyptian and Sumerian religions through Indian, Greek, Jewish, and Christian traditions to modern ecofeminism. Primavesi's book is a collection of essays that, despite having a foreword by Lovelock, also does not offer a treatment of the science of the Gaia theory, though it comes closer to what I am attempting here than Ruether.[2]

DOI: 10.4324/9781003110750-1

FIGURE 0.1 James Lovelock and Timothy Lenton, Global Systems Institute, University of Exeter, July 2019. Author photo.

A third book which might be said to treat of the Gaia theory from a religious perspective is Bruno Latour's Edinburgh Gifford Lectures, which I attended in 2013, and which were later published as *Facing Gaia*, although the dominant *topos* of the lectures is not Gaia theory but the political implications of climate science. The disciplinary perspective of the lectures was not religious studies or ethics but Science and Technology Studies in which discipline Latour is a pioneer.[3] Where religion is treated, it is more for illustrative and metaphorical purposes, and to draw analogies between the claim contemporary humans are influencing the climate of Gaia in new ways and earlier religious conceptions of human interactions with earthly forces and heavenly divinities.

The core claim of this book is that Lovelock and Lynn Margulis' idea that the Earth and her creatures are active agents in the generation of conditions which make the Earth habitable for Life is a revival, and scientific restatement, of long-held traditional religious conceptions of the co-agency of all Beings, and the Earth herself, in making and sustaining the habitats and resources humans enjoy and utilise for food, for dwellings, and for cultural activities such as making music and performing religious ceremonies. The belief that the Earth and her creatures are a living composite was represented in ancient times by mythic and artistic representations of beings which included large mammals, the Earth, the moon, sun and stars, invisible forces such as the wind and demons or spirits, and the gods: all of these beings and their places of habitation the Greeks knew as the *cosmos*. Ancient representations of this pantheon of Beings among whom humans lived include prehistoric cave art in which large mammals such as cows and deer are represented as anthropomorphic or quasi-divine figures. In classical Greece and medieval Christendom, this pantheon was known as the Great Chain of Being. It reached from the habitation of the gods beyond the moon to the habitation of sea creatures at the bottom of the ocean.[4] Humans dwelt in the midst of this chain of being, 'little lower than angels' as the Psalmist has it, and were neither its centre nor its pinnacle.

In religious traditions including the Indian Vedas, Greek Pythagoreanism, Platonism, Judaism, and Christianity, the awareness of early or 'indigenous' humans that they were surrounded by a vast array of beings who sustained their life was encoded into religious pantheons of the gods from whom all beings were said to emanate. Hinduism and Christianity also held that the life principle that runs through all being – Vishnu in the Vedas and Logos from Heraclitus to the Greek New Testament – was Incarnate in Krishna and Christ respectively. I argue in this book that this idea of the emanation of all life from the divine, and of *creatio continua* as the upholding of all life through the divine breath, energies, or Spirit, together represent the original cosmology of what Karl Jaspers called the Axial Age traditions, and that it is restated in scientific discourse in the Gaia theory.[5] For people of faith to appropriate the Gaia theory as I propose here is therefore not a *pagan* move as some argue but a revival of core aspects of their traditions which modern secular science and philosophy have diminished.

The core 'takeaway' of this book for environmental concern is that humans are not alone in their efforts to restore the Earth from the ecological damages industrial activities have inflicted. Instead, contemporary humans need to recover a sensibility for, and revive the capacities of, all Beings – the Earth, mammals, fishes, plants and soils, and the divine Life force working through them – to restore the Earth to the state of species abundance which filled most of the Earth's surface until the colonial and industrial eras.

This book, like many academic books, is in part a composite of previously published papers but the research for this book did involve much new writing which, because of exceptional global travel restrictions from 2020 to 2022, I had few opportunities to try out on other academics. In Chapter 1, I commence with a discussion of 'deep time,' the Anthropocene, and what I call 'ancestral time.' I start the book here because the first modern scientist and philosopher to claim, as Lovelock and Margulis did in the 1970s, that the Earth is a System which has generated conditions suitable for Life was the eighteenth-century Scots Enlightenment philosopher James Hutton. In his *System of the Earth* Hutton proposed that volcanoes, soil erosion, the weight of the oceans and the Earth's molten core between them represent a living system which is responsible over very long time periods for the protrusions of rocks and soils from the sea in which terrestrial creatures evolved, and that these systemic processes were therefore responsible for the evolution of Life, including humans. Hutton's greatest impact was from his claims as to the very long time that these Earth System processes took to bring the Earth to its present state. Hutton's theory ultimately displaced anthropocentric conceptions of the Earth's history with a geocentric theory in which Earth processes, rather than human generations, became the prime markers of natural history. Some propose that this new geocentric creation story teaches modern humans humility and promotes greater care for life on Earth than the Jewish-Christian creation story which some say underwrites human dominion. But I argue that the Genesis myth of the exile of Adam and Eve from an original forested garden, and hunter-gatherer way of life, is better understood as a warning to early-Holocene humans about the risks of overreach in their newly acquired agricultural powers of plant selection and animal domestication that the new era of climate stability facilitated. I also argue that the apocalyptic idea of an Anthropocene epoch, though advanced to promote greater environmental concern, instead tends to promote scientistic Prometheanism rather than environmental care. Against environmental apocalyptic, I argue that an intergenerational ethic in which past generations and future generations are both honoured in the ways in which humans take up with the Earth is the most ecologically appropriate temporality.

In Chapter 2, 'In Borneo,' I look at the most significant intergenerational feature of the ecological crisis – since it is potentially irreversible – which is the threat of species extinctions. I investigate biodiversity decline and its connection with industrial projects, and alleged 'climate repair' via 'biofuels,' through the lens of the island of Borneo which has the oldest and most biodiverse rainforest on Earth but which has been under sustained assault from industrial development over the last fifty years. Before the large-scale destruction of its dipterocarp forests, Borneo evidenced the shaping role of humans in the distribution of species over tens of thousands of years and provides an important corrective to the modern scientific claim that only since industrialism have humans begun to significantly influence life on Earth. The reality is that indigenous people have in the main benignly influenced life on Earth for tens of thousands of years and are in part responsible for the extraordinary species diversity of places such as Borneo. The chapter concludes with a case study of a highland 'peace park' established informally by the nomadic Penan tribe along with Dayaks and Iban in order to resist the efforts of industrial loggers and palm oil investors to convert so much of Borneo into industrial plantations for the production of cooking oils and biofuels.

In Chapter 3, I situate the contemporary biodiversity crisis in the longer story of what moderns call economic 'development' which is a euphemism for the continued spread of industrial practices, and of post-colonial European (and latterly North American and Chinese) expropriation of ecosystems, human labour, lands, and species in Africa, Asia, and Latin America. I show how the increasingly concentrated oligarchic capture of Western capital and financial systems by a small group of wealthy families, investment companies, and 'philanthropic' private wealth trusts – in concert with governmental and intergovernmental agencies including the European Commission, the US Federal Government, the IMF, UN agencies, the World Economic Forum, and the World Bank – is gradually turning ever more areas of the Earth into investment zones where humans and other beings are treated only as 'resources' for exploitation in industrial projects which homogenise cultures and biodiversity in a radically evil and single-minded quest for rates of return on investment. And I show how the attempt to revise the ideology of development with the allegedly 'Earth friendly' ideology of 'sustainable development,' far from turning back the anti-ecological and antihuman direction of industrial economic development, is accelerating it by giving yet more power to globalist agencies to capitalise, homogenise and transform diverse local and regional forms of production and exchange – cultural as well as economic – into a centrifugal economic network in which wealth is accumulated in the capital cities of Asia, the Americas, and Europe at the cost of local ecological, economic, and social resilience.

In Chapter 4, I look at the role of mechanistic and reductionist modern science in systematically reducing biodiversity and the evolving energetic powers of Gaia in nurturing ever rising levels of species diversity in Earth history. I examine the attempts of scientists and technologists to wrest from Gaia the power of photosynthesis itself, and the processes of reproduction of plants, animals, and insects. The development of the genetic engineering of life down to the cellular level is playing an ever-increasing role in systematically reducing diversity and increasing human control and power over the processes of photosynthesis that produce nutrients for all life on Earth including humans. Often presented as means to increase the food supply, or the dependability of food availability, field research I participated in on genetically modified organisms reveals that the spread of GMOs in the Americas and Asia has deleterious effects on ecosystems, on human food cultures, and on human health. It also promotes oligarchic control by a small number of seed and chemical companies and investment agencies of industrial agriculture, and so underwrites the tendency discussed in Chapters 1 and 2 to concentrate wealth in fewer hands, a concentration which seriously undermines the resilience and sustainability of human communities as well as ecosystems and nonhuman species.

In Chapter 5, I consider a second case study of scientific reductionism and control of life in the shape of the science of genetic engineering of human life as evidenced in the genetic engineering of bat-originated coronaviruses into chimeric forms capable of infecting humans. I investigate the published scientific history of gain-of-function engineering of SARS viruses which strongly supports the theory that SARS-CoV-2 is a lab-engineered organism that escaped from various laboratories, including the Wuhan Virology Institute in China. The scientific record of gain-of-function engineering of bat viruses commences with peer reviewed papers on research funded by the US National Institute of Health in the virology laboratory of Ralph Barac at Chapel Hill, University of North Carolina in 2004. These led to over 5000 registered patents for newly engineered coronaviruses, and related experimental therapeutics. Despite its laboratory origins SARS-CoV-2 was nonetheless nothing like as dangerous as widely alleged, and millions of alleged 'COVID-19' deaths were misattributed due to concerted but unhelpful changes to death recording in many countries, combined with government payments to doctors and hospitals for attributing deaths of patients to COVID-19, and the widespread adoption of 'treatments' for COVID-19 which exacerbated rather than resolved symptoms. The use of the resultant pandemic to institute new forms of biopolitical and genetic engineering of human health, and new systems of technological surveillance and control of the most intimate aspects of human life, underwrites the critique which I present in this book of scientific materialism and its malign effects when turned towards systematic technological control of plants, animals, and persons, and the urgency of discovering a more God-conscious and Earth-sensitive cosmology.

In Chapter 6, I describe the modern rediscovery of the Earth as a System in the form of the Gaia theory. I give a brief overview of how Lovelock, while working for NASA, together with micro-biologist Margulis, developed a theory of the connectedness of all matter and organisms, and their symbiotic roles in generating over billions of years an Earth which is capable of regulating its own atmospheric chemistry and climate and in so doing has created an Earth which is habitable by Life. I discuss how mainstream biologists attacked the theory because it indicated there was a teleological – goal-oriented – convergence of matter and organisms towards Life, and how Lovelock then revised his original statement of the Gaia theory, by means of scientific studies in cybernetics and atmospheric chemistry, in such a way as to eschew biological teleology, and make the theory more consistent with atheist, mechanistic, anti-teleological science. With Margulis, I argue that the Gaia hypothesis in its original form was a new scientific natural theology which included the claims that: the Earth is alive; that all life and all matter are connected; and that there is a convergence of matter and organisms on Earth towards the evolution of biodiversity, conscious living organisms, and a planet habitable for them, from the dawn of time. I argue that the Gaia theory replaced the deistic and mechanistic tendencies of early modern and modern science with a theory of Life which best fits the surprising evolution of biodiversity, and of conscious mammals, on a large lump of molten lava and congealed rock hurtling around a solar system, and a galaxy, that is otherwise empty of Life.

In Chapter 7, I explore the convergence between Lovelock's Gaia theory and prior traditional sacred cosmologies and ontological theories from indigenous beliefs and the Indian Vedas through classical Greek, early Christian, and medieval cosmologies. I argue that the Gaia theory is not so much a new discovery as the recovery, in modern scientific discourse, of the original ontology which was held in almost all human cultures before the modern, and according to which humans dwell in the midst of networks of Beings which stretch from the depths of the Earth and the bottom of the ocean to the tops of mountains, the atmosphere and clouds all the way to the moon, sun, and stars and include material, organic, and spiritual beings such as demons, angels, and gods. This 'Great Chain of Being' was not the anthropocentric cosmology often ascribed to Christian and Greek predecessors of modern Europeans but rather one which situated humans in the midst of networks of co-agency, and which requires humans to fit themselves and their cultures, lives, and practices to the given cosmological order in order to maintain and not destroy the world.

In Chapter 8, I present a case study from the island of Bali in which I describe the attempt of European agricultural engineers to improve on Bali's 1000-year-old rice growing system using modern Green Revolution techniques. The science-inspired Bali Irrigation Project ultimately failed and those who sponsored it were forced to admit that they could not improve on the symbiotic system of rice growing, water control, and pest management that had evolved in Bali in close interaction with the ceremonies and water temples of the Balinese, and which produced rice more efficiently than any science-based approach could replicate. I use this and other case material to show how Gaian ontology is reflected in symbiotic cultural and production relationships between humans and their Earthly habitats. I argue that enabling the co-agency of all beings on the Earth to co-manage and co-create resilient and biodiverse habitats will enable contemporary humans to revive the co-agency of Gaia and other creatures in repairing the divinely originated capacities of Gaia to heal the damages wrought by industrial humans.

Notes

1 Rosemary Radford Ruether, *Gaia and God: An Ecofeminist Theology of Earth Healing.* San Francisco, CA, Harper, 1993.
2 Anne Primavesi, *Sacred Gaia: Holistic Theology and Earth System Science.* Foreword by James E. Lovelock. London, Routledge, 2000.
3 Bruno Latour, *Facing Gaia: Eight Lectures on the New Climate Regime.* Trans. Catherine Porter, Cambridge, Polity Press, 2017.
4 Arthur O. Lovejoy, *The Great Chain of Being: A Study of the History of an Idea.* Cambridge, MA, Harvard University Press.
5 On the traditional Christian understanding of *creatio continua* as the continuous divine sustaining of creation see further Emil Brunner, *The Christian Doctrine of Creation and Redemption, Dogmatics, Vol. 2.* Philadelphia, PA, Westminster Press, 1952, 149–50: on the axial age see Karl Jaspers, *The Origin and Goal of History.* London, Routledge & Kegan Paul, 1953.

1

FROM DEEP TIME TO ANCESTRAL TIME

Abstract

In the eighteenth century, Scots polymath James Hutton invented a new theory of the 'System of the Earth' and argued from empirical observations of rock strata in Scotland and Northern England that the Earth evolved the conditions for the emergence of life over a huge span of time. After Hutton, deep time cosmology gradually displaced humans and gods in the modern scientific history of life. But this modern history conceives of the Earth as an autonomous machine and accompanied the Promethean re-engineering of life on Earth since the dawn of the Industrial Revolution. Industrial technologies are now so invasive globally that they have turned industrial humans into a geological force, and industrialism is even said to have birthed a new geological 'epoch' – the Anthropocene. In the light of this, some argue that a 'new creation story' based on deep time can help to recover more modesty and humility in human uses of the Earth than that offered by the machine cosmology, or by predecessor religious creation stories. Evidence is presented that the predecessor stories have more cultural and spiritual power in resisting industrial re-engineering of the Earth than the new creation story.

Philosophy Outdoors

A long-weekend 'Outdoor Philosophy' course in Scotland in June 2014 comprised a group of environmentalists who included five academics and six executives of environmental organisations. Among them were Professor Robert Watson who collaborated with James Lovelock, and led the first experiment in iron fertilisation in the Southern Ocean to estimate the potential of this geoengineering technology to draw down anthropogenic atmospheric carbon.[1] Myself and Andrew Dobson had published research monographs on ecological and climate

DOI: 10.4324/9781003110750-2

politics.[2] Kate Rawles and Ed Gillespie had written travelogues of continent- and planet-spanning journeys by bicycle and boat/train to highlight their perception of the urgent need for a civilisational turn towards 'low carbon' travel in the light of the dominant scientific theory of recent planetary warming, which is that it is caused by anthropogenic carbon dioxide from burning fossil fuels.[3]

The weekend short course was based in a small Scots baronial-style country house on the West Coast of Scotland. The main activity was sea kayaking around the Caribbean-like seas, kelp forests, and coral sand islands off the coast around Arisaig on the edge of the Inner Hebrides. The sea kayaking trips were powerful as ways to engage the nonhuman world since in a kayak you effectively sit in the ocean and feel its movements as you paddle along. You become more aware of all the elements, including the wind and the sun as well as the ocean. And your view of other creatures – arctic terns, sea eagles, seals, kelp forests, white-sand coral shorelines – is multi-sensory and powerfully involving. Daily sea kayaking expeditions were interspersed with activities, discussion, and meditation on topics from deep time to environmental politics and plastic waste (Figure 1.1).

Holding the course for the most part out of doors has the purpose of 'doing philosophy' while connecting with nature by setting reflective activities principally in non-anthropogenic environments. The first morning began with a ritual exercise in which the participants were invited to place in chronological order

FIGURE 1.1 Sea Kayaking off Arisaig, Inner Hebrides, Scotland, June 15, 2014. Author photo.

laminated cards depicting the principal geomorphological events, from the Big Bang to the inventions of agriculture and the steam engine, in the approximately 3.5-billion-year history of Earth on a 40-metre climbing rope laid out on the grass. The participants mostly got the order right, and in some cases even the appropriate temporal intervals. All the human events were bunched up on the last few centimetres of the rope, while the time of the pre-human forces of volcanoes, stars, species, rocks, gases, and biota stretched into the distance to the furthest edge of the large lawn. The course leader, environmental philosopher Kate Rawles, concluded the exercise with a reflection in which she suggested that the tiny part of the timeline on which humans had existed palpably represented the insignificance of the human to the history of life on Earth. This ritual reflection on Gaian deep time should enable us to overcome our anthropocentric perspective by reframing the dominant role humans have played in recent centuries in shaping Earth's habitat: moving beyond anthropocentrism should promote humility towards other than human life and motivate a more ecologically situated set of attitudes and practices than those displayed in consumer societies (Figure 1.2).

The deep time exercise was pioneered by Australian deep ecologist John Seed in the Tasmanian rainforest as part of a larger ritual called the 'council of all beings' in which participants 'step aside from their human identity and speak on behalf of another life-form.'[4] The ritual is designed to enable 'evolutionary remembering' of the life histories of volcanoes, rocks, rainforests, oceans, and even the Big Bang

FIGURE 1.2 Deep Time Exercise, Outdoor Philosophy Retreat, Arisaig House, Scotland, June 14, 2014. Author photo.

long before human life began. The aim is to overcome contemporary alienation between nature and culture by stimulating the species memory of humans as mammals, and before that as cosmic elements: 'as the fog of amnesia disperses, there is a transformation in your relationship to other species, and in your commitment to them.'[5] The ritual reflects religious beliefs in the interdependence of all life, and the 'ecosophy' of the Norwegian deep ecology philosopher Arne Naess. For Naess, fostering the ability of modern humans to imaginatively extend their self-identity to nonhuman creatures and environments is the principal means of restraining the impacts of modern technologies and lifestyles on the ecosphere and redirecting humans towards a more Earth sensitive way of living.[6] Seed contrasts evolutionary remembering that situates human life in the deep time history of life on a 4-billion-year-old Earth with the biblical creation story which promotes 'the idea that humans are the crown of creation.'[7]

The ritual is an example of the 'New Genesis' genre of science-informed creation myth and ritual inspired by the North American Roman Catholic theologian, or 'geologian,' Thomas Berry. For Berry the ecological crisis is as much a crisis of worldview as of behaviour. To transform worldviews he argues that the Christian creation story, based as it is on the history of successive human generations, should be replaced with a new sacred myth of human and Earth origins based upon scientific knowledge of geological or 'deep time' and evolutionary history.[8] For Rawles, as for Berry, experiential engagement with the scientific story of the formation of the Earth and of biological life is able to reconnect contemporary humans with 'the value of the natural world, and our relationships with it' on a planet at risk from the ecological impacts of excessive consumption and of human beings living in a disconnected relationship to their environment.[9]

James Hutton and the System of the Earth

The West Coast of Scotland is an appropriate place to reflect on deep time since the Lewisian Gneiss rocks of the Scottish Hebrides, at 3 billion years old, are among the oldest visible on the Earth's land surface.[10] And it was in Scotland, and on the basis of observations of Scotland's ancient rocks, that eighteenth-century Scottish farmer and geologist James Hutton invented geological or 'deep time.' Hutton was the first scientist to realise that soil erosion, combined with pressure from the ocean, creates new layers of rock which over geological time are pushed up through the visible land surface by subterranean heat and geomorphic rocks. He argued that life on Earth had evolved as a result of the interaction between the molten core of the Earth, the Earth's surface, including its oceans, and the gaseous atmosphere which together made what he called the 'System of the Earth,' which was the theory he first presented at the nascent Royal Society of Edinburgh in 1785.[11]

Hutton was a landowner and farmer in the Scottish Borders. He also studied the science of mineralogy and he came up with the idea that rocks were primarily formed from eroded soils that, under pressure from the ocean, turn into sedimentary rock, and that may later be pushed up or overlaid by extrusions of magma – which becomes geomorphic rock such as granite and basalt when cooled – from the Earth's hot inner core.[12] The evidences for his theory came first from his observation of soil erosion on his farmland in the Scottish Borders, where he observed how soils are carried by rivers to the ocean. Over time, the weight of the oceans leads to the formation of sedimentary rocks. His second source of evidence was the veins of granite and other geomorphic rocks that were visible extrusions around older sedimentary rocks at various places in Scotland. Hutton theorised these veins were solidified molten magma which had been thrust from deep below the Earth's surface. Hutton found the strongest evidence for his theory at Siccar Point on the North Sea coast south of Edinburgh, where Old Red Sandstone (Devonian and 350–400 million years old) meets Limestone (Silurian and 420–440 million years old) at right angles, creating a 'T' formation.[13] Hutton argued that geological 'unconformities,' such as that at Siccar Point, and others he observed at Glen Tilt, on the Isle of Arran, and in the Lake District in Cumbria, are not instances of exceptional events, as was generally believed but examples of the uniform operation of physical laws that operate in the present and have always so operated and by which the Earth evolved to its present state. What were once called geologic 'unconformities' indicate the Earth as presently constituted is a living 'Earth System' which over deep time has been *uniformly* created, and is *still being* created, by living processes which over long time periods change its appearance and constituent parts (Figure 1.3).

FIGURE 1.3 'Geological unconformity' in Glen Tilt, Scotland. June 8, 2016. Author photo.

Since these geomorphic processes could be seen to occur in the present, Hutton judged that their incremental effects must have taken many millions of years to reproduce the arrangements of rocks presently visible on Earth. This meant that Hutton needed to retheorise historical time and, against the reliance of his scientific contemporaries and predecessors on an intergenerational estimate of the beginning of time from the biblical record of the creation and succeeding generations, Hutton argued that only a vastly long temporal history could have achieved the present formation of sedimentary and geomorphic rocks on the Earth, and he concluded his magnum opus, *A New Theory of the Earth*, with the observation that in the geological history of Earth, as evidenced in the 500-million-year-old rocks of the British Isles, he could 'find no vestige of a beginning, no prospect of an end.' For Hutton, Father Time, in combination with the living Earth as a system of rocks and gases, soils and plants, had generated the environments humans dwell in and observe, and so it is 'vain to look for anything higher in the origin of the Earth' such as a divine Creator.[14]

Geological versus Biblical Time

Hutton's thesis was received with scepticism in his lifetime since it contradicted the established view that the Earth was no more than 6,000 years old. The dating of the Earth at 6,000 years old originated in a theological interpretation of the first Hebrew creation story in Genesis 1 which described the creation of the Earth as having taken place over 'six days.' Another Hebrew text, Psalm 90.4, proposed that 'one day is as a thousand years' from a divine perspective. So Jewish theologians writing in the Talmud in the second century of the Christian Era proposed that each of the six days of creation as described in Genesis 1 correlates to a thousand years. This chronology was adopted by Christian theologians Julius of Africanus and Eusebius of Caesarea in the second and third centuries CE.[15] It was later adopted by early modern theologians, including Martin Luther in Germany, and the English Bishop Ussher. Luther and Ussher proposed only slight variations to Julius' chronology to keep the 'six millennia' extending beyond the then present into the near future and this chronology was accepted by most scientists even up to the time of Hutton.[16]

Hutton was also criticised for sponsoring atheism, since his theory indicated that the Earth in its present form is not the work of a divine Creator but the product of systemic and interactive mechanical processes including vulcanism, soil formation, atmospheric change, plant development, oceanic ingress and regress, and climate heating and cooling. Not until the publication in the mid-nineteenth century of Charles Lyell's *Principles of Geology*, and its influential adoption by Charles Darwin in *On the Origin of Species*, was Hutton's chronology accepted, and opposition to it gradually dispatched.

The Abyss of Time and Sacred Chronologies

Hutton's deep time chronology not only displaces divinity from Earth history beyond its primeval beginning; it also displaces humanity. This produced a novel bifurcation between natural and human history, and between traditional cosmologies and modern scientific cosmology since for most of the vastness of geological time the Earth after Hutton appears to be uncaused by divinity, and empty of gods and even humans.[17] Although Nicolai Copernicus had argued that the Earth was not at the centre of the universe, as Ptolemaic cosmology held, Copernicanism did not initially challenge humanity's central place in Earth's history. Christian chronology from the second century of the Christian era had mapped human intergenerational history onto Earth history with only slight variations in calendrical enumeration for more than 1,500 years before Hutton. It was therefore universally believed in Christendom that there were three ages of the Earth: the first was the era from the Creation to Christ (BC); the second was the era from Christ to the present and near future (AD); and the third was the eschaton which would herald the Last Judgement, and the return of Christ to the Earth to inaugurate a literal 'new heavens and new earth,' as indicated in eschatological interpretations of the Revelation of St John the Divine, the last book of the Christian bible.

After Hutton, the history of the Earth is said to be divided into a much longer set of eras that had nothing to do with God or humans, and which extend back over roughly 3.5 billion years. In all of these eras, apart from the Holocene, which roughly corresponds to BC and AD combined, humans were either completely absent or very epiphenomenal. However, it is still the case that the Christian division of human history and Earth history until the present into two eras – Before Christ and Anno Domini – remains culturally dominant. The two best-known 'eras' in popular culture are still eras in which human history and Earth history are aligned, and the division between them remains the estimated date of the birth of Jesus Christ. The principal concession to contemporary secularism is that the BC and AD nomenclature have been revised to BCE and CE where 'C' stands for 'Common' rather than Christ and E for Era. There are alternative faith-based calendars, and most influentially, the Muslim calendar. But the BC/AD two-era chronology is so influential that it is represented in astronomic clocks, digital computers, and most printed and online encyclopaedias and history books. The enduring role of the birth of Christ in dividing human history into two calendrical eras may explain why a significant minority of individuals leave school in Western Europe and the United States still believing the Earth is only 10,000 years old. Nonetheless for most scientifically educated people Hutton's deep time chronology is well established and human history and Earth history no longer coincide in cultural consciousness.

Hutton's deep time chronology, by making humans so evidently peripheral to the network of beings which generated and sustained life on the Earth humans now experience, while it may promote humility at the insignificance of humans in Earth history, may also underwrite the refusal to acknowledge that humans since the eighteenth century, under the dominant influence of industrial technologies, began a trajectory towards what some call the new epoch of the

Anthropocene in which human effects on the Earth are so invasive as to affect the behaviour of the Earth as a System, and risk driving it beyond critical thresholds towards a new post-Holocene instability that threatens the current array of diverse species, and even humans, with major, even cataclysmic, changes to their habitats.[18] But recognition of this possibility is influenced in a problematic way by the scientific argument that humans had so little influence through almost all of deep time. If humans were and are so insignificant in Earth history, how is it that now all of a sudden, they have become an Earth-changing geological force? This is why for deep time futurists, the recent modest warming of planetary temperatures since the Industrial Revolution represents a mere blip in the history of the planet, and decisions about fossil fuel use or deforestation have minimal import for the future of life on Earth, and the Earth herself, in this longer view.[19]

Advocates of science-based myths and rituals of the 'New Genesis' genre envisage that scientific accounts of the Earth System and its history function in a way that is analogous to religious cosmologies and sacred stories by generating 'the desired values and sense of connection' with nature and underwriting a civilisational turn from ecological destruction to restoration.[20] But as Bruno Latour argues, it is the tendency of scientists to set apart scientific 'factishes' from human cultures, meanings, and values that promotes the bifurcation of nature from culture and the disconnect between 'moderns' and the environment.[21] This bifurcation had cultural implications that are not fully appreciated by those who propose that participation in a deep time ritual, or a 'new universe story' that resituates human consciousness in deep time, will facilitate greater identification between humans and the rest of the 'natural' world. A deep time ritual, while it may provoke wonder at Hutton's 'abyss of time' also generates a sense of the epiphenomenal character of human history as compared to the history of life on Earth. If the passing of human generations, and the birth of children and grandchildren, is peripheral to Earth history, it may be said to be unreasonable to argue, as many now do, that humans are capable of significantly influencing the course of natural history or the Earth's atmosphere. Hutton's deep time chronology underwrites the refusal, which is particularly prominent in Anglo-Saxon cultures, to acknowledge that industrial humans' impacts on ecosystems may potentially destabilise the Earth as a System. The claim that geological time provides the basis for a new sacred universe story that promotes greater care for the Earth than traditional intergenerational sacred stories may therefore be erroneous.

There is a presumption among advocates of the 'new universe story' that myths based on science are more 'earth-friendly' and produce more pro-environmental behaviours while traditional religious creation stories are anthropocentric and anti-environmental. This presumption emanates not from contemporary evidence but from Lynn White Jr.'s *Science* essay in which he argues that the invention of new technologies by medieval Catholic monks indicated that the Christian doctrine of creation de-animated and instrumentalised the Earth.[22] But the accumulating contemporary evidence on environmental attitudes, practices, and values in the United Kingdom, North America, and Europe indicates that the population group most likely to advocate maximal human domination of 'nature,' or the environment, is not Christians but atheists, and especially people with a higher level of *scientific* education than the general population.[23]

The claim that geological time provides the basis for a new sacred universe story that promotes greater care for the Earth than the pre-modern and Christian intergenerational story from Adam and Christ to the present may therefore be erroneous precisely because it promotes the new Baconian reading of time according to which the Scientific Revolution conferred on humans powers to recreate and reorder life on Earth so as to recreate Paradise on Earth, and even to bring about their own immortality. On this evidence, creating new sacred myths around scientific information, and discarding 'old' creation stories, and their association with intergenerational rather than deep time, may not have the effect intended since it will remove cultural sources of resistance to the domineering and mechanistic approaches to Earth, life, and matter that modern science and technology, and the ideology and practices of scientism, together advance.

The New Time of the 'Anthropocene'

Instead of looking back into deep time to stimulate ecological consciousness, some Earth scientists argue that the best way to provoke care for the future is to remap Hutton's geological history onto the recent and near future history of humanity. Paul Crutzen, Will Steffen, and others propose that scientists name the recent intergenerational history of increased anthropic influence on the Earth system as a new geological epoch which they call the Anthropocene. The term 'Anthropocene' for a successor geological epoch to the Holocene was coined by Russian scientists in the 1960s and popularised by Paul Crutzen at a meeting of the International Geosphere-Biosphere Programme near Mexico City in 2000. In a plenary discussion Crutzen said: 'Whenever you go somewhere in the world and make measurements, you cannot avoid having to deal with humankind,' and that this is what is meant by the term Anthropocene.[24] Crutzen, with Eugene Stoermer, wrote up the idea in the newsletter of the IGBP and the term snowballed from there, initially in the natural sciences and subsequently in the social sciences and humanities. To date, thousands of scholarly articles, 50 books, and numerous media reports have been written on the Anthropocene. But despite this outpouring, the Subcommission on Quaternary Stratigraphy of the International Union of Geological Sciences did not ultimately find for the Anthropocene as the successor epoch to the Holocene but describe it instead as a 'geological event.'[25]

Crutzen initially identified the commencement of the Anthropocene epoch with growing atmospheric emissions from coal burning triggered by James Watt's invention of the condensing steam engine, the emissions of which are discernible in raised levels of carbon in air bubbles trapped in the polar ice pack generated from the eighteenth century. Watt's engine was so efficient compared to its predecessor that it enabled the large-scale extraction of coal, whose burning in sky-darkening quantities sparked the Industrial Revolution and anthropogenic climate change. The Earth has undergone a modest rise in temperature of 0.8 degrees Centigrade since 1850 and, on the basis of the claim that industrial-era greenhouse gas emissions are the only cause, Earth scientists argue that this means that humanity will remain a major environmental force for many millennia.[26]

In a subsequent paper, Crutzen and Will Steffen argue for the later date of 1950 for the commencement of the Anthropocene. From 1950, in a period sometimes called the 'Great Acceleration,' industrial activity across the planet evidences a huge increase as shown in exponential increases in a range of indices of human activity including: nuclear isotopes in air, soil, and water from atmospheric testing of nuclear weapons; atmospheric carbon and methane from fossil fuel extraction and burning; boreal and tropical deforestation; industrial ocean fishing; nitrogen fixing in agriculture; intensive animal husbandry; production of concrete, steel, copper, and aluminium. On this account, the 'period of the Anthropocene since 1950 stands out as the one in which human activities rapidly changed from merely influencing the global environment in some ways to dominating it in many ways.'[27]

Identification of the date of the event(s) thought to have set the Anthropocene in motion depends upon the choice of facts, or factishes, that may be said to characterise the emergence of this new epoch. Some confine the human marker of the Anthropocene to biogeochemistry and hence to Earth System effects and especially the Atmosphere. Others use the term to speak about measurable global human impacts not only on the atmosphere, climate, and cryosphere, but also on minerals, nitrogen, oceans, rocks, rivers, soils, and species. The widespread clearance of forests since the colonial era, and the substitution of grazing animals and animal feed crops such as soya and other monocultural plantations in areas such as the Amazon; the hoovering up of the majority of living biomass in the ocean by industrial trawlers; the displacement of nitrogen from fossil fuels into soils and waters subsequent to the invention of the Haber Bosch process for making artificial nitrogen, as used in artificial fertiliser and explosives; these and other processes have together significantly changed the chemistry and resilience of the Earth, significantly reducing the life-generating and Earth-healing capacities of such core planetary components as the atmosphere, forests, soils, oceans, and rivers. Other significant examples of Earth-System–wide industrial pollution in the last 100 years are the diffusion of plastic and microplastics in the oceans and soils; the atmospheric spread of desert dust and soot from anthropic activities; and the presence in soils and waters of new radioactive substances from nuclear fission.

In this broader definition, the human behaviours that generate the Anthropocene include not only such practices as fossil-fuel burning and deforestation, which have rapidly accelerated since 1950, but also agricultural and horticultural plant selection, forest-felling, fishing and hunting, mining, damming rivers, deep-sea fishing, and the movement of species to non-native regions. Adopting this broad definitional approach to the Anthropocene, William Ruddiman and others argue that elevated methane levels from early agriculture are discernible in sedimentary layers from 5,020 Before Present. This would take the dating of the Anthropocene closer to biblical chronology. Simon Lewis and Mark Maslin prefer the date of 1610, in recognition of the significance of the beginning of the 'Columban Exchange' in which millions of Europeans settled the New World and took with them European species, while at the same time displacing, and killing, tens of millions of American cultivators and hunters. They argue that the notable atmospheric cooling of the Northern hemisphere from 1610 to 1850 was a consequence of the greening of North America which followed the disappearance of so many native land managers and the regrowth of forests.[28] The principal stratigraphic evidence of this dating is the mixing of Old and New World plants, pollens, and species from the 1600s which show up in soil samples. This dating also fits well with the historiography of Fernand Braudel for whom 1492 was the origin of a new era of globality in human history, in which human cultures, minerals, pollens, and species were gradually mixed up between and across continents.

The range of possible starting points for the Anthropocene indicates that the dating of this claimed new epoch depends as much upon 'values' as 'facts.' For those who use it to highlight the dangers to human- and other-kind from anthropogenic influence on the climate, Crutzen's emphasis on atmospheric carbon from fossil-fuel burning as the definitive historical marker is correct. For those who believe that industrial humanity's influence on the climate is not so cataclysmic, then the role of pre-industrial agrarians in increasing atmospheric methane, or of early modern colonials in the movement of species, may seem equally significant. For those who find the roots of the ecological crisis in the industrial revolution and the ongoing displacement of naturally evolved ecosystems with scientific and technological systems, the date of 1748 makes the best sense and if talk of an Anthropocene is to be a permanent feature of the geological record the industrial revolution dating makes the best sense. While there has been an acceleration in industrial practices since 1950, I present evidence in the next few chapters that it is the invasive nature of these practices on natural systems which is the root of the modern ecological crisis and of the destabilisation of ecosystems such as to reduce their habitability to other species and often to humans also, and especially indigenous humans whose ancestral lands are still being raided and destroyed by industrial development agencies and projects.

The Anthropocene as Spectacle

The claim that humans are living in an anthropic epoch, the Anthropocene, is not only a matter of historical or scientific judgement. It involves a departure from nineteenth-century evolutionary science and the related rise of environmentalism. In essence it involves the claim that nature and culture are irrevocably 'mixed up,' and that what humans call the 'environment' is human as well as natural. Back of this claim is the claim that this mixing up of nature and culture is a novel situation, and that at some previous point in history humans were more *disconnected* from and *less influential* over other species and ecosystems.

Against these claims, Jared Diamond argues that human influence has been consistently ecologically destructive for thousands of years. However, many of his examples of the causes of earlier 'collapse' are contested, such as the Easter Island episode which Diamond blames on religious practices whereas the weight of evidence suggests the island collapsed under the influence of visits from European ships in the colonial era: another contested example is the Mayan Civilisation where the weight of evidence indicates this civilisation also collapsed as a result of invasions from outside, not crop failures.[29] The most significant pre-modern large-scale ecological collapse occurred in the Levant and North Africa as a result of the Roman Empire's soil-depleting and water-catchment–wrecking wheat-growing practices.[30]

What happened in the late Roman Empire is indicative of the way in which that powerful empire disrupted predecessor collective action institutions for managing commons such as grasslands, water catchments, forests, coastal fishing grounds, lakes, and rivers. Garrett Hardin argues that the uses of such commons are inevitably tragic but Elinor Ostrom and other scholars of the history of commons governance institutions have demonstrated that they are the longest enduring forms of human institution on the planet: in some places they have been effective means for regulating human–habitat relationships sustainably for up to two millennia, including, for example, water management on the island of Bali where this book was written.[31] These pre-modern and pre-industrial institutions of 'commons governance' were ignored by colonial-era settlers who in the main overrode local customs and traditions in the New World, Africa, and parts of Asia when they introduced European-style farming, mining methods, plantations, and so on. Hence it was colonialism, and the associated introductions of industrial agriculture and other industrial resource-harvesting techniques, which were the real causes of the 'tragedy of the commons' in colonised domains: Garret Hardin's classic essay of that title failed to consider the significance of pre-settler and pre-colonial commons governance arrangements and yet it remains highly influential because it fits with the broader modern environmentalist and misanthropic narrative that humans are intrinsically in conflict with the health of ecosystems and the diversity of species.[32]

I argue in this book that it is the machine-dominated culture of industrialism, including industrial agriculture, fishing, mining, and urban development, together with the associated cosmological picture of the Earth and of nonhumans as autonomous machines, which are the principal roots of the modern ecological crisis, not an intrinsic conflict between humans and 'wild nature.' It is industrialism which has disconnected billions of people from 'nature' and reduced the presence of the nonhuman in the lives of industrial urbanites to mere spectacle. The alleged *crises* of nature – culminating in the apocalyptic announcement of the 'Anthropocene' epoch – are also presented to urbanites as spectacles in which they are but passive participants and in relation to which they have no agency other than as 'consumers' since they do not directly produce food or other material and cultural essentials of daily life in direct engagement with ecosystems.

Rather than challenging the cultural disconnect between modern industrial citizens and the natural environment, the announcement of the Anthropocene shares another key feature of modern environmental discourse, which is its apocalyptic character. The 'Anthropocene' is ecology and geology *as* modern spectacle. As Guy Debord insightfully argues, the modern mechanical reproduction of media-replicated visual spectacles is 'the sun that never sets over the empire of modern passivity. It covers the entire surface of the globe, endlessly basking in its own glory.'[33] Spectacles are presented primarily as mechanically reproduced images: and as Walter Benjamin argued in a classic essay, the industrial capacity to reproduce images was the most significant cultural marker between the industrial and the pre-industrial cultures, turning moderns into 'flaneurs,' walking observers of culture rather than its active co-producers.[34]

Spectacular images of the Anthropocene include plastic submerged in soils; plastic waste floating in the ocean or blanketing Pacific Island beaches; floods inundating urban settlements; drone films of vast grid-like urban areas such as the desert city of Phoenix; documentaries charting the industrial pillage of the oceans or the moonscapes created by tar sands development; photographs of the Blue Planet seen from space or even from the dark side of the moon. Debord argues that industrial societies do not create spectacles by accident. Rather spectacles – whether of buildings being blown up, or wars in the oil rich Middle East, or the bodies of the homeless being buried in Manhattan by prisoners in orange jump suits, or burning forests in Northern California and New South Wales – present reality as mechanically repeated images for those for whom reality has ceased to be an agential experience of embodied engagement with the material processes of production in interaction with the network of Beings which composed the world humans inhabit.

The Industrial Origins of Environmental Apocalyptic

Before the industrial revolution, most humans in every culture derived their living by their own ecological agency and through direct interaction with animals, plants and soils, water catchments, and weather systems. It is industrialism which has denied these agential connections with other Beings to billions of people and through which all people in predecessor cultures to industrial modernity learned day-to-day and intergenerational care for species, soils, and water catchments. As if to underwrite this disconnect, the first mode in which environmentalism emerged in response to the industrial revolution was that of aesthetics and the spectacle, in which the new and fatal disconnect between humans and nature was not so much resisted as underlined.

The eighteenth- and nineteenth-century nature romantics of Britain, Germany, and North America may reasonably be said to be the originators of modern environmentalism, and they were inspired in part by biblical poetry and prophetic prose in the stances they took in attempting to resist industrial developments – and especially large dam projects – in the mountainous landscapes they celebrated in the American West, the English North West, and Germany.[35] But in the main, the romantics failed to critique or resist the reduction of modern life for millions to industrial serfdom in death-dealing factories and life-shortening noxious cities, though there are important exceptions to this, and especially William Blake and John Ruskin.

Blake and Ruskin are arguably the originators of the modern rhetorical strategy of 'environmental apocalyptic' which is evidenced in the scientific announcement of the Anthropocene, the climate crisis, the extinction crisis, and so on. Among the earliest environmental apocalyptic texts are Blake's 'Jerusalem' with his famous references to the 'dark satanic mills' of England's life-wrecking industrial cities. Blake also percipiently connected industrialism and colonialism in his poem 'Albion' in which he prophetically showed how the oppressive reach of Britain's empire across the world originated in the injustice visited on the peoples of Britain's shires through the purloining of their forests and common fields by aristocratic landowners.[36]

Another important early environmental apocalyptic text which also identified empire and industrialism as the roots of the portentous change in human–nature relations is John Ruskin's *Storm-Cloud of the Nineteenth Century*. In these lectures Ruskin described his observations over 50 years, as recorded in his journals, of new weather conditions – including a persistent 'tremulous' 'plague-wind,' a 'lurid' 'smoke-cloud,' and changes in the colour and quality of sunsets – in the skies around his house above Coniston Water in the Lake District, which he blamed on the industrial pollution of 'Manchester's devil darkness' and its 'manufacturing mist.'[37] Ruskin noted that the smoke besmirched the British Empire; instead of being an Empire on which the sun never set, it was becoming one on which the sun never rose. He called for smoke controls on industrial chimneys, and for citizens to be housed in smoke-free zones or greenbelts away from the harmful effects of industrial pollution. But he was equally concerned at the aesthetic change that industrial pollution was bringing about, for it was endangering the whole tradition of landscape, cloud, sky, and weather painting that J.W. Turner, Constable, and other nineteenth-century artists had developed to such a fine art. For Ruskin, an anthropic sunset or atmosphere was not merely a new 'fact' to be accepted but a reality that was to be resisted and that indicated a moral and spiritual, as well as a scientific and technical, pathology. Hence Ruskin linked the apocalyptic portent of humanly induced weather changes with a call to moral and spiritual renewal, drawing on the words of Christ when he stilled the storm on the Sea of Galilee, and on the words of the late Hebrew Prophet Malachi:

> Whether you can affect the signs of the sky or not, you can see the signs of the times. Whether you can bring the sun back or not, you can assuredly bring back your own cheerfulness, and your own honesty. You may not be able to say to the winds, "Peace; be still," but you can cease from the insolence of your own lips, and the troubling of your own passions. And all

> that it would be extremely well to do, even though the day were coming when the sun should be as darkness, and the moon as blood. But, the paths of rectitude and piety once regained, who shall say that the promise of old time would not be found to hold for us also?—"Bring ye all the tithes into my storehouse, and prove me now herewith, saith the Lord God, if I will not open you the windows of heaven, and pour you out a blessing, that there shall not be room enough to receive it.[38]

Ruskin treated the changes in the weather as a portentous 'sign of the times' which indicated that industrial humans, having neglected and degraded both nature's beauty and goodness *and* the dignity of humans, will one day be called to account in a heavenly court. Ruskin uses environmental apocalyptic as a call to moral, spiritual, and social renewal. He calls the industrialists and wealth accumulators who were driving the injustice and pollution of the factory-filled cities to recover the old paths of justice and piety, to return to God: to return the land to its former beauty and its inhabitants to health. In moral and spiritual renewal, Ruskin envisaged that humanity, and more especially the English, would find it in their power to restrain the civilisational rush towards a coal-fuelled industrial economy which, though it profited the wealthy few, brought ill health and misery to millions, while besmirching the beauty of the Earth and skies for future generations of artists and nature lovers.

Blake and Ruskin's origination of environmental apocalyptic differed from the apocalyptic tenor of other nature romantics in their emphasis on the connections between human injustice – the theft of ancestral lands and shared forests, fields, and moorlands by aristocrats and industrialists – and the destructive alienation of moderns from the nonhuman world. Instead of commending the protection of nature as 'wilderness' from industrial predation, both Blake and Ruskin envisioned the end of empire and the scaling back of industrialism as the way out of the human–nature rupture. For them the rupture was an apocalypse, a revelation not of an intrinsic alienation between humans and nonhumans but of the alienation that emanated from the theft of the commons from the people, and their corralling in the besmirched and toxic air and factories of industrial cities.

The Misanthropic Anthropocene

As with Ruskin's storm-cloud, Crutzen's announcement of the Anthropocene is intended to generate a sense of grave crisis, a *Kairos* event, and so to create a shift in temporal consciousness.[39] The Anthropocene is a rhetorical device for presenting a new and portentous global horizon of risk, not merely to the appearance of the skies and the health of city dwellers, as in the case of the polluted skies and Satanic mills Ruskin and Blake complained of, but to the whole Earth System. But whereas Ruskin's 'signs of the times,' and Blake's apocalyptic vision of Albion, proposed that a different future for humanity was possible provided humans recovered humility and care for the Earth and turned away from the industrial nightmare of control and destruction of life, and tread more gently on the Earth, Crutzen uses the *Kairos* event of the dawn of the Anthropocene to call for a significant ramping up of research and development by scientists and technologists of the industrial engineering technologies to centrally and globally coordinate interventions in the Earth System, including active geoengineering of the atmosphere to reduce solar radiation. For Crutzen, the term 'Anthropocene' indicates not only that the relatively stable Holocene epoch is ending but also that humanity has an enhanced Promethean power over, and hence collective responsibility for, the state of the Earth and her future.

The contemporary announcement of a new geological epoch or event, the Anthropocene, is in the same genre of environmental apocalyptic begun in modernity by Blake and Ruskin. But it takes apocalyptic rhetoric in a new *misanthropic* direction. Instead of identifying industrial imperialism, and the planet-wide deployment of scientific and technological engineering projects which displace older less invasive forms of human dwelling in and harvesting from the environment, the Anthropocene narrative suggests that people in general are the problem, and that only when science and technology is allowed to take over control of all human–nature relations and uses will the bad Anthropocene be turned around. Hence for climate scientist H.J. Schellnhuber, the dawn of the Anthropocene represents a second Copernican Revolution.[40] The Copernican Revolution revealed that the Earth, and hence humans, were not at the centre of the universe but instead that the Earth revolves around the sun. The Copernican turn decentred humanity from the cosmos, and seminally shifted human cosmology such that humans gradually diminished in their awareness that their activities, and the activities of other Beings, including the stars, the sun and the moon, trees, plants, oceans, rivers are all part of a grand interconnected network of all Beings. The Anthropocene narrative, by contrast, puts humanity back at the centre of the network of Beings that constitutes the Earth System, and suggests, as did religious traditions in former times, that humans have a shared responsibility *with* the other Beings for the present and future evolution and trajectory of life on Earth.

Against Scientistic Prometheanism

Hans Jonas was the first modern philosopher to propose that the novel horizon of ecological risk opened up by modern technologies requires the recovery of an intergenerational ethic which includes not only duties to humans but duties to care for the Earth and ecosystems so as to pass them on to future generations in such ways as not to reduce their life chances.[41] Some might consider this an overly anthropocentric argument, and most contemporary environmentalists – including environmental philosophers – pose the ethical dilemmas raised by industrial technologies and large-scale degrading and re-engineering of ecosystems and species via appeals to, and new approaches to valuing, the beauty, diversity, and given evolved order of the nonhuman. But when Jonas argued for the recovery of a new cultural and moral awareness of human responsibility, and a recognition of ethical limits to human agency over the biosphere, he intended to promote a new moral sensitivity for humanity's relationship with life on Earth and the conditions for Earth's persistence.

If, however, we look at the rhetoric of natural scientific advocates of the new epoch of the Anthropocene we see that far from calling for restraint, they argue that heightened human agency over the Earth requires *greater* technological control over natural systems by scientists and increased scientific management over humans themselves. On this account, humanity should form clear intentions about desired planetary and climate states, and use the technological instruments of Earth System engineering, climate modelling, global 'meteorological governance,' and mass social engineering of human behaviours to bring these about.

Advocates of recognition of the Anthropocene epoch in some respects recall the anthropic cosmological assumptions of the pre-Copernican age since they argue that humans in the Anthropocene are central to the future of life on Earth. But they recast pre-Copernicanism in the form of an anthropic *scientistic* cosmology in which the human brain, computers, and machines controlled by centralised networks of elite scientists, economists, corporations, and bureaucrats determine and direct the future interactions of humans with nonhumans, and the desired end states of nonhuman and human habitats and societies.

Scientism is the modern ideology that is gradually displacing religions – indigenous religions and world religions – and successor modern philosophies such as Kantian idealism and existentialism. In essence, scientism is the view that only knowledge derived from observations of the behaviours of empirical objects, and the generation of theories from such observations, is true. Since the time of Francis Bacon, the influence of scientism has grown inexorably and its stated goal, as laid down by the British Royal Society, is that natural science should lead all other branches of human knowledge and understanding including philosophy and the other liberal arts, as well as practical disciplines such as human health, house building, food growing, and child rearing.[42] Scientism gives power to natural scientists and to the models and theories they develop on the basis of data models built around observations of empirical objects. The construction of data models from empirical observation is well described in the discipline of science studies of which Bruno Latour remains the best-known practitioner. In a series of anthropological studies, Latour records and describes the cultures scientists inhabit and through which they construct scientific 'facts' or 'factishes,' by which term Latour is suggesting scientific facts take on the role of religious fetishes in pre-scientific cultures. Since the time of Francis Bacon, who influentially delineated the modern scientific method, and via the spread of the industrial empires of Britain, France, and the Netherlands, and latterly of the United States, the Soviet Union, and China, science and its application to the control and re-engineering of the nonhuman through industrial techniques has overspread the Earth, and systematically eradicated traditional forms of knowledge including 'primitive' art, dwelling construction, food growing, herbalism and healing and their association with place-specific knowledges and languages, and religious traditions.

Natural scientists treat the Earth as a uniform space, and they do not allow that different techniques may be more appropriate in different places because they believe that the 'laws' of nature are uniform. This belief in the uniform operation of natural laws across all time and space was challenged by Einstein's discovery of quantum mechanics but it continues to shape the homogenising impacts of industrial techniques on the Earth.

Another major difficulty with scientism is that scientists believe that the natural world contains no given 'ends', or *telos*. They believe that the arrangements of atoms, cells, and living Beings that humans encounter on Earth arose randomly, and without the influence of mentality or mutually shared experiences. But if this is the case then the only legitimate source of ends are those of scientists and if a scientist determines that the best forest is one in which she can more easily count the number and size of trees, and so predict the rate of return, or the best fishing boat is one equipped with radar and a freezer capable of hoovering up and storing a whole species of fish from its spawning ground, there are no scientistic grounds for challenging these beliefs for in both cases the 'ends' imposed by humans are rational, while challenging those ends because evolved diverse forests and oceans are more beautiful lacks empirical validity.

Martin Heidegger and C.S. Lewis are among the more prominent philosophical voices from the mid-twentieth century, along with Jacques Ellul, who first questioned the totalitarian direction towards which scientism was driving both nature and humanity. Heidegger's famous analogy of nature being treated as a 'standing reserve,' as in the technical transformation of a river, via an engineered dam, into a huge store of water which humans can turn on or off at will, finds significant echoes in C.S. Lewis' account of the 'abolition of man' in which he suggests that heightened technical powers over nonhuman life, once conferred on humanity's rulers, will soon be turned on humans themselves, who will ultimately find their very humanness – which is to say their conscious free agency as ensouled beings – also at risk of abolition.[43]

The Hazards of Crisis Temporalities

Philosophical warnings from the mid-twentieth century of the dangers of scientism were very prescient. The 'pandemic' of 'COVID-19' in 2020–2 was said to be such a serious threat to human health that extreme social control measures were imposed. These unique social controls led to billions of people being put at risk, not from COVID-19 but from hunger, denied health care, social isolation, sedatives used to 'calm' care home residents, and side-effects from novel genetically modified drugs that did not stop transmission of the virus.[44] Crisis temporalities have a poor history and are typically utilised to promote totalitarianism. In the case of COVID-19 in most domains the crisis was used to shower public wealth on pharmaceutical companies and other large corporations and investors by an unprecedented quantity of central-bank money creation while millions of small businesses collapsed. This was combined with the emergency-use authorisation of expensive and hazardous genetically modified organisms in novel gene-editing drugs administered as an alleged prophylactic against SARS despite cheap, generic, and safe drugs already being widely available for this purpose.[45]

The extent to which the global health crisis of an allegedly untreatable disease was used to impose totalitarian levels of regulatory and technologically enforced control in many countries around the world highlights the dangers of the rhetorical strategy of the announcement of the 'Anthropocene' as a novel crisis in human–Earth relations which only *more* science, *more* industrial engineering, *more* technological surveillance, and State and corporate controls, of human decisions to travel from A to B, or to have children, or to grow food, or to eat certain foods can allegedly resolve.

Against the scientistic reading of the rise of human influence over the Earth System as an invitation to more intentional Earth System control, Ruskin argued that industrial humans, and more especially the English, should pull back from their increased influence on the atmosphere by moral and spiritual change, economic restraint, and turn away from wealth creation via industrial carnage: 'We have been taught a religion of pure mercy, which we must either now betray, or learn to defend by fulfilling.' A 'religion of pure mercy,' were they to recover it, would enable the English once again to make their country 'a sceptered isle,' and free it from the besmirching ugliness and gloom of industrial pollution, and ugly death-dealing slums, which together were driving many in Ruskin's day to depart Britain's shores and seek freedom and opportunity in the colonies.[46]

In contrast to Ruskin, Crutzen, and others use the apocalyptic temporality of the Anthropocene – the claim that humans stand on a precipice of imminent Earth System collapse – to argue for a more conscious and deliberative embrace of human influence by the global research and engineering community towards large-scale environmental management. This suggests the cause of the disease is also the cure. And yet the *core* scientific claim involved in the crisis temporality suggested by the onset of the Anthropocene is that there are *limits* to the extent of human interference in the atmosphere, oceans, rocks, soils, species, and waters, and that if humanity exceeds these it risks challenging the boundaries of the 'safe operating space' for human history. Johan Rockström and others use the language of 'safe operating space' to describe the limits to the Anthropocene, and they observe that already humanity is exceeding its safe space for influence on soils and oceans and that it is in danger of doing so in the atmosphere as well.[47] As with Jonas' philosophical argument for greater intergenerational responsibility, the deployment of the language of limits is intended to heighten a sense of responsibility for the future state of the Earth between present and future generations, and between asymmetric resource users. But if there are 'safe' operating limits which humans are in the process of exceeding, this underwrites the need for restraint among those who use excessive amounts of resources, and who direct the operations of huge extractive and engineering corporations, not only including fossil fuel corporations but industrial fishing, deforestation, mining, genetic engineering, and even climate engineering. The Earth needs a breathing space not from humans in general, but from the technologically empowered industrial agencies – governmental and corporate – whose managers use science and technique to capture quantities of power over other people, and species, unprecedented in human history, and at the same time to turn the labour and lands of other peoples into accumulated wealth again at levels unprecedented in human history.

Chronos, *Kairos*, Mechanical Time, and Cosmic Time

The crisis-driven temporality of the announcement of the 'Anthropocene' is of a piece with the revolutionary turn that occurred in human affairs from the onset of the Industrial Revolution and related political upheavals in Europe and in the European colonies. The invention of the clock, and the way in which the controlled time of the clock displaced the time of the rising and setting sun from the late Middle Ages, played a crucial role in this revolution, as Lewis Mumford and E.P. Thompson have shown.[48] The monastic invention of mechanical timepieces to reliably predict the hourly times of prayer for monks changed the human relationship to time in the following centuries. This gradually led to a bifurcation in perceptions of time, and in particular a disconnect between clock time and the experience of time through encounters with the Earth and other Beings, which has reached new extremes in modernity.

Traditionally, time was measured by the sun, and the movements of the moon and stars, together with their roles in bringing darkness and light and changing seasons. Religious calendars in many cultures, and this is still the case in Bali to this day, mapped ritual events, including rituals concerning such activities as food planting and harvesting, hunting, house building, raising children, and burying the dead, around the lunar calendar. So different events are timed according to what is the right or *fitting* time – which is what Aristotle meant when he talked about time as *Kairos*. A Balinese example – I wrote this book in Bali – is the putting of a child's feet for the first time on the ground. Balinese believe that children new-born are souls from heaven, and they cannot bear too much Earthliness in their early weeks, so they are not supposed to touch the ground in their first few months. There is a ceremony known as *Otonon* for putting the feet of the child on the ground for the first time which takes place after six months of the child's birth as this is considered the fitting time when the reincarnated soul of the child has sufficiently adjusted herself to being in the new environment of Earth and the child is becoming more human and less like a god or goddess. The precise day on which the child's feet touching the ground ceremony is performed also depends on the Balinese lunar calendar so both the measure of the rising and setting of the sun, and the phases of the moon, play roles in determining this most significant of ceremonies in the first years of a child's life.[49]

This Balinese example of the connection between a significant rite of passage and the human experience of time as mediated by the sun and the moon has analogies in other traditional cultures. One of the older evidences of human time reckoning is from Scotland and relied on observations of the phases of the moon and dates from the eighth century BCE.[50] More recently, medieval records of the building of York Minster, as investigated by anthropologist Kevin Birth, show that the hours of the day in which masons and joiners worked in the European Middle Ages varied according to the seasons, and hence also the rate of pay. Craftsmen were expected to work for longer hours in summer than in winter, but their rates of pay did not vary between the seasons. Hence although the time a particular operation would take was longer in the winter than the summer, because of fewer daylight and hence fewer working hours, the daily rate of pay remained the same.[51]

The power of clock time – or time as *chronos* – reached new heights in the transformation of human productive powers as 'labour time' in the Industrial Revolution in which time as measured by clock hours turned 'chronic' in its influence on human affairs, and 'proper time,' or *Kairos* time, diminished in power and significance.[52] *Kairos*, the Greek deity of rhetoric, was the word used by Aristotle and Plato to describe a conception of time as fitting or just or appropriate to particular tasks, rituals, or events.[53] Again this is close to how time was practised in the European Middle Ages. The 'just price' payable for tailoring a suit or a pair of shoes in the Middle Ages was measured according to the time it took to make the suit, the time involved in training the craftsman, and the time needed to purchase and maintain a workshop and home.

Chronos on the other hand measures time as duration.[54] It counts passing moments which moderns now experience via the control of so much of life by digital clocks, and the constant presence of digital time in wearable human devices, in homes and offices. The mechanical measurement of *chronos* in modernity increasingly separated work time from the time and seasons of the Earth, and from the time needed to learn a craft. Then via usury and debt-based money creation, *chronos* was essentially weaponised in early modern Europe by central bankers who monopolised money as interest-bearing debt. In so doing they began a process which gradually stored up more and more of the labour time of humans, and the 'resources' of the Earth, into the accumulated wealth of the elite class of banking families, royal families, and owners of large corporations who still own much of the world to this day. Aristotle called making money from time by lending money at interest *chrematistics* and argued that it was contrary to the human or common good, a position that was also adopted by the Catholic Church until the Reformation.[55] But after the Reformation time *as* money was increasingly normalised by central bankers and this began the process of turning both the deep time in which the Earth evolved its wondrous wealth of minerals and biodiverse ecosystems, and the time of human lives measured as timed work and timed debts (mortgage in English means debt until death), into accumulated wealth.

Chronos and the Monetisation of Time

Wealth accumulation via the monetisation of time was further advanced in the twentieth century by the development of the scientific management of work through the 'time and motion' approach initiated by Frederik Taylor.[56] Taylor used stopwatches and film cameras to measure and observe the bodily movements of factory workers performing specific repeated tasks in industrial processes, and used formulae derived from these measurements to determine rates of pay. These techniques were used to increase the pace at which industrial work is performed to the point that work increasingly resembled the adapted service of the human body and mind to machines. By the end of the twentieth century scientific management had taken over most crafts and professions in North America, Europe, and much of Asia, often to the point that craftspeople such as bakers and even architects are increasingly deskilled by the machines, or computer-aided design software, that they are required to work with and according to time-measured repetitive tasks.[57]

Time as *chronos* in debt-based money creation,[58] together with industrial time, produced a deep alienation between the products of work and the worker or craftsperson, as Karl Marx argued, drawing on Aristotle.[59] But the monetisation of mechanically measured work time also produced a deep alienation between human craft, production, and work and the Earth herself. The forms and techniques of modern production increasingly subject both human body and mind, and other species and ecosystems, to such invasive forms of re-engineering that the networked agencies which gave rise to life on Earth, and ultimately to conscious human life, over the aeons of deep time are unravelled and both human life and the life of the Earth are increasingly submitted to the priorities of bankers and investors in seeking to maximise rates of return on investment of debt-based money. Time as *chronos* has in other words become chronic in the extreme in the escalating and interrelated crises of being human and of nonhuman being which characterise the present moment.

Prometheanism and the Age of Revolutions

If the Anthropocene is in the main associated with the Promethean claim that the Earth and humans need to be subjected to even *more* industrial scientific control, then the announcement of the Anthropocene, far from recovering time as *Kairos*, and so fitting human culture and production to the time of the Earth, risks accelerating the lack of fit of industrial civilisation to the Earth. By contrast, when John Ruskin protested the industrial invasiveness of 'Manchester's devil's darkness' on the Earth and on his fellow humans, he appeals against the invasive temporality of industry and usury to the temporality of *Kairos*. Ruskin invokes the New Testament *Kairos* language of 'signs of the times' in resisting the ugly and anti-human as well as anti-natural storm-cloud of industrialism.

Since Ruskin's time, the military industrial complex, and now the medical industrial complex, has used crisis after crisis to capitalise human interactions not only through industrial production and scientific management but via war. War is among the most powerful means for bankers to accumulate wealth since it requires huge amounts of government spending, and large-scale borrowing from central bankers. War is also a powerful device for deepening the industrialisation and centralised scientific management of human societies, and this was especially true of the Second World War, in the course of which European and American governments grew hugely in both expenditure and in the size of their bureaucracies.

Wars also involve hugely invasive impacts on ecosystems, and even the Earth System since, as Peter Sloterdijk argues, modern total wars – and especially the First World War – were the origin of the intentional modification of the atmosphere to poison combatants: hence the modern science of meteorology – both weather prediction and weather modification – began in war.[60] Recent forms of 'war'-related crises requiring ever more invasive industrial management and technological control of human life and Earthly habitats include such fake wars as the 'war on terror,' and now an alleged 'war on viruses,' a project which has been as destructive of human and planetary wellbeing – estimates suggest there are 1.5 billion plastic masks in the oceans from two years of this 'war' – as those of the 'war on terror,' the 'Cold War,' or 'World Wars.'

After more than two centuries of large-scale European and global wars since the initial break with organic and traditional societies represented by the Industrial Revolution, there is accumulating evidence that crisis-driven apocalyptic rhetorics do not promote societal efforts to address the causes of crisis. Instead they result in huge misdirection of societal resources from the genuine sources of human and planetary harmony, peace, prosperity, and resilience towards destructive activities such as the global weapons industry, the scale of which dwarfs all other industrial activities; the fossil fuel industry; industrial food production; and the manufacture of vast quantities of superfluous consumer and fashion items. Furthermore, apocalyptic 'climate porn' and 'extinction porn' are associated with denial that there is anything that can be done, even if the problems are real, because the collapse of 'nature' or the 'climate' or 'biodiversity' are said to be too immanent and engender passivity: individuals are overwhelmed by apocalyptic messages and spectacles and are led by them to believe there is nothing they, or even groups of people working together, can do positively to change things. Modern resort to crisis temporalities does not therefore recover a *Kairos* temporality in which humans are encouraged to live more fittingly and harmoniously with each other and the Earth. Instead, it deepens the seminal break between humans and the Earth which originated with the invention of a debt-based usurious banking system in the seventeenth century and the promotion by that system of ever more invasive and homogenising industrial 'development' of human activities, and of species and ecosystems, across the face of the Earth.

Against Revolutionary Time and for Time as the Accretion of Memory

In my last book I argued that a sense of place, and the association of a sense of place with the sacred, is something which is productive of pro-environmental behaviours or, to use a more philosophical register, of an 'ethics of care' towards nonhumans and humans alike.[61] I also noted that what makes places sacred is the action of nonhuman forces, *combined* with human habitation, over considerable periods of time. In other words, a sense of the sacred, and of what is profane and therefore 'forbidden,' or at least in need of constraint, is the outcome of a build-up of experiences, or 'chains of memory,' in particular places over long – which is to say intergenerational – periods of time.

The idea that time itself is a property of the evolutionary process of life on Earth originated in many respects with James Hutton's deep time theory. But for Hutton, as for Darwin, Huxley, Spencer and the other key theorists of evolution, the role that time played was simply that of duration. According to modern biological sciences, and geophysics, the autonomous mechanisms of the Earth produced life, and ultimately humans, not by an inner convergence or teleological orientation to life, including complex Beings such as the higher Primates and humans, but by the random operation of the laws of nature over extremely long periods of time. But in an important revision to this mechanistic cosmology Henri Bergson argued that the depth of time over which life evolved towards the evolution of the most conscious beings that ever lived – human beings – together with the highest levels of biodiversity in geological history, are the outcomes of the cumulative temporal memory of *experiences* which accrues not only within each conscious being but between generations of conscious beings, and at all levels of material and organic life.[62] Bergson argued that memory resides in networks of beings, cells, matter, and organisms and that this explains how embryos, even when damaged at birth, still 'know' what form they should ultimately take and achieve that form, or how caterpillars become butterflies, neither of which processes is explicable via conventional mechanical and materialist understandings of cellular, genetic, or organic behaviours.[63]

Bergson's theorisation of the accretion of memory of experiences over considerable time periods as the means of the creation of life has profound implications for human cultural development, and in particular for the revolution and crisis-ridden phase of development which has occurred since the scientific and the industrial revolutions. Empirical studies of customary governance, commons governance, trade guilds, and other pre-industrial and pre–nation-state forms of organisation indicate that humans grew in their capacities for collective and peaceable organisation over very long time periods. But on every continent the huge social forces unleashed by capital accumulation via land expropriation, and by the invention of joint-stock–holding corporations for deep mining and trans-ocean colonial projects, together with central banks, and new technologies – including gunpowder, chronometers, and long-distance navigation – released new forms of social power which were exercised at a distance and at scale from the seventeenth century, first by the East India Company as so lucidly chronicled in William Dalrymple's *The Anarchy*.[64]

From the mid-eighteenth century the process of consolidation of land, labour, oceans, and societal structures was advanced by the invention of central banking, and from the nineteenth century by oil companies and trading dynasties including the central-banking Rothschilds dynasty, the Rockefeller oil dynasty, Jardine Mattheson's opium-war dynasty, Goldman Sachs and Morgan Stanley Wall Street dynasties, and in the late twentieth century the technological dynasties of Gates, Bezos, Jobs, and Zuckerberg. Today just one investment company – Blackrock – controls more stock than any company in history and with that control it can manipulate markets, natural resource use, and value exchange on every continent.

The forces unleashed from the nineteenth century by new forms of centralised banking and investment, together with European colonial expansion, were combined with the new technologies associated with the scientific and industrial revolutions. These initially led to huge increases in capitalist production and accumulation. But this was at the cost of greatly increased poverty and immiseration for millions of people subjected to these forces, whose lives were shortened by the 'devil's darkness' of industrial cities of which Ruskin complained. These immiseration tendencies were further enhanced by the use of industrial technologies from the nineteenth century to unleash what Carl von Clausewitz first described as 'total war' in the Napoleonic Wars, and then in the First and Second World Wars, the Cold War, Communist totalitarianism, and more recently the fake 'war on terror' from 2001, and now the 'war on pathogens.' I will return to a fuller analysis of this pattern of events later but for now my point is to highlight the extent to which the modern form of revolutionary temporality is uniquely crisis-ridden and relies on radical disruptions in the human and ecological experiential chains of memory which for Bergson are the origin of the creative evolution of life towards higher levels of complex consciousness and biodiversity.

The implication of this very brief overview of the industrial era 'age of revolutions' is that to recover forms of life and social organisation which enable humans to care for each other and ecosystems in such ways as to foster the ongoing evolution of memory and experience towards higher levels of consciousness and care requires rejection of the apocalyptic, crisis-ridden temporality of instantaneous 'real-time,' and the distancing and virtual spectacle associated with its representation in contemporary ecological, health, and social 'crises.'[65] Instead it requires a more conservative temporal timbre, as first suggested by Edmund Burke in his *Reflections on the Revolution in France*, in which a longer-term intergenerational temporality – what I call 'ancestral time' and 'descendant time' – is recovered.[66]

This temporality was the norm among indigenous peoples who dwelled in the same land area intergenerationally and honoured not only their ancestors but the rich mix of species who lived in these areas over thousands of years. This temporality is also evident in the long endurance of ancient traditional civilisations and religions, and especially those of China and India which are among the longest living traditions in all history. Christianity and Islam are of course much less ancient as traditions but in certain forms – and especially Orthodox Christianity – they also are characterised by customs and traditions that have remained relatively constant in form and meaning over more than 100 human generations.

Temporality as a form of consciousness and experience is under-investigated both by natural scientists, who have mainly ignored Bergson's powerful insights, and by social scientists. Lack of proper investigation of temporality in the social sciences is in part responsible for the common presumption among scientists that 'crisis' rhetorics are capable of producing the kinds of changes environmentalists propose are needed in human ethics, behaviours, and forms of social organisation if what they call the 'Anthropocene' is not to result in the extinction of significant numbers of species and even to threaten human life itself, at least in vulnerable areas prone to flooding or drought.

Ancestral and Descendant Time

To address the gap in social scientific investigation of temporal rhetorics in relation to the ecological crisis, I led a collaborative research project at the University of Edinburgh from 2013 to 2017, funded by the UK government's Arts and Humanities Research Council, in which we explored temporality as a register among groups of environmentally concerned churchgoers, students, and citizen organisations in Scotland, England, and in parts of the United States. The research revealed that apocalyptic rhetoric, and in particular the classic climate and extinction rhetorics that soon, very soon, the Earth will become unstable and mass extinctions, floods, droughts, and famines will render it increasingly uninhabitable for humans, is ineffective in promoting pro-environmental behaviours and was more likely to produce passivity and denial. Our findings were summed up in the comments of one respondent on the futility in terms of behavioural change of appeals to future apocalyptic crises:

> We don't really have a concept of the future – we can't conceive of the world being a seachange different so when you try and explain that sort of thing to people they find it confusing and upsetting and not very enlightening. People are much better at dealing with problems that are here and now so I find that sort of tactic – you know the sky's falling in – does not work.[67]

More positively, the research revealed that more effective environmental rhetorics linked with conscious ecological responsibility and practical actions arise from the temporality of intergenerational care for children and grandchildren

> Saving the planet, good stewardship, climate change; yeah that is the bottom line for me, which comes to the same thing. It is about what we are doing to the earth, what we are handing on to our children. Thinking ahead a couple of generations what will it be like, what are we doing, what basis are we doing that? You feel these economic arguments and you know

> it will be really difficult to do it – but try not doing it and see what it is like in fifty years' time, I think you'll find the economy starting to creak a bit.[68]

Rather than ancestral I think of this form of temporality as 'descendant time' – a concern for the condition of the Earth and human society as it will be experienced by children and grandchildren, and other species, who are already alive. It occurred in a number of the interviews I conducted in the course of the research and was more positively associated by respondents with practical actions in the present to address the roots of ecological crisis than appeals to apocalyptic crises in the future.

This ethnographic finding engages intriguingly with philosophical and theological accounts of intergenerational time. In the midst of the cataclysm of the rise of Nazism in Germany in the 1930s, theologian Dietrich Bonhoeffer argued that the central question posed by civilisational crisis is not 'how I am to extricate myself heroically from the affair, but how a coming generation is to live.'[69] Bonhoeffer, like Martin Heidegger, Hannah Arendt, and Hans Jonas, had a distinctively existentialist sense that the modern revolutionary condition – scientific, industrial, political, and technological – gave rise to the extremes of industrial servitude and re-engineering of life that led to the Third Reich. The revolutionary condition is novel because scientific and industrial technologies have conferred on modern humans capacities to change both the human condition and the ecological landscape in ways that potentially disrupt the bodies as well as the habitats of future generations. These new powers which can turn a river into a 'standing reserve' of energy or kill the million-year-old life of farmland soil with artificial chemicals, or empty the oceans of 90 per cent of all life, potentially also involve change to the human future and so have intergenerational implications.

Intergenerational Care and Honouring the Dead

Whether we call it 'ancestral time,' 'descendant time,' or 'intergenerational time,' the conception of temporal responsibility for the condition of life and life's habitats – including soils, forests, rivers, and the oceans – that passes from one generation to another has a long history in indigenous and Asian cultures, and in the West until modernity. Wendell Berry is one of the strongest advocates of this conception of intergenerational responsibility in North America and in numerous of his essays he shows how chemical agriculture and industrialism have wrecked the soils and waterways of North America and spread a culture of irresponsibility and lack of care for the condition of the land which future generations will inherit.[70] Berry's analysis explains why the Amish have become the most successful farmers in North America because by refusing to adopt most industrial technologies their farms have healthier soils, and fewer monetary debts, so they are more financially resilient and own more fertile land than their chemical-farming neighbours, many of whom are driven by the chemical-farming treadmill to sell out to banks and big farming corporations.

The research I and my colleagues conducted in the United Kingdom on religious temporalities and environmental attitudes indicates that temporalities that promote intergenerational care for the Earth are connected neither with the slow, and mostly pre-human, temporality of 'deep time,' nor with the apocalyptic rhetoric of crisis associated with the 'Anthropocene.' The temporality most associated with Earth care is one that links human generations in shared concern for the heritage left to them by their ancestors and the inheritance they will leave for their descendants.

Religions in their origins nurtured such intergenerational connections, as evidenced in the fact that many of the earliest recorded religious rituals and structures concern the burial of the dead. Often, burial structures such as the ancient stone structures on Orkney in Scotland were built in the midst of human dwellings and built in such a way as to honour the other beings in the cosmos, including in the case of the Neolithic Orkney burial sites, the seasonal movements of the sun, the moon, and the stars in the sky.

Robert Pogue Harrison argues that the sense of ancestral connection between the living and the dead memorialised in such structures is a strong feature of pre-modern cultures and that respect for the dead and care for the land and the waterways that the dead bequeath to the living were traditionally strongly associated.[71] Moderns on the other hand have mostly forgotten the religious rites and traditions that connect the living and the dead. In the formerly Christian West, most moderns tend to bury their dead apart from their living spaces, or to cremate them, so removing any organic trace of the deceased from the Earth. And in modernity, the events associated with departing and death are also in the main hidden away from normal living places, in hospitals, or end-of-life care homes. If such a seminal temporal process of transformation of consciousness and organic life as the departure of the soul from the body is no longer considered central to the rituals and practices of modern industrial cultures it is perhaps not surprising that these cultures also manifest a deep disregard in the ways they treat the Earth, and for the condition in which future souls will inherit it.

Hebridean Sabbaths and the Hallowing of Time

A fine exemplar of the capacity of traditional religions to resist modern industrial disrespect for Mother Earth, and for the habitations of living souls, arose in the Hebrides in Scotland in the 1990s. The waters and coastline around the Scottish Inner and Outer Hebrides represent Western Europe's most extensive and biodiverse coastal wilderness environment. They are inhabited by a largely Protestant people who have adhered more actively to rituals that celebrate the traditional doctrines and stories of the Old and New Testaments than any other people in the United Kingdom or Europe, attending church at rates 20 times those of most West Europeans.[72] There is evidence, arising from a case study of a proposed superquarry in the Hebrides on which I report elsewhere, that the old creation story, and its association with the hallowing of the seventh day – the Sabbath – as a day on which no work should be performed, and hence is opposed to 24/7 industrial time, played a significant role in successful resistance to the proposed superquarry.[73] Time, in other words, and especially *sacred* time, may play a larger role than previously thought in resistance to the scientific and technological re-engineering of life on Earth which is the root of the modern ecological crisis. It is then ironic that advocates of the New Genesis story discussed above envisage that this rare biocultural region will be better protected from the impacts of consumerism by the substitution of a cosmopolitan and scientistic universe story composed by a North American 'geologian' for the traditional creation story and the related hallowing of time and place, that has served the Hebrides so effectively as a source of resistance to the extractive and biodegrading economy of the British mainland.

In the next chapter I present a case study based on field work on the island of Borneo in Southeast Asia which also involves the role of indigenous people in conserving their environments and their contest, in so doing, with scientism and industrialism. Borneo has remnants still of the oldest and most biodiverse tropical rainforest on Earth and I show how 'ancestral time' in the indigenous traditions of Borneo resulted in the thousands of years old indigenous custodianship of the unique biodiversity of the island, in contrast to the temporalities of industrial and 'climate' time, and which between them have advanced, and provided ideological cover, for an island-wide extent of ecological destruction that amounts to ecocide.

Notes

1 P.W. Boyd and A. Watson, 'A Mesoscale Phytoplankton Bloom in the Polar Southern Ocean Stimulated by Iron Fertilization.' *Nature* 407 (2000) 695–702.
2 Andrew Dobson, *Listening for Democracy: Recognition, Representation, Reconciliation.* Oxford, Oxford University Press, 2014; Michael S. Northcott, *A Political Theology of Climate Change.* London, SPCK, 2014.
3 Kate Rawles, *The Carbon Cycle: Crossing the Great Divide.* Isle of Lewis, Two Ravens Press, 2012; Ed Gillespie, *Only Planet: A Flight Free Adventure Around the World.* London, Wild Things Publishing, 2014.
4 John Seed, Joanna J. Macy, Peter Fleming and Arne Naess, *Thinking Like a Mountain: Towards a Council of All Beings.* Gabriola Island, BC, New Society, 1998.
5 Seed, Macy, Fleming and Naess, *Thinking Like a Mountain*, 32.
6 Arne Naess, *Ecology, Community and Lifestyle.* Trans. and Ed. David Rothenberg, Cambridge, Cambridge University Press, 1989.
7 Seed, Macy, Fleming and Naess, *Thinking Like a Mountain*, 19.
8 Thomas Berry, *The Great Work.* New York, Random House, 1999, 63–5.
9 Rawles, *The Carbon Cycle.*
10 Malcolm Rider and Peter Harrison, *Hutton's Arse: 3 Billion Years of Extraordinary Geology in Scotland's Northern Highlands.* 2nd Edition, Edinburgh, Dunedin Academic Press, 2019.
11 James Hutton and John Playfair, *James Hutton's System of the Earth.* New York, NY, Hafner Publ. Co, 1785.
12 John Repcheck, *The Man Who Found Time: James Hutton and the Discovery of Earth's Antiquity.* London, Simon & Schuster, 2003.
13 John Playfair, *Biographical Account of Dr James Hutton in the Works of John Playfair.* Edinburgh, William Constable, 1822.

14 James Hutton, *A New Theory of the Earth*. Edinburgh, William Creech, 1786.
15 Ronald Hendel, *The Book of Genesis: A Biography*. Princeton, NJ, Princeton University Press, 2019.
16 James Fuller, 'Before the Hills in Order Stood: The Beginning of the Geology of Time in England.' *Geological Society London, Special Publications* 190 (2001) 15–23.
17 Dipek Chakrabarty, 'The Climate of History: Four Theses.' *Critical Inquiry* 35 (2009) 197–222.
18 Will Steffen, Johan Rockström, Katherine Richardson, et al., 'Trajectories of the Earth System in the Anthropocene.' *Proceedings of the National Academy of Sciences of the United States of America* 115 (2018) 8252–9.
19 Curt Stager, *Deep Future: The Next 100,000 Years of Life on Earth*. New York, Thomas Dunne Books, 2011.
20 Lisa H. Sideris, 'Science as Sacred Myth? Ecospirituality in the Anthropocene Age.' In Ricardo Rozzi, et al. eds. *Linking Ecology and Ethics for a Changing World: Values, Philosophy, and Action*. Dordrecht, Equinox, 2013, 147–62.
21 Bruno Latour, *On the Modern Cult of the Factish Gods*. Durham, NC, Duke University Press, 2010.
22 The most influential essay in which sacred creation stories are proposed as the dominant factor in environmental behaviours is Lynn White Jr., 'The Historical Roots of Our Ecologic Crisis.' *Science* 155 (1967) 1203–7; for a contemporary theoretical overview of the possible sources of pro-environmental behaviours, including social, psychological, and economic as well as cultural factors, see Anja Kollmus and Julian Agyeman, 'Mind the Gap: Why Do People Act Environmentally and What are the Barriers to Pro-environmental Behavior?' *Environmental Education Research* 8 (2002) 239–60.
23 Bernadette C. Hayes and Manussos Marangudakis, 'Religion and Environmental Issues within Anglo-American Democracies.' *Review of Religious Research* 42 (2000) 159–74.
24 Paul J. Crutzen, reported oral comments, cited Michael S. Northcott, 'Going Gently into the Anthropocene.' In Celia Deane-Drummond, Sigurd Bergmann, and Markus Vogt eds. *Religion and the Anthropocene*. Eugene, OR, Cascade Books, 2017, 19–34.
25 P. Gibbard, A. M. Edgeworth, W. F. Ruddiman, et al., 'A Practical Solution: The Anthropocene is a Geological Event, not a Formal Epoch.' *Episodes* (2021). https://doi.org/10.18814/epiiugs/2021/021029.
26 Paul J. Crutzen, 'Geology of Mankind: The Anthropocene.' *Nature* 415 (2002) 23.
27 Will Steffen and Paul J. Crutzen, 'How Long Have We Been in the Anthropocene Era?' *Climatic Change* 61 (2003) 251–27.
28 Simon L. Lewis and Mark A. Maslin, *The Human Planet: How We Created the Anthropocene*. New Haven, CT, Yale University Press, 2018, Ch. 5.
29 Jared Diamond, *Collapse: How Societies Choose to Fail or Succeed*. New York, NY, Viking, 2005; for well evidenced critiques see the essays in Patricia McAnany and Norman Yoffee eds. *Questioning Collapse: Human Resilience, Ecological Vulnerability, and the Aftermath of Empire*. Cambridge, Cambridge University Press, 2010.
30 On ecological factors in the collapse of the Roman Empire see Clive Ponting, *A New Green History of the World*. London, Penguin, 2007.
31 See further Elinor Ostrom, *Governing the Commons: The Evolution of Institutions of Collective Action*. Cambridge, Cambridge University Press, 1990.
32 For a fuller discussion of Hardin see Michael S. Northcott, *The Environment and Christian Ethics*. Cambridge, Cambridge University Press, 1996, 191–2.
33 Guy Debord, *The Society of the Spectacle*. Trans. Ken Knabb, Canberra, Hobgoblin Press, 2012.
34 Walter Benjamin, 'The Work of Art in the Age of Mechanical Reproduction.' In Hannah Arendt ed. *Illuminations*. London, Fontana, 1968, 214–18.

35 See further Michael S. Northcott, 'The Romantics, the English Lake District, and the Sacredness of High Land: Mountains as Hierophanic Places in the Origins of Environmentalism and Nature Conservation.' In Hans-Günter Heimbrock ed. *Ecotheology: Contemporary Essays in Religion and Nature: Festschrift for Sigurd Bergmann*. Leiden, Brill, 2021, 74–90.

36 For discussion of William Blake's critiques of the industrial revolution, empire, and ecological destruction, and the reasons for the neglect of Blake as an environmental prophet, see Northcott, *A Political Theology of Climate Change*, 287–301.

37 John Ruskin, *The Storm-Cloud of the Nineteenth Century: Two Lectures Delivered at the London Institution February 4th and 11th, 1884*. Published online by Project Gutenberg, 2006, http://www.gutenberg.org/ebooks/20204 accessed June 1, 2021.

38 Ruskin, *Storm-Cloud, Lecture 1,* 8.

39 On the distinction between time as Chronos and Kairos see John E. Smith, 'Time, Times and the 'Right Time'; "Chronos" and "Kairos".' *The Monist* 53 (1969) 1–13.

40 H.J. Schellnhuber, '"Earth System" Analysis and the Second Copernican Revolution.' *Nature* 402 (1999) C19–23.

41 Hans Jonas, *The Imperative of Responsibility: In Search of an Ethics for the Technological Age*. Chicago, IL, University of Chicago Press, 1984.

42 On scientism see Tom Sorell, *Scientism: Philosophy and the Infatuation with Science*. London, Routledge, 1991; Bruno Latour, *We Have Never Been Modern*. Trans. Catherine Porter, Cambridge, MA, Harvard University Press, 1993.

43 Martin Heidegger, *The Question Concerning Technology and Other Essays*. Trans. W. Lovitt, New York, NY, Harper & Row, 1977, 3–35; C.S. Lewis, *The Abolition of Man*. New York, NY, Macmillan, 1947; Jacques Ellul, *The Technological Society*. Trans. John Wilkinson, New York, NY, Vintage Books, 1964.

44 For scientific references and a fuller discussion of COVID-19 see Chapter 5.

45 There are many clinical trials showing the effectiveness of hydroxychloroquine or ivermectin, combined with vitamin D, zinc, and vitamin C, in successfully alleviating early symptoms of COVID-19 and in reducing mortality in late-stage COVID-19 in hospitals: see for example Juliana Cepelowiz, Michael S. Sherman, Naaz Fatteh, et al., 'ICON (Ivermectin in Covid Nineteen) Study: Use of Ivermectin Is Associated with Lower Mortality in Hospitalized Patients with Covid-19.' *The Lancet* (2020). https://ssrn.com/abstract=3631261; and Hashim A. Hashim, Mohammed F. Maulood, Anwar M. Rasheed, et al., 'Controlled Randomized Clinical Trial on Using Ivermectin with Doxycycline for Treating COVID-19 Patients in Baghdad, Iraq.' *medRxiv* (2020). https://doi.org/10.1101/2020.10.26.20219345.

46 John Ruskin, *Lectures on Art*. New York, NY, John Wiley & Son, 1870, 28.

47 Johan Rockström, Will Steffen, Kevin Noone, et al., 'Planetary Boundaries: Exploring the Safe Operating Space for Humanity.' *Ecology and Society* 14 (2009) 32.

48 Lewis Mumford, *Technics and Civilisation*. New York, NY, Harcourt Inc., 1934; E.P. Thompson, 'Time, Work-Discipline and Industrial Capitalism.' *Past and Present* 38 (1967) 56–97.

49 Fred B. Eiseman Jr., *Bali: Sekala and Niskala: Essays on Religion, Ritual, and Art*. Denpasar, Turtle Publ., 2011, 93–5.

50 Vincent Gaffney, Simon Fitch, Eleanor Ramsey, et al., 'Time and a Place: A Luni-Solar "Time-Reckoner" from 8th Millennium BC Scotland.' *Internet Archaeology* 34 (2013). https://doi.org/10.11141/ia.34.1.

51 Kevin Birth, *Objects of Time*. London, Routledge, 2013.

52 Kevin K. Birth, '"Hours Don't Make Work": Kairos, Chronos, and the Spirit of Work in Trinidad.' In Kevin K. Birth ed. *Time Blind: Problems in Perceiving Other Temporalities*. Chambery, Switzerland, Palgrave Macmillan, 2017, 39–54.

53 Birth, *Time Blind*, 49.

54 Birth, *Time Blind*, 50.

55 Aristotle, *Politics*, 1. 10, 1258a 27-34; for a fuller discussion of Aristotle's distinction between chrematistics and oikonomia see Herman Daly and John B. Cobb, *For the Common Good: Redirecting the Economy Toward Community, the Environment, and a Sustainable Future*. Boston, MA, Beacon Books, 1994.
56 Frederik W. Taylor, *Scientific Management: The Early Sociology of Management and Organizations*. New York, NY, Harper and Bros., 1947.
57 See further Richard Sennett, *The Corrosion of Character: The Personal Consequences of Work in the New Capitalism*. New York, NY, W W Norton, 1998.
58 On the origins of most modern money in debt see Richard Werner, 'Can Banks Individually Create Money Out of Nothing? – The Theories and the Empirical Evidence.' *International Review of Financial Analysis* 36 (2014) 1–19.
59 Jonathan E. Pike, *From Aristotle to Marx: Aristotelianism in Marxist Social Ontology*. London, Routledge, 2019.
60 Peter Sloterdijk, *Terror from the Air*. Trans. Amy Patton and Steve Cocoran. Cambridge, MA, MIT Press, 2009, 77–9.
61 Michael S. Northcott, *Place Ecology and the Sacred: The Moral Geography of Sustainable Communities*. London, Bloomsbury, 2015.
62 Henri Bergson, *Creative Evolution*. Trans. Arthur Mitchell, New York, NY, Random House, 1944.
63 Henri Bergson, *Matter and Memory*. Trans. N.M. Paul and W. Scott Palmer. New York, NY, George Allen and Co., 1912.
64 William Dalrymple, *The Anarchy: The Relentless Rise of the East India Company*. New York, NY, Bloomsbury, 2019.
65 Eric Hobsbawm rightly connects the revolutionary timbre of both the French political revolution and the British industrial revolution as complementary origins of the revolutionary and intergenerationally disruptive character of industrial modernity in his magisterial *The Age of Revolution 1787–1848*. London, Vintage, 1996.
66 Edmund Burke, *Reflections on the Revolution in France, and on the Proceedings in Certain Societies in London Relative to that Event. In a Letter Intended to Have Been Sent to a Gentleman in Paris*. London, J. Dodsley, 1790.
67 Respondent 1, Melrose, Scotland, May 11, 2015.
68 Respondent 6, Dunscore, Scotland, June 26, 2015.
69 Dietrich Bonhoeffer, *Widestand und Egebung*. Ed. Eberhard Bethge, Munich, Kaiser, 1952, cited and translated by Rachel Muers, 'Pushing the Limit: Theology and Responsibility to Future Generations.' *Studies in Christian Ethics* 16 (2003) 36–51.
70 For his now classic monograph on the intergenerational irresponsibility of modern chemical and industrial agriculture see Wendell Berry, *The Unsettling of America: Culture and Agriculture*. San Francisco, CA, Sierra Club Books, 1977.
71 Robert Pogue Harrison, *The Dominion of the Dead*. Chicago, IL, University of Chicago Press, 2003.
72 Alastair McIntosh, *Island Spirituality: Spiritual Values of Lewis and Harris*. South Locks, Isle of Lewis, Islands Book Trust, 2013.
73 Northcott, 'Place, Religion and Resistance to Corporate Power.' In Northcott, *Place Ecology and the Sacred*, 87–103.

2

IN BORNEO

Abstract

The highly volcanically active Malay Archipelago is a place where the Earth is felt to be alive, as in the city of Yogyakarta which lies to the South of Java's most active volcano. The largest tropical forest island in the archipelago is Borneo which was a key site in the scientific discovery of the Wallace/Darwin theory of evolution, and it remains one of 25 globally significant biodiversity hotspots. Since the colonial era, Borneo's ancient and biodiverse forests have been largely destroyed for industrial timber extraction and replaced with oil palm plantations. Indigenous people have been driven off their ancestral lands despite their critical evolutionary role in sustaining the biodiversity of Borneo, although their ecological role was neglected by Wallace and is actively resisted by State governments. The roles of the misguided claim that palm oil is a 'climate friendly' biofuel and of international finance in destroying Borneo's forests are discussed, and two contemporary conservation projects in Borneo are discussed which recover the agency of indigenous people and of other animals and plants in protecting and reviving the biodiverse ecosystems of Borneo.

Earthly Agency in the Malay Archipelago

Borneo is the seventh largest island on earth and, along with Sumatra, Java, and Peninsula Malaysia, forms an area known as Sundaland, which is one of 25 global 'biodiversity hotspots' where the long history of species evolution generated the largest assemblies of distinct species in Earth history.[1] The region's exceptional richness in species was first revealed to science by the Victorian traveller Alfred Russell Wallace, who, with Charles Darwin, originated the modern theory of species evolution. Wallace also made a seminal

DOI: 10.4324/9781003110750-3

discovery in the history of Earth System science when he found that the Malay Archipelago is divided by a geological fault-line, running between the islands of Bali and Lombok. On the north side of the line he found predominantly Eurasian plants and animals and, on the south side, he found predominantly Austronesian plants and animals. This discovery helped Wallace to develop the first scientific account of the causes of species distribution and evolution in a paper which he sent to his friend Darwin who, on receiving it, hastily wrote up his own related theory and presented his findings in a joint presentation, with Wallace's, at a meeting of the London-based Linnean Society in Wallace's absence in 1858.[2]

Though the two sets of findings about species variety, and the reasons for their perpetuation, were originally presented as one, a close reading of the two papers indicates that Wallace's account of evolution differs from Darwin's in two crucial respects. First is the seminal role of the Earth as a system, or what James Lovelock later called 'Gaia,' in Wallace's theory. For Wallace the geological and maritime division between Sundaland, and what later came to be called Wallacea, played a critical role in the distribution of species. Scientists now know, as Wallace did not, that this division is explained by plate tectonics, and the role of vulcanism in moving huge land masses on the Earth's crust. But even without this knowledge, Wallace's account of evolution, as indicated in the significant role of the bio-geological 'Wallace line,' gives more agency to the Earth, or Gaia, than does Darwin's in the distribution and evolution of species.[3]

The second difference between Darwin's and Wallace's theory concerns Wallace's greater emphasis on what has come to be known in the social sciences as 'actor-network' theory (ANT): ANT plays a large role in the oeuvre of Bruno Latour, and in his ecological writings he argues that the failure of humans to include other animals as agents in their networks is a crucial cause of the ecological crisis.[4] Both in his field observations, and in his theoretical writings, Wallace emphasised the seminal role of *groups* of organisms in species types in the causes of the distribution and preservation of species while Darwin's theory focused more on the behaviours of individual organisms.[5] Wallace summarised this role, in what he called Sarawak's Law, as follows: 'Every species has come into existence coincident both in space and time with a pre-existing closely allied species.'[6] In other words, and against Darwin's account of species evolution through competition, Wallace found that the distribution of species in the Malay archipelago was such that groups or networks of species of similar kinds were usually found in the same localities, and this suggested that symbiosis between species, rather than competition, had played the major role in their evolution and distribution. Wallace's theory of evolution contrasts with Darwin's because it suggests a degree of networked agency within the nonhuman world that is responsible for the exceptional biodiversity which characterises the assemblies of beings to be found in places such as the forests of Borneo.

Wallace's theory of evolution suggests that what moderns sometimes call 'living things' – which in Darwin's theory are not supposed to have goals or purposes let alone get together and organise themselves so as to have *shared* goals or purposes – have together fostered the exceptional biodiversity he found in Borneo. And, as Wallace also observed, life has done this many times over since the fossil record already in his day indicated there had been a series of events in geological time in which large numbers of species were extinguished.[7] Since Wallace's time, global surveys reveal that the geological record is punctuated by five 'mass extinction events' in which between 30 and 50 per cent of marine and terrestrial species were extinguished.[8] And yet life came back each time with even greater varieties of species than before, culminating in the present geological era in which the greatest range of species that has ever existed on Earth now exists. It is as though the Earth, as Holmes Rolston puts it, is a 'generator of value' which, after each extinction wave, successively generated higher values each time the Earth was subjected to cataclysmic change.[9] But how can this be, since modern philosophers after David Hume argue that only humans generate and impute 'values' such as aesthetic beauty and moral goodness, and to believe that 'things' also do this is magical thinking which rational Enlightened moderns must reject?

That 'life' came back from the dead after five climatic cataclysms five times over with even greater diversity, and that Wallace was among the first scientists to reckon with this, perhaps explains why he was drawn to the Victorian offshoot of Eastern philosophy known as 'spiritualism,' even though this harmed his scientific reputation because scientists are not supposed to believe that 'things,' when alive, and even less when dead, have purposeful, let alone collective, agency. But absent some account of collective memory, or spirit, working within Earthly processes, and the networks of beings that generated the huge range of species he discovered, and so many of which he was the first modern scientist to collect and name, it was hard for Wallace to envisage how these assemblies of beings had composed themselves in all their Malayan and Austronesian magnificence.

The agency of the Earth or Gaia, and of Gaian networks of beings, are the guiding concerns of this present work.[10] The 'Anthropocene' is said to represent a new geological epoch in which one species – *Homo sapiens* – has acquired so much agency, through science and technology, over other species and ecosystems that the evolution of life on Earth is for the first time in Earth history dependent on the actions and intentions of one species. The distinctive condition of the Anthropocene is that wherever humans go, as Crutzen observed when he first made the case for this new geological epoch, they find their own influence – even in Arctic ice cores and deep rainforest.[11] Modern humans are asymmetrically suppressing the agency of all the other beings which in prehistory have composed, and even today continue to compose, an atmosphere, soil bacteria, photosynthetic processes and assemblies of beings in deep oceans, hilly water catchments, forests, savannah, and coastal regions which are the sources of humanity's capacities to breathe, eat, drink, dwell, and make distinctively human habitats such as cities and monocultural fields of corn and rice and wheat. As Bruno Latour argues, ignorance and neglect of the agency of the beings who co-assembled Earth *before* humans evolved, and who continue to sustain a habitable planet, are centrally implicated in the ecological crisis.[12]

The mentality that drives the most influential human agencies – which are State and multi-state globalist agencies, and public and private economic corporations[13] – is strongly influenced by the modern *scientistic* belief, going back to the seventeenth-century occultist and philosopher Francis Bacon who first described the modern scientific method, that humans are the only conscious agents capable of imposing collective purpose on Earth. For Bacon the ability to craft inventions and re-engineer the elements, and even Life itself, are the markers of being human and set humans apart from all other beings.[14] This scientistic belief – that humans are the only rational willing and cooperating agents capable of imposing order on Earth – renders moderns peculiarly insensitive to the agencies of the others whose evolved habitats and practices collectively produced landscapes and ecosystems whose extraordinary beauty and biodiversity remains a cause of awe and wonder.

As a counter to the modern belief that only humans have collective agential powers to act towards shared purposes and goals, James Lovelock and Lynn Margulis' Gaia theory proposes that the Earth herself has agency in the generation and sustaining of life. Although this theory is reminiscent of beliefs in the agential powers of other beings in indigenous and world religious traditions, it is a novel claim in modern natural science. In Indo-European religions, including the Vedas, classical Greek myths including of the goddess 'Gaia,' and even the Genesis creation story as I show in Chapter 7 below, the Earth is given a formative role as the generator of nonhuman life.

While the 'Gaia theory' is most associated with the work of Lovelock, whom I engage more fully below, earlier scientific conceptions of the Earth as a System, as we have seen, go back to the eighteenth-century Scottish farmer and vulcanologist James Hutton.[15] The nineteenth-century German scientist Ernst Haeckel also came close to something like the Gaia theory when he coined the word 'ecosystem' to describe particular Earth habitats, and the organic processes which sustain them as living systems, and so in effect created the science of ecology.[16] Lovelock's Gaia theory is nonetheless new to modern science, and remains controversial, because it suggests that the Earth herself, as well as the beings that dwell on Earth, together composed the aqueous and gaseous conditions that facilitated the evolution of life, including oxygen-breathing mammalian life, in the 'critical zone' which is the life-sustaining surface of the earth in which biota and more complex organisms are able to live.

Lovelock perhaps made this theory – which is now widely accepted but more often called Earth System science or 'biogeochemistry' – harder for modern scientists to embrace by naming it after the Greek goddess Gaia, as suggested by his friend, the novelist William Golding. His intention in so doing was not only to make the theory more memorable but also to give it cultural 'legs' by linking it to premodern and indigenous religious traditions which ascribe living agency to the Earth. This book is an attempt at a dialogue between the Hindu, Hellenic, and medieval Christian religio-cultural sensibility of the agency of the Great Chain of Beings, which includes the Divine origin of all things emanating divine forces or spirits; angels; humans; and all other creatures, including Gaia/Earth, and the scientific theory of Gaia/Earth as a living system or network of agencies. My core claim is that Gaian science will only acquire greater influence over how *Homo industrialis* lives in the precious and inter-galactically unique 13-kilometre-deep 'critical zone' from deep ocean trenches and terrestrial soils to the upper atmosphere, when religions and traditions recover their ancient and deep-rooted beliefs and rituals concerning the Earth, and the living networks of agential beings with whom humans share the Earth, and for the care and conservation of which they have distinctive responsibilities to the divine ground of all Being.[17]

Gaian Agency, Gunung Merapi, and the Geomorphic City of Yogyakarta

Recognition of the Earth as a network of living agents is far stronger in the cultures and religious traditions of Indonesia than it is in England where Darwin lived and from studies of which he developed the principal lines of his theory. Indonesia is situated on the liveliest part of the 'ring of fire' of active volcanoes which surrounds the Pacific Ocean and where the most active current movement of continental plates meet and push up against each other. The terrestrial impacts of this constant movement, in the form of volcanic eruptions, earthquakes, and tsunamis, creates the cultural sensibility of the Earth as a living assembly of agents whose powers may overwhelm human powers at any moment. In Indonesian culture active volcanoes are considered living, quasi-divine, agencies. The most active of all, Merapi, is situated just 30 kilometres north of the Javanese city of Yogyakarta where Universitas Gadjah Mada is situated. Merapi is honoured as a living being not only by animist followers of the old Javanese shamanic religion of *Kajawen*, but also by the Muslim Sultan of Yogyakarta who annually makes a pilgrimage to the slopes of Merapi where offerings of rice, fruit, and meat are made along with prayers in old Javanese.[18]

Yogyakarta, and the large royal palace complex known as the Kraton around which the city grew up, is situated in a notable cosmographic relationship to Merapi since it sits halfway between the top of the volcano and the coast of the Southern Ocean. The slope of the land built up over many centuries of eruptions gently reaches from the foot of the volcano to the sea, serrated by deeply gouged riverbeds which are the contours down which lava principally flows during eruptions. The Kraton in the 'middle place' effects a cosmologically mediating role between the sea, and particularly the 'Queen of the South' who is thought to dwell in the ocean, and Merapi. The inhabitants – over 300,000 people live and farm on the fertile slopes of Merapi – believe that eruptions are a sign of disturbance in the ocean between the Queen of the South and other spiritual forces[19] (Figure 2.1).

This is a spiritual analogy of the geophysical reality since Merapi's frequent eruptions are indeed a result of disturbances in the Earth's crust off the coast of Java where a deep trench indicates the meeting point of the Pacific tectonic plate and the Eurasian plate. The political and spiritual role of the governor of Yogyakarta – the Sultan – is therefore also a cosmo-geologic role since his residence, and the rituals performed there by him and his household, and on the slopes of Merapi and at

FIGURE 2.1 Gunung Merapi seen from the campus of Universitas Gadjah Mada. Author photo.

the coast, are intended to mediate between the spiritual powers that reside in the depths of the ocean, and at its edge with the land, and Merapi herself. The result is a landscape and a religio-cultural 'humanscape' which are both among the liveliest on earth, with frequent eruptions replenishing the soils around the mountain and providing a steady supply of building material in the form of rock and ash which is incorporated in the cement, and walls, of many of the dwellings of the more than 4 million inhabitants of the city and its environs.[20]

The religio-cultural humanscape of Yogyakarta is host to the second largest Buddhist temple complex in Southeast Asia – Borobudur – and to the large Hindu temple complex of Prambanan, as well as to thousands of mosques and suraus, a smaller number of churches, and a few Chinese temples. The result is a city and a region which is both geophysically active and religiously 'alive,' with constant dialectic interactions between the Earth and human culture; between religious traditions ancient and modern; and between religious traditions and the late modern quasi-religious tradition of consumer capitalism which is overlaid on the city and whose dominant forms include motor scooters, Indonesian television, social media, cultural tourism, a huge hotel and hospitality sector, extensive batik, furniture and instrument making, and other workshops, and a hugely productive farming area. Indonesia's liveliest volcano produces a human religio-cultural space which, like the constantly moving flux of the volcanically shaped landscape in which earth tremors are almost daily events, is also constantly evolving and changing. As with the Earth around Merapi, in Yogyakarta also the past is layered under, and sometimes peeks up above, the present: and the post-industrial future – in the form of the digital economy, social media, and so on – also takes shape juxtaposed with a surrounding economy of rural peasants still owning their own land, growing their own food, and engaging in religious practices which are also a dialectical mix of indigenous, world religious, and modern.

Journeys into Highland Borneo

The tropical island of Borneo is roughly half-Indonesian and half-Malaysian and played a seminal role in my ecological awakening during a five-year working residency in Southeast Asia in the 1980s. That encounters with the Earth and particularly 'wild' or 'remote' places – wild and remote relative to modern urban life that is – play a crucial role in the development of ecological perception, and what is now called 'environmentalism,' is recognised by the romantics going back to Samuel Taylor Coleridge and William Wordsworth, and in the romantic-inflected scientific writings of Alexander von Humboldt. Humboldt draws particular attention to the Tropics, or what he calls the 'torrid zones,' as places of discovery of nature's laws because:

> (T)he regions of the torrid zone not only give rise to the most powerful impressions by their organic richness and their abundant fertility, but they likewise afford the inestimable advantage of revealing to man, by the uniformity of variations of the atmosphere and the development of vital forces, and by the contrasts of climate and vegetation exhibited at different elevations, the invariability of the laws that regulate the course of the heavenly bodies, reflected as it were, in terrestrial phenomena.[21]

Walking and staying in tropical forests; swimming in tropical rivers; climbing jungle-covered hills; hearing the early morning calls of lemurs and hornbills; seeing epiphytic plants, including orchids and ferns, growing in the branches of huge trees whose buttressed roots and leaves form a forest floor teaming with life in the form of myriad species of beetle, centipede, millipede, leech, mosquito, ant; watching butterflies of different hues flit from leaf to flower; hearing the cacophonous night sounds of a tropical forest including the ever-present cicadas rubbing their wings; and seeing fireflies flickering in the dark: this is to experience a diversity of life forms constantly interacting and rapidly evolving in ways that it is rare to experience in Boreal forests, and particularly those of the Northern hemisphere most of which have been managed and reconstituted into either bare land – as in the hills of Scotland in my native island – or into Continental European and North American forest monocultures – and especially fast-growing Sitka spruce and pine forests – favoured by scientific foresters from the eighteenth century onwards.[22]

I taught at Seminari Theologi Malaysia, Kuala Lumpur between 1984 and 1989 and it was in those years that I first encountered tropical forests, and saw with my own eyes, heard with my ears, and felt under my feet both the incredible aliveness and diversity of tropical forests, and at the same time the devastating effects of clear-cutting of these forests by the tropical timber industry, and their subsequent conversion into the monotonous monocultures of oil palm plantations. Tropical forest clearance exposes thin red earth to soil erosion from heavy rains which rapidly form deep gouges in the thin forest soil stripping the Earth's surface back to red bare earth, while rainwater falling on denuded hills is no longer retained by forests and drained into subterranean aquifers, but instead pours off the hills in torrents that frequently flood human towns and farmlands in the plains below.

At the same time as the forests are cleared, the clouds that usually form and condense water from the oceans above the forests dissipate: this natural 'air conditioning' system is in effect 'turned off' and consequently human tropical settlements in deforested regions once cooled by forests are not only more exposed to flooding but are also too hot for comfortable human habitation. In city apartments and town houses built of cement, it becomes so hot that it is not possible to sleep comfortably without fans or air conditioning whereas in the forest all that is needed is a mosquito net or screened windows and the night temperature is perfectly comfortable for sleeping. This local form of 'climate change,' combined with the denuding of the huge range of plants and species that live in tropical forests, awoke me at first hand to the two dominant features of the larger ecological crisis, which is to say, biodiversity decline and *local* climate change.

Among many exposures to ecological change from destructive human interventions in Southeast Asia, the most seminal was a boat trip I took with my family along the Rajang River in Sarawak, North Borneo, in 1987 with two of my students. After flying into the bustling Chinese town of Sibu, where we ate among other things baby ferns which are a distinctive local delicacy, we travelled up-river in a flat-bottomed metal boat with a huge in-line six cylinder engine and a long shaft propellor balanced on the rear gunnel. This delicate arrangement propelled the substantial boat at some speed while it allowed the propellor to be lifted to the surface of the water when traversing shallow rapids. The river is wide and gave the impression of being surrounded by tropical forest on both sides though, as we learned later, forestry and oil palm companies leave a small area of intact forest on the riverbanks even while devastating the forest just a few metres from the water's edge.

We stayed for some days in the logging town of Belaga and visited two riverside Dayak longhouses, which are 100-metre-long communal houses on stilts in which each family has its own private dwelling attached to a wide and long common hall running the length of the house where children are nurtured and play, food is prepared, baskets and fabric are woven, meals eaten, and cultural and religious ceremonies are held. We also went one day on foot to a temporary bamboo jungle shelter of the kind commonly constructed by the nomadic Penan people. We were taken by local people to their 'swidden' farming areas in which wild rice, vegetables, and fruits are grown, and enjoyed meals, sitting on the floor cross-legged with our small children, of these deliciously cooked and freshly picked foods.[23] The longhouses showed signs of Western cultural influence, including posters of Michael Jackson and television sets, but the people sustained many of their traditional cultural and religious practices, including communal cooking, child rearing, and ceremonial performances, as well as crafts such as weaving.

In Belaga we met a doctor who worked for a Christian mission agency and who recounted her visits by helicopter to Dayak and Iban communities living in the highland interior. She described how children were suffering from malnutrition because the fish, mammals, and plants their parents used to gather and hunt were no longer available. The fish in the rivers had died as the quantity of soil displaced by logging activities had filled the rivers with silt, depriving the fish of oxygen. The mammals and plants the communities had once hunted and gathered, as had their ancestors for thousands of years, had disappeared because of the clear-cutting practices of the logging companies. It is possible to remove the largest trees from tropical forests without destroying forest ecosystems and in such a way as to permit the forest to regenerate. However, in Sarawak and Sabah, and in Kalimantan, the Indonesian part of Borneo, the forest is cleared not only to remove tropical timber for sale but to convert the land into oil palm plantations, coal mines, and rice fields.

Often when the largest and most valuable trees have been extracted, the remaining forest is burned as this is a more economical way of clearing the land than removing the remaining trees and undergrowth before draining the land and terracing it in preparation for oil palm plantation (Figure 2.2). So extensive

FIGURE 2.2 Recently deforested peatland drying out in preparation for an oil palm plantation, Mount Ophir, State of Johor, West Malaysia, December 2019. Author photo.

is the burning of forests that the cities of Sumatra, West Malaysia, Borneo, and Java – including Medan, Padang, Singapore, Malacca, Kuala Lumpur, Penang, and Jakarta – are shrouded for days or even weeks at the height of the dry season in August and September by harmful smoke and smog. In the worst year on record for smog – 1997 – most of Southeast Asia was shrouded in smoke and this large-scale pollution incident occasioned a sudden downturn in regional stock markets which precipitated a financial crash throughout Southeast and East Asia that impacted Korea and the Philippines as well as Myanmar, Cambodia, Vietnam, Laos, Thailand, and the Malay Archipelago.[24]

Tropical Deforestation and the Spread of Biofuel Crops

The lowland tropical forests of Borneo and Sumatra – Sumatra being the other large island currently being cleared for oil palm in the Malay Archipelago – grow mainly on top of very deep peat and in some places the surface fires have led to subterranean peat fires which are impossible to extinguish. When the peat is not burned, it is dried out by drainage channels which are a common feature of oil palm plantations since the oil palm is not a native plant and needs drier soils than those sustained by tropical rainforests. The burning of the forest, together with the drying out of the methane-rich peat, produces vast regional greenhouse gas emissions. The quantity of greenhouse gas emissions from fires from land being cleared for oil palm plantations is so vast that Indonesia, where most of the burning and clearing now goes on, is the eighth-largest global greenhouse gas emitter.[25]

Oil palm is widely used in industrial products including cooking oil, vegetable shortening, detergents, cosmetics, tea lights, and candles. It is also used to make liquid fuels – or 'biofuels' – for vehicular engines. Biofuels are made from several plant monocultures including oil palm, soya, rape, corn, sugar beet, sugar cane, and wheat. Oil palm is the largest source of biofuels in the tropics. The huge land area globally already devoted to such plants, together with the widespread use of full-spectrum pesticides and herbicides on plantations, is a major driver of tropical forest destruction, and hence of biodiversity decline and species extinction.

The global market for biofuels is advanced by the claim made by, among others, the Malaysian and Indonesian governments and the European Commission, that biofuels are forms of 'renewable energy' which have lower 'carbon footprints' than fossil fuels despite the clear association between oil palm cultivation and tropical forest clearance.[26] But far from reducing greenhouse gas emissions from motorised vehicles, the substitution of fossil fuels with biofuels actually increases them. This is because the greenhouse gases emitted by the combination of tropical forest clearance, peat, and other fires, the subsequent drainage and drying out of soils, and the production process for biofuels, are higher from palm oil and other biofuels than from conventional crude oil and natural gas wells.[27] Recognition of this led the European Parliament in 2017 and 2018 to call on the European Commission to remove palm oil from European biofuels by 2020 although the governments of Malaysia and Indonesia, the principal global producers of palm oil, resist this proposal.

Corruption in the Forests

Oil palm plantation companies are closely connected with the governments of Malaysia and Indonesia. In Malaysia, the party that has governed almost continuously since independence in 1957 – the United Malay National Organisation – has extensive holdings of shares in plantation companies, while a Malaysian government plantation agency – FELDA – is a major producer of palm oil. This helps explain why, when Malaysian cities including Penang, Ipoh, Kuala Lumpur, Malacca, and Johor Bahru are shrouded in extremely unhealthy levels of particulates from forest fires in Sumatra and Borneo, the government makes little fuss since the fires are set in many cases by subcontractors acting for Malaysian-owned companies.

The conversion of so much of Borneo from ancient and highly biodiverse primary forest to oil palm has occasioned an ecological disaster on an unprecedented scale with thousands of endemic species now at risk of extinction from loss of habitat.[28] Behind this disaster there is systemic political corruption, extensive criminal activity, and destruction of the habitats of native peoples and wildlife. In Sarawak, the people most affected by the destruction of the forests are the nomadic Penan people. Though under the Malaysian customary law, or *adat*, the Penan ought to be considered as owners of most of the highland forests of Sarawak since they have lived in them for generations, this customary ownership is not recognised by the Sarawakian government, and there is a similar state of affairs in Indonesia which also refuses to recognise that those whom it calls 'remote peoples' can be legal owners of land.

The Sarawak State Government was led from 1983 to 2014 by Chief Minister Abdul Taib Muhamad and he declared the government he led as 'trustee' of the forests of native peoples for 60 years. 'Taib,' as he is known, also dominates the forest trade in Sarawak and amassed a vast personal and family fortune by a range of devices. These included granting licenses to log forests to companies owned by Taib's family; receiving payments from other logging companies for logging concessions; and the receipt of export 'fees' to a Taib-family company registered in Hong Kong to whose port most of Sarawak's timber is shipped, for onward shipping to timber-importing nations in the region, and most notably Japan, Taiwan, Korea, and China. Finally in the wealth chain, much of the land, once logged, is turned into oil palm plantations that in many cases also become assets for members of the Taib family, whose product ends up in biofuels as well as in cooking oil, cosmetics, detergents, and processed foods.[29]

The destruction of the forests led to extensive protests and resistance by the nomadic Penan people and other indigenous groups, including logging blockades. Indigenous peoples also raised claims in Malaysian courts for recognition of their land rights based upon extensive mapping projects undertaken by the Iban and Penan with help from the Swiss Art for Rainforests Foundation and the Bruno Manser Fund. Bruno Manser was a Swiss adventurer who lived extensively with the Penan in the 1990s and campaigned on their behalf internationally. In response the Sarawakian government *outlawed* land surveying and mapping other than by government surveyors, and Manser was declared an illegal immigrant and eventually disappeared in mysterious circumstances in the year 2000 on land belonging to Samling, the largest Sarawakian timber company.[30] Individual Penans attempting to protest or raise legal actions against logging companies have also been killed, while Penan women, and especially young girls, were often raped by timber company employees.

The interaction in Borneo between criminality, corruption, primary forest exploitation, and wealth accumulation is facilitated by banks, investment agencies, and a range of technologies.[31] Deutsche Bank, Goldman Sachs, and other international financial agencies provide investment capital for new tropical forest 'conversion' projects, and the financial means to offshore illicit funds from such developments including ultimately into real estate in North America, Australia, and the United Kingdom.

The Ambiguous Role of Climate Science in the Destruction of Borneo

In addition to the role of international finance in funding and legitimating these activities, there is the role of international science to consider. Science is on both sides of the story of the destruction of Borneo. On the one hand, climate science has been used by American and European legislators to promote biofuels as a response to global climate change, and this has provided capital opportunities for primary forest conversion to biofuel production. On the other hand, climate scientists and conservationists have used satellite imagery and local oral testimony to chart the local and global climate impacts of tropical forest conversion and to challenge the claims that biofuels are more 'climate friendly' than fossil fuels.

Under the auspices of the UN Framework Convention on Climate Change, nations agreed to monitor and publish inventories of national greenhouse gas emissions from their territories. But such inventories are sites of mediation and manoeuvring by governments in pursuit of national interests. The Malaysian government's inventory counts oil palm plantations as 'forest': in official statistics 64 per cent of the land area of Sarawak is 'forest' while less than 10 per cent of primary or secondary tropical forest remains.[32] Labelling oil palm plantations as 'forests' reduces the public visibility of the gradual conversion of the forests of East and West Malaysia from tropical rainforest to oil palm, though it is evident to anyone who flies over these lands since oil palm plantations now stretch as far as the eye can see in lowland and parts of upland Sabah and Sarawak, as in large parts of Kalimantan, Sumatra, and West Malaysia.

The nexus of science and technology engaged in the measurement of climate change, and in generating policy responses to it – including 'renewable' energy – is scientistic and globalist. It fails to look holistically at the whole-Earth impacts of the 'solutions' it promotes to climate change, and promotes the destruction of the biodiversity of particular ecosystems in particular places in the name of the global problem of greenhouse gas emissions, which in reality are made worse locally and globally by such 'renewable energy' projects as biofuel monocultures substituted for tropical rainforests which store carbon, are home to a huge array of diverse species, and which cool the local climate.

Indigenous Resistance to 'Environmentality'

Juxtaposed against globalist scientism are the traditions, customs, and beliefs of native peoples whose ancestral forests are destroyed to grow biofuels. The common assumption among governments, corporations, science-informed foresters, and many Western 'conservationists' is that indigenous people are not good guardians or managers of forests and need to be excluded from living in or using them to 'preserve them' for posterity. Hence when Yosemite in California was made a National Park, among the first acts of the US Federal Government National Parks Authority was to evict the indigenous people who had lived there for thousands of years and managed the environment. Only after they were excluded did the National Park Authority realise that these evicted humans had provided 'ecosystem services' for the 'park' – such as setting fires and limiting deer numbers through selective hunting – and that these services had subsequently to be provided by National Park employees if the park was to be maintained in its 'pristine' state.

Seven per cent of the land area of Borneo is designated National Parks from which indigenous peoples are also formally excluded.[33] But outside of that area the government claims 'trusteeship' of much of the rest of the forest – both lowland and highland – under the ruse that it stewards the land on behalf of the people with the aid of science-informed conservation practices. And once indigenous people are excluded from their ancestral lands, most of the 'State-stewarded' forest is logged and converted into biofuels plantations. In Sarawak the only areas that have been saved from destruction outside of the gazetted boundaries of National Parks, which constitute only 6.6 per cent of the forested area of Sarawak, are areas protected from deforestation by the non-violent actions of native peoples, and in particular the Penan. Using logging-road blockades, warning posters, and bamboo structures, while at the same time registering land claims in Sarawak courts, the Penan successfully saved 163,000 hectares of upland rainforest in the Baram District. Penan villages in this area have collaborated to form the 'Penan Peace Park.' The ancestral lands and their ecosystem and living creatures are intact and now designated by the Penan as a tribally governed nature reserve though it is not included on official maps of nature reserves in Sarawak. As a result, the Penan are able to continue to hunt and fish there and plant small gardens and fruit trees.[34]

The case of the Penan highlights the issue of what Arun Agrawal calls 'environmentality.' This is the tendency of modern environmental, atmospheric, and species-conservation science and policy-making to be used to justify governance regimes on peoples, creatures, and habitats in which agency is denied to both human and nonhuman inhabitants.[35] The increasingly technocratic tendencies evidenced in the COVID-19 pandemic, and the broader centrally driven policies of the United Nations' 'Agenda 21' and 'Agenda 2030,' including their globalist and anti-ecological free trade proposals, and 'Sustainable Development Goals,' are part of this larger pattern in which multinational corporations and agencies use the ecological crisis to close down the historic agency and freedoms of ordinary people *and* species.

Religion, Ecological Perception, and Conservation Biology

The limited success of the small Penan tribe – which numbers around 20,000 – in protecting some of their ancestral forests is an example of 'environmentalism from below' and of the 'moral force of indigenous politics.'[36] By resisting the post-colonial regime of a corrupt government and crony capitalists, and forcibly blocking logging roads, the Penan asserted their agency as carers of the Earth. In so doing they resisted the consensus in West and East Malaysia, and in Indonesia, that indigenous peoples only acquire recognition as citizens, and possessors of land rights, when they come out of the forests and live like 'moderns' in permanent houses and adopt the cultures and religious traditions of the nations that claim to own their forests. Courtney Jung argues that through contestation and resistance to such a consensus indigenous people become agents, subjects, in the public sphere and hence *contestation* – and not only democratic governance and the law courts – is an essential source of justice in societies where minorities are excluded from full participation in normal legal and political processes.[37]

The Penan are not what moderns call 'environmentalists' since they are neither students nor agents of environmental science. Instead, they resist it when powerful interests invoke science to collude with crony capitalism in the theft and destruction of their forest homes. The Penan's resistance is an example of what Joan Martinez-Alier calls the 'environmentalism of the poor.'[38] The Penan are not protesting the destruction of the rainforests because they are a 'sublime wilderness' unmodified by humans as John Muir described the tropical forests of Latin America when he visited them. Nor are they principally protesting the impacts of this destruction on endangered species such as orangutans or pygmy elephants. They resist logging and the spread of biofuel plantations because the forests are their home, their livelihood, their indigenous wealth, and the modern turn of their habitat into liquid fuel, or 'capital,' leads to their environmental exclusion from the lands of their ancestors in a process analogous to the spread of dispossession and penury which accompanied the Enclosures of common fields and forests in England from the seventeenth century in the name of 'agricultural improvement.'[39] The Enclosures set in train a global process of land theft and de-peasantisation which spread from England to the colonies, and was also replicated in parts of mainland Europe. In the post-colonial era, the same process has continued, now directed by nation States in partnership with agricultural and energy corporations, and globalist agencies such as the United Nations and private wealth Foundations, and under the aegis of promoting global free trade, scientific progress, and 'sustainable development.'

The origins of the idea that humans have a responsibility to protect habitats for their own wellbeing *and* that of other animals are untraceable since they lie in what today are called 'indigenous cultures' which were pre-literary. But that indigenous peoples from early in the history of the evolution of *Homo sapiens* developed cultural practices which limit their capacities as a 'keystone species' to prey on other animals may be surmised from evidence assembled by anthropologists in their studies of ecological knowledge and practices of contemporary hunter gatherers. This is not to say that indigenous peoples did not hunt some species to extinction. There is good evidence of a correlation between the extinction of certain large vertebrates and the spread of humans into their habitats, including the disappearance 8,000 years ago of woolly Mammoths in Siberia and North America, and the disappearance of the large flightless Moa birds in New Zealand 700 years ago.[40] But indigenous people have a record of living long term in ecosystems in which the vast majority of species endure alongside them and this record is at odds with the record of industrial moderns whose modes of habitat modification, land use, hunting, food growing, and dwelling express a systematic tendency to create conditions favourable to a very small number of species, so reducing overall species diversity.

Indigenous Traditions and the Guardianship of Species

North American Anthropologist Roy Rappaport pioneered the study of the ecological function of indigenous religion when he investigated the pig-keeping and sacrificial customs of a tribe in Papua New Guinea, and found that their religious practises enabled them to keep the numbers of their domesticated animals within the carrying capacity of their ecosystem habitat.[41] Analogously Tim Ingold charts the depth of ecological perception among sub-arctic indigenous peoples, as expressed, for example, in their hunting and food-sharing rituals. He shows that indigenous peoples in the sub-arctic regions 'develop their skills and sensitivities' as foragers, herders, and hunter gatherers 'through histories of continuing involvement with human and nonhuman constituents of their environments' and that there is no clear divide between human 'mentality' and nonhuman beings and things.[42] Instead mentality emerges from the relational nexus in which indigenous people live and dwell, human and nonhuman.

A related, and converse, observation can be made of modern urban humans for whom inanimate things, more often than beings, are crucial parts of their mentality, especially in the acquisition of life skills: the most obvious example is the 'smart' phone, though in fact this is increasingly an animate object given its capacities for aural, haptic, and visual mimicry of embodied communication.

Other 'formative' relational objects for modern urban humans include clocks, cutlery, clothes, computers, communication devices such as radios and televisions, and mechanical vehicles. Such technological artefacts play roles in developing and extending human mentality through the skill set that enables them to co-construct, communicate and dwell within their material and social worlds. But mechanical 'things' are not 'alive' in the way organisms are, and increasingly they are made as 'black boxes' whose inner workings are hidden to the user and non-repairable, and so are resistant to negotiation.[43]

Unlike moderns, for indigenous people, the extent to which the 'things' that co-construct their worlds are other living beings means that their acquisition of skills and artefacts for dwelling, making, socialising, and sharing in the world involves myriad acts of negotiation with other living beings. Hence indigenous cultures, while they have, like all human cultures, impacted on some dominant species to the point of extinction, nonetheless manifest capacities for co-dwelling with other beings in ways that do not destroy their habitats and capacities to nurture their progeny. The traditional ecological knowledge of indigenous peoples, including their evident skills in managing habitats in ways that better preserve the biodiversity and resilience of ecosystems than modern agricultural and industrial habitat management, was for centuries disregarded or dismissed as anti-scientific superstition by colonists, missionaries, anthropologists, and scientists. As British explorer Robin Hanbury-Tennyson puts it in an account of an expedition to Sarawak to map its geography and plant biodiversity in 1978, the centuries-long trusteeship exercised by the indigenous peoples of Sarawak. What

> makes the forests so valuable and worth turning into a park is that there have been people looking after it in a symbiotic relationship for ages. And what is more, as we were increasingly to learn on our expedition, their knowledge and understanding of it far exceeded our own superficial scientific analysis.[44]

In a subsequent talk to the Royal Geographical Society, Hanbury-Tenison

> paid tribute to the immense scientific knowledge of the local tribes people, including Penan and Berawan, who he said knew far more about the species, uses and life-pattern of the creatures of the jungle than any of the distinguished foreign scientists who had joined the expedition. … Yet, he pointed out, they had been mocked and discriminated against in local schools at the time and their special knowledge and intelligence was woefully undervalued. [45]

He also lamented that,

> When we went in [to the jungle] in 1978, 95% of Sarawak's low-land rainforest was intact and now forty years later only 5% of low-land rainforest

> remains. So we failed in our mission to explain to those in authority that it was counter-productive to destroy the rainforest.

It was not religion that destroyed Sarawak's ecosystems and biodiversity but the rapacious greed of one man and his family. That the destroyer is a Muslim raises the larger question of whether what are called 'world religions,' including the Abrahamic faiths of Christianity, Islam, and Judaism, have the same capacities as indigenous religions to enable their followers to conserve ecosystems and species through relational and traditional knowledge of the fellow beings who inhabit them.

That the Abrahamic faiths, and their 'de-animating' and rationalising legacy, particularly in the modern scientific method and the European Enlightenment, paved the way for the modern technological quest to control and dominate the Earth was first proposed by Horkheimer and Adorno, and followed up by Lynn White Jr. in a famous essay discussed in Chapter 7. But in weighing the evidence for that argument in Borneo, it may be pointed out that the colonial dynasty of 'Rajah' James Brooke were nominal Christians, and they allowed the trusteeship of indigenous people in Sarawak to continue almost without disturbance.[46] It is a similar story in Kalimantan, or Indonesian Borneo. Under the nominally Protestant Dutch administration of Kalimantan most of the highland and lowland forests were conserved intact and their indigenous inhabitants left relatively unmolested to dwell, forage, and hunt in the forests until the end of the colonial era. It was the advent of Indonesian independence in 1945, and the post-colonial and Islamic administrations of Presidents Sukarno, and Suharto, and their successors, which began the systematic destruction of the forests of Kalimantan, for coal and timber extraction, and the conversion of much of the island into oil palm plantations.

Post-colonial Resource Extraction and Protected Areas

The approach taken by post-colonial administrations to the environment, in Borneo, and more broadly in Malaysia and Indonesia, reflects the same tension as that of colonial administrations between conservation and 'primitive accumulation' through 'natural resource' extraction. The new post-colonial elites were in a rush to accumulate capital with which to build the cities and service infrastructure which colonial administrations in the main had failed to provide outside of important trading centres such as Batavia, Malacca, Kuching, Singapore, and Penang. Post-colonial administrations adopted a two-pronged strategy. On the one hand they pursued resource extraction over extensive parts of the land area newly available to them for exploitation and primitive accumulation, while on the other they sought to consolidate and expand protected areas that had been first set aside by colonial administrations.

These protected areas originated in three ways, which reflected the history of protected areas in the Old and New Worlds. The first were areas of forest or savannah that were set aside as hunting grounds for Europeans and where birds or mammals that are particularly suitable for recreational hunting are to be found. In the colonial era such land areas might be either leased by a group of private individuals or set aside by colonial administrators from peasant agriculture or indigenous use. The majority of 'big game' parks in Africa go back to colonial times, and in Indonesia and Malaysia it was also the case that among the first areas to be preserved from deforestation and conversion into agriculture were areas with relatively large populations of animals suitable for hunting.[47]

The second root of protected areas is European nature romanticism, and originated in projects in England, Germany, the Netherlands, and North America to protect landscapes identified as 'sublime,' and especially mountainous regions, by nature romantics – first among whom is Samuel Taylor Coleridge who coined the term 'mountaineering' and is recognised as the originator of the modern practices of mountain climbing and hill walking. With his friends, including William Wordsworth, John Ruskin, and Thomas Carlyle, Coleridge was a member of the first nature protest organisation set up to defend the Thirlmere valley in what is now the English Lake District from being turned into a reservoir by the Manchester City Corporation.[48] Subsequent campaigns led to the setting up of the English National Trust, which is one of the oldest nature conservation organisations, and in whose name parts of the Lake District, and many other highland areas in England, and later Scotland, were set aside for recreational use and aesthetic enjoyment. John Muir took up the same approach when he campaigned for Yosemite to become a National Park in the late nineteenth century, and the idea of protected areas spread to other European colonies including Indonesia and Malaysia.

The third root of protected areas is 'scientific' forestry which was invented in nineteenth-century Germany where forests were managed as plantations for the provision of masts and lumber for war and trading ships, and for city dwellings. The practice of monoculture was introduced into silviculture in Germany in the nineteenth century as it was believed to be more 'efficient' than mixed forests in growing timber for construction and military purposes, and the practice was widely adopted in Europe and European colonies including in Southeast Asia.[49]

In the Indonesian archipelago there was recognition in the second half of the nineteenth century among Dutch foresters that certain highland areas of mixed forest should be set aside as protected forests from monocultural forests such as teak and rubber tree plantations which led to the Batavia administration setting these areas aside in a forest ordinance in 1858.[50] In Borneo, the colonial administration of the Brookes did not see the need to do this since they left the forests largely undisturbed and governed, as they had been for centuries, by indigenous Dayaks, Ibans, and Penan. Only riverside settlements and coastal areas were seen as important for the model of trading entrepôt which Brooke promoted to fund his colony, in partnership with the East India Company.

The independence achieved by Indonesia and Malaysia in the post–Second World War era gave rise to post-colonial environmental governance regimes in which, as already indicated, there was a new nationalist and economic imperative to accumulate wealth for the new nations and their elites. Hunting areas, forests, and especially biodiverse or 'sublime' ecosystems, such as highland rainforest and coral reefs, were designated as national government protected areas. But in this way, post-colonial environmentality rejected the traditional, or '*adat*,' guardianship of these areas by indigenous people which, alone among colonial administrations in the region, the Brooke 'rajahs' had sought to honour. This has meant that the integration of indigenous ways of living with the preservation of intact and biodiverse ecosystems has been displaced on a huge scale throughout the Malay-Indonesian archipelago, and not only in Borneo.

Contemporary 'nature conservation' in Borneo, and in Malaysia and Indonesia more broadly, is focused on the protected areas that have been so designated in the post-colonial era even as nationalist governments have stripped bare ecosystems in large parts of the archipelago. As with conservation biology elsewhere, this has meant that conservation has tended to focus on the twin priorities of attempting to prevent the extinction of particular endangered species, and restoring the particular habitats that endangered species need in order to endure despite the considerable shrinkage of their range areas.[51] But this reductionist approach focuses exclusively on scientific practices in 'protected areas' in which non-scientific human agency is problematised even although the historic record shows that indigenous humans have a much better track record of species and ecosystem conservation than scientistic humans who, when they claim to 'protect' areas formerly inhabited by indigenous people, typically reduce species diversity and especially when protected areas involve evictions of indigenous people from their lands. Modern scientific conservation also dismisses the possibility that not only indigenous people but nonhuman species are capable of expressing agency in their own conservation and the conservation of their own habitat, something I discuss more fully in Chapter 8. But not all conservation projects dismiss the agency of indigenous people, and one project I visited in Borneo employs local people in habitat restoration and works carefully to honour and enhance the agency of other primates in the area it manages.

Agential Encounters in the Forest

I had the occasion in 2008 as Guest Professor of the University of Malaya to visit a species-conservation project in Sabah – the smaller of the two Malaysian States in Borneo – under the auspices of the Biology Department of the University of Malaysia, Sabah, where I had given a lecture on religion and ecology. The Department of Biology arranged for me to visit the Kinabatangan Forest Reserve, which is an area of mostly secondary forest on the upper reaches of the Kinabatangan River, and site of an orangutan conservation project run by HUTAN. HUTAN was founded by French primatologists Marc and Isabelle

Ancrenaz in 1998. Working with the Sabah Wildlife Department, and Universiti Malaysia Sabah, HUTAN established and maintains the Kinabatangan Orangutan Conservation Project (KOCP) in the upper reaches of the Kinabatangan River. KOCP has as its core objective 'to restore harmonious relationships between people and the orangutan, which in turn will support local socio-economic development compatible with habitat and wildlife conservation.' The project aims to achieve this by employing and educating local residents and indigenous people who still live in the area in an orangutan conservation and habitat restoration project.

The principal work of the project has focused on efforts to protect and enhance the mostly secondary forest in which Orangutans live in the Kinabatangan River catchment.[52] This area was logged more than 50 years ago in a process in which the largest trees are extracted and floated downriver to sawmills for sale as hardwood timber, or to be thinly sliced and steamed into plywood sheets for the construction industry. Logged forests are often subsequently burned and cleared for agricultural plantations. But this forest had been left as 'secondary' jungle in which sapling trees, bamboo, and other species had continued to grow. Orangutans are fewer in number in logged forest as the trees are smaller and weaker, so not so robust for nesting for these large animals, and there is less fruit to eat. The HUTAN project objective is to demonstrate that orangutan numbers can be enhanced in secondary forest provided it includes a good range of trees of sufficient size in which the animals can nest, and of fruit trees that match their preferred diet. This is achieved by selective planting of fruit trees in formerly logged forest. By increasing potential orangutan habitat, the project aims to help prevent the extinction of orangutan, whose numbers in Borneo and Sumatra – its remaining wild habitats – have shrunk to just over 100,000 individuals in Borneo and 14,000 in Sumatra, due principally to habitat loss from logging and land clearance principally for oil palm plantations.

KOCP employs 40 local people in managing a project which includes a research station, and an eco-tourism project. Project scientists and guides chart the presence of nests and know most of the individuals who live around the station. The project is increasing scientific knowledge of these animals while also enhancing the orangutan survival chances through habitat restoration. To this effect the project employs four full-time tree planters. Orangutans eat around 300 species of tree and they are essential to the germination of some of the tree species whose seeds expand and absorb moisture and nutrients from the faeces of orangutan after being eaten and deposited on the forest floor. To mimic this evolved process, the seeds of these tree species are therefore premixed by human hands with collected Orangutan faeces before planting. The areas where the planting is done are protected with electric fences or else Borneo pygmy elephants – another endemic and endangered species – will eat the saplings. The tree planters are all local women. Experience shows that women are better than men at tree planting: they are more careful to clear the ground, especially 'Lalang' – long razor-sharp grass which grows abundantly in forest clearings

– where the sapling is planted, and to dig and water it in properly, and so their plantings are more successful.

KOCP employs local people both because they know the area and live nearby but also because conflicts between local people and orangutans grow as the habitat and range area of orangutans, and of available forest for indigenous human inhabitants, declines from the spread of palm oil plantations in which neither orangutan nor indigenous people can live. The KOCP also use indigenous guides for their eco-tourism project. Such links between the project and local cultures are designed to promote indigenous resource management so that local people, instead of being excluded from deriving material benefit from protected parks and conservation projects, are actively engaged in employment and educational opportunities linked to species conservation.

I went into the Kinabatangan Forest Reserve on a morning in May 2008 in the hope of seeing orangutan in a restored natural environment. I was however to be frustrated in that goal, albeit rather delightfully frustrated, by the rare presence along the river that day of a group of pygmy elephants that were passing through the district. When I got off the small boat on which we travelled to the tree planting area from the Project's riverside lodge, our guide soon explained that orangutans were making themselves scarce because the herd of elephants – around 20 in all – were moving through the area. Consequently, our guide decided to call on two others from the project who are local elephant guides in order that we might have the opportunity of walking safely through the area. In due course our elephant guides came and led our small group of three visitors to a small tributary river across which we saw through the trees a group of elephants on the other side. We stood still and the group trumpeted, turned, and ran off into the woods. We held our ground. A few minutes passed in the intense and humid midday heat of the jungle. Then the elephants returned again to the side of the stream across from us. At this point they wagged their heads and trunks from side to side, and our guide whispered that this meant, in elephantine communication, 'it's OK' and 'they might be friendly.' Then one of the group, a six-year-old female about a metre and a half in height and a metre in width, left the herd and approached us across the stream. As she drew near, she turned and began to walk backwards, and then raised a back leg in the air and wiggled it. At this one of our guides moved forwarded and spanked her on her large rump. She ran away a few metres and undertook the same procedure again, with the same response. She then came again, and this time swished her tail which our guide then held and swished in turn. Eventually she began to approach us frontally and we found our small group being chased through the undergrowth, effectively playing 'catch' with an elephant. After about 10 minutes, though it seemed a lot longer, the rest of the herd began to take an interest in joining in, and one of the large males, tusks and all, standing around 3 metres tall, began to walk towards the younger juvenile with the apparent intention of joining the game. And at that point our guides suggested we back off as things could get a little too interesting (Figure 2.3)!

FIGURE 2.3 Pygmy elephant, Kinabatangan River Conservation Area, July 2008. Author photo.

After this primeval meeting in Sabah, I reread the second creation story in Genesis 2 in which Adam is said to have been invited by God to give names to his fellow creatures. Were the animals slaves of the divine will who could do no other, which is the suggestion of a fourth-century Syrian bishop and theologian, Severian of Gambala, and of Augustine of Hippo? Or did they, as did the elephants in Borneo, approach Adam naively but volitionally, as a fellow mammal? The elephants did not know the humans they saw in the forest as *Homo industrialis*, who burns their habitats down, or kills them with a gun. They discerned in us fellow primates, communicative agents, wanderers who crossed one of the routes of their nomadic lives. They recognised us. They chose to play with us. They sought us out as companion animals. I read the Genesis story of primeval naming – which is the origin story of all scientific taxonomies – in a new way after this encounter. I saw that it is reasonable to suppose that it was memories of such naive meetings, and primitive companionship unmarked by fear between humans and other animals, that were recorded in this primeval story of species identification and naming in Eden.

For me this meeting in Borneo was a spiritual experience. The most striking aspect of the whole encounter was the agency expressed by the juvenile female in crossing the river and coming to engage us in play. It was not just that she was not afraid of us. I have encountered flightless birds on small islands off the South Island of New Zealand where they have been reintroduced, having been bred in captivity by the New Zealand Department of Conservation (DoC), after DoC scientists have eradicated introduced mammals such as rats and possums from the islands.[53] But these flightless birds express no fear of humans because they have no evolved concept of 'human' or even 'mammal,' since New Zealand had no mammals until Polynesians and then Europeans arrived with rats, dogs, and possums, and so these flightless birds do not register humans as meaningful subjects. The elephant, on the other hand, clearly registered us as fellow mammals and made the decision, as a free mammalian agent, that we had the potential to be playmates, and so engaged us.

Despite global concern at the rising tide of extinctions, conservationists are not stemming the tide of extinctions caused by marine and terrestrial habitat destruction of the kind that is occurring still in Borneo. The research question, which led me to the ontological claims I go on to unfold in this book, is 'why not?' In an earlier paper I argued it was because of the economic and political power of corporate agents, and especially multinational and government agencies such as fossil fuel and chemical companies, in crowding out the agency of conservation agencies and environmental regulators.[54] But the problem is deeper than that. The proposal I present in this book is that globalism, industrialism, and scientism are imperial forces spreading their monolithic and often misanthropic as well as anti-ecological tentacles into every human and nonhuman habitat on Earth. Against these oligarchic post-colonial forces, I propose in what follows that the Gaia theory, and the 'original ontology' of indigenous peoples and world religions of which the Gaia theory is a scientific revival, have the potential to revive human consciousness and spiritual sensitivity towards the Life-generating and sustaining agencies of Gaia and the agencies of the myriad other beings whom Gaia supports, and who in turn sustain Gaia in habitable form. This approach has come to me in the course of digging deeper into the role of science itself, in partnership with the modern State, modern corporations, and under the aegis of the modern doctrine of development, in systematically reducing biological and social diversity. It is to the nature of these partnerships, and the doctrine of human development, including the ambiguous new variant of 'sustainable development,' to which I turn in the next chapter.

Notes

1 Norman Myers, Russell A. Mittermeier, Cristina G. Mittermeier, et al., 'On the Definition of a Biodiversity Hotspot.' *Nature* 403 (2000) 853–8; Mark de Bruyn, Björn Stelbrink, Robert J. Morley, et al., 'Borneo and Indochina are Major Evolutionary Hotspots for Southeast Asian Biodiversity.' *Systematic Biology* 63 (2014) 879–901.

2 Charles Darwin and Alfred Wallace, 'On the Tendency of Species to Form Varieties: And on the Perpetuation of Varieties of Species by Natural Means of Selection.' *Proceedings of the Linnean Society* (August 1858) 45–62.

3 Jane R. Camerini, 'Evolution, Biogeography, and Maps: An Early History of Wallace's Line.' *The History of Science Society* 84 (1993) 700–27.

4 Bruno Latour, 'The Parliament of Things.' In Bruno Latour, *We Have Never Been Modern*. Trans. Catherine Porter, Cambridge, MA, Harvard University Press, 1993, 142–5.
5 M.B. Fagan, 'Wallace, Darwin, and the Practice of Natural History.' *Journal of the History of Biology* 40 (2007) 601–35.
6 Alfred Russell Wallace, 'On the Law Which Has Regulated the Introduction of New Species.' *Annals and Magazine of Natural History, 2nd Series* 16 (1855) 184–96.
7 Wallace, 'On the Law Which Has Regulated the Introduction of New Species.'
8 Jessica H. Whiteside and Kliti Grace, 'Biomarker Records Associated with Mass Extinction Events.' *Annual Review of Earth and Planetary Sciences* 44 (2016) 581–612.
9 Holmes Rolston III, *Conserving Natural Value*. New York, NY, Columbia University Press, 1994.
10 This approach is informed by the work of Bruno Latour on actor-networks and the relevance of this approach for thinking about ecological crisis: Bruno Latour, *Reassembling the Social: An Introduction to Actor-Network Theory*. Oxford, Oxford University Press, 2005.
11 Paul Crutzen, 'The "Anthropocene".' *Phys. IV France* 12 (2002). https://doi.org/10.1051/jp240020447.
12 Bruno Latour, 'Agency in the Time of the Anthropocene.' *New Literary History* 45 (2014) 1–18.
13 Michael S. Northcott, 'Artificial Persons against Nature: Environmental Governmentality, Economic Corporations and Ecological Ethics.' *Annals of the New York Academy of Sciences* 40 (2011) 1–14.
14 Francis Bacon, *Novum Organum*. Ed. Joseph Devey, New York, NY, P. F. Collier, 1902, 249; and Michael S. Northcott, *A Political Theology of Climate Change*. Grand Rapids, MI, Wm. B. Eerdmans, 2013, 63.
15 Although Hutton was the first to coin the phrase 'system of the earth,' reference should also be made to the French scientist La Place, and the German explorer Friedrich von Humboldt, both of whom also advanced a new scientific and cultural conception of 'nature' as a totality of relations rather than as Newtonian empty space for myriad discrete entities.
16 Ernst Haeckel, *The History of Creation: Or the Development of the Earth and Its Inhabitants by the Action of Natural Causes, Volume 1*. Trans. E.R. Lankester, London, Kegan Paul, Trench and Trubner, 1899.
17 On the critical zone see H. Lin, 'Earth's Critical Zone and Hydropedology: Concepts, Characteristics, and Advances.' *Hydrology and Earth System Sciences* 14 (2010) 25–45: on the Divine as 'ground of being' see Paul Tillich, *Biblical Religion and the Search for Ultimate Reality*. Chicago, IL, University of Chicago Press, 1955, 31–6, 81–3.
18 On *Kajawen* and Merapi see Adam Bobbette, 'Cosmological Reason on a Volcano.' In Adam Bobbette and Amy Donovan eds. *Political Geology: Active Stratigraphies and the Making of Life*. Chamberry, Switzerland, Palgrave Macmillan/Springer, 2019, 169–200.
19 On religious interpretations of volcanic eruptions in Java see further Judith Schlehe, 'Reinterpretations of Mystical Traditions: Explanations of a Volcanic Eruption in Java.' *Anthropos* 91 (1996) 391–409.
20 On the spatial axis of Merapi, the Kraton, and the South Sea see Werner Hennings, 'On the Construction of Space and the Construction of Places: Java's Magic Axis.' In Jurg Wassmann and Katharina Stockhaus eds. *Experiencing New Worlds*. New York, NY, Bergahn Books, 2007, 125–45.
21 Alexander von Humboldt, *Cosmos: A Sketch of the Physical Description of the Universe Vol. 1*. Trans. E.C. Otte, New York, NY, Harper & Row, 1858, 34.
22 James C. Scott, *Seeing Like a State: How Certain Schemes to Improve the Human Condition Have Failed*. New Haven, CT, Yale University Press, 1997, Ch. 4.

23 Swidden farming has been practised sustainably for thousands of years and involves the clearing of small patches of land in the forest, mostly close to river banks where longhouses are also situated, for rice, vegetable, and fruit crops: traditionally a swidden patch is farmed for six years and then left for the forest to regrow, or it may be replenished by animal grazing before being used again: see further Evelyn Hong, *Natives of Sarawak: Survival in Borneo's Vanishing Forests*. Penang, Institut Masyaraka, 1987.
24 David Swinbanks, 'Forest Fires Cause Pollution Crisis in Asia.' *Nature* 389 (1997) 321.
25 World Resources Institute, 'Climate Data Explorer: 2017'. At www.wri.org/our-work/project/cait-climate-data-explorer accessed September 3, 2020.
26 *Directive 2009/30/EC of the European Parliament and of the Council of 23 April 2009*. https://www.eea.europa.eu/policy-documents/directive-2009-30-ec accessed October 10, 2020.
27 L. Pesqueira and P. Glasbergen, 'Playing the Politics of Scale: Oxfam's Intervention in the Roundtable on Sustainable Palm Oil.' *Geoforum* 45 (2013) 296–304.
28 E.B. Fitzherbert, Matthew J. Struebig, Alexandra Morel, et al., 'How Will Oil Palm Expansion Affect Biodiversity?' *Trends in Ecology & Evolution* 23 (2008) 538–45.
29 Lukas Straumann, *Money Logging: On the Trail of the Asian Timber Mafia*. Zurich, Bergli Books, 2014, 107–9.
30 Straumann, *Money Logging*, 167.
31 Anna L. Tsing, *Friction: An Ethnography of Global Connection*. Princeton, NJ, Princeton University Press, 2005.
32 Jason Hon and Shibata Shozo, 'A Review on Land Use in the Malaysian State of Sarawak, Borneo and Recommendations for Wildlife Conservation Inside Production Forest Environment.' *Borneo Journal of Resource Science and Technology* 3 (2013) 22–35.
33 Hon and Shibato, 'A Review on Land Use.'
34 Straumann, *Money Logging*, 270.
35 Arun Agrawal, *Environmentality: Technologies of Government and the Making of Subjects*. Durham, NC, Duke University Press, 2005.
36 Courtney Jung, *The Moral Force of Indigenous Politics: Critical Liberalism and the Zapatistas*. Cambridge, Cambridge University Press, 2008.
37 Jung, *Moral Force of Indigenous Politics*, 264–5.
38 Joan Martinez-Alier, *The Environmentalism of the Poor: A Study of Ecological Conflicts and Valuation*. Cheltenham, Edward Elgar, 2003.
39 See further E.P. Thompson, *Customs in Common*. London, Merlin Press, 1991; for fuller discussion of the environmental impacts of Enclosure see Michael S. Northcott, *The Environment and Christian Ethics*. Cambridge, Cambridge University Press, 1996, 66–94, 291–300.
40 Paul S. Martin, *Twilight of the Mammoths: Ice Age Extinctions and the Rewilding of America*. Berkeley, CA, University of California Press, 2005.
41 Roy A. Rappaport, *Pigs for the Ancestors: Ritual in the Ecology of a New Guinea People*. 2nd edition, Long Grove, IL, Waveland Press, 1984.
42 Tim Ingold, *The Perception of the Environment: Essays on Livelihood, Dwelling and Skill*. London, Routledge, 2000.
43 This is why in the film *The Matrix* it is suggested that the only ones who are able to engage the Matrix as agents, and knowing subjects, are those who know how to code and can therefore 'get inside' the software simulacrum and challenge it with its own logic.
44 Robin Hanbury-Tennyson, *Finding Eden: A Journey into the Heart of Borneo*. London, I B Tauris, 2017, 64.
45 'The Garden of Eden Was Turned to Desert - Hanbury-Tenison on Sarawak,' *Sarawak Report*, September 30, 2017, at https://www.sarawakreport.org/2017/09/the-garden-of-eden-was-turned-to-desert-hanbury-tennison-on-sarawak/ visited January 21, 2022.

46 'On the colonial stewardship of North Borneo under the Brookes see Steven Runciman, *The White Rajahs: A History of Sarawak from 1841–1946.* Cambridge, Cambridge University Press, 1960.
47 Peter Boomgaard, 'Oriental Nature, Its Friends and Its Enemies: Conservation of Nature in Late-Colonial Indonesia, 1889–1949.' *Environment and History* 5 (1999) 257–92.
48 See further Michael S. Northcott, 'The Romantics, the English Lake District, and the Sacredness of High Land: Mountains as Hierophanic Places in the Origins of Environmentalism and Nature Conservation.' In Hans-Günter Heimbrock and Jorg Persch eds. *Eco-Theology. Essays in Honor of Sigurd Bergmann.* Leiden, Brill, 2021, 74–90.
49 See further chapter 4 in Scott, *Seeing Like a State.*
50 Boomgaard, 'Oriental Nature'; see also Michael S. Northcott, 'Religion and Ecology in Indonesia After Covid-19.' In Zainal Bagir, Michael S. Northcott, and F. J. S. Wijsen eds. *Varieties of Religion and Ecology: Dispatches from Indonesia.* Zweigniederlassung Zürich, ALIT VERLAG GmbH & Co. KG Wien, 2021, 183–210.
51 On the modern emergence of nature conservation practices and conservation biology, see further Michael E. Soulé with Robert L. Peters, *Collected Papers of Michael E. Soulé: Early Years in Conservation Biology.* Washington, DC, Island Press, 2014.
52 Felicity Oram, Mohamed Daisah Kapar, Abdul Rajak Saharon, et al., '"Engaging the Enemy": Orangutan (*Pongo pygmaeus morio*) Conservation in Human Modified Environments in the Kinabatangan Floodplain of Sabah, Malaysian Borneo.' *International Journal of Primatology* (2022). https://doi.org/10.1007/s10764-022-00288-w.
53 Sabrina S. Taylor, Ian G. Jamieson, and Doug P. Armstrong, 'Successful Island Reintroductions of New Zealand Robins and Saddlebacks with Small Numbers of Founders.' *Animal Conservation* 8 (2005) 415–20.
54 Northcott, 'Artificial Persons Against Nature,' 104–17.

3

DIVERSITY AND DEVELOPMENT

Abstract

In the 500-million-year history of multicellular organisms, Life has realised five speciation explosions, each of which was preceded by a wave of extinction caused by a sudden climatic shock. These climate shocks, caused in the main by extra-terrestrial objects, set the earth on a trajectory towards greater biodiversity over deep time. This trajectory was mirrored in human linguistic and religio-cultural diversity until the modern era. In the last 700 years Europeans developed scientific and industrial technologies which enabled them to begin eradicating species diversity in their own domains, and to colonise other continents where they spread the same process, eradicating indigenous species and peoples. In the nineteenth century, Europeans developed a philosophical and social scientific ideology – the doctrine of development – with which to justify the European colonisation of other peoples' habitats, and the erosion of native species, and indigenous cultural traditions and languages. This ongoing process was redefined by the United Nations as 'sustainable development,' but the trajectory is unchanged.

Extinction and Diversity

The Earth, or Gaia, has undergone five waves of species formation – or speciation – each of which was preceded by a wave of extinctions caused by sudden climatic events which killed billions of living beings, many of which as species did not survive into the succeeding speciation era. The best-known extinction wave was caused by a huge asteroid which is buried in the Chicxulub crater beneath the Yucatán Peninsula in Mexico.[1] Dust from its hit led to decades of global cooling and decimated the dinosaurs, of which just a small remnant survived into the human era. Humans evolved after the fifth speciation wave which also produced the greatest species diversity in Earth history. One of the dinosaurs which survived

DOI: 10.4324/9781003110750-4

is the giant leatherback turtle. A remarkable beast, the leatherback regularly laid its eggs in deep nests which it digs with its back flippers in the night on beaches of the South China Sea and the Pacific Ocean. Sightings of the leatherback turtle on the shores of the South China Sea in Southern Thailand and West Malaysia were common until the 1960s and attracted many tourists to the area. But in West Malaysia the leatherback is now extinct, firstly because the eggs were harvested for sale as food in local markets, and then for use in a Malaysian government turtle hatchery which, while it was intended to conserve and increase the numbers of leatherback turtles that survived to adulthood, hatched the eggs in such a way that they all came out as one gender. Conservation scientists thought they could improve on the agency of the creature in her own reproduction. They failed, and now no female leatherbacks lay eggs in West Malaysia (Figure 3.1).[2]

The leatherback turtle is not at risk globally as it still has more isolated beaches in its vast range area that individuals use to lay their eggs. But although this gentle giant survived the huge climate cataclysm of 65 million years Before Present caused by the Yucatán asteroid, its extinction in West Malaysia indicates that it is not immune to a novel cause of species extinctions, namely *Homo industrialis*. Mass extinctions are Gaian processes which over deep time have converged on greater species diversity and the evolution of primates, including humans. But modern humans, by the scale of their agricultural and industrial modifications of terrestrial and marine habitats, their domestication, harvesting, hunting, and

FIGURE 3.1 Leatherback turtle nest markings, West Sumbawa, Indonesia. Photo Esther Dispozici, June 2020. Author Photo.

movements of species, and changes in atmospheric and soil chemistry, are said to be provoking a global wave of species extinctions sometimes referred to as the 'sixth extinction.'[3]

The third-century BCE Greek philosopher Euphorion is the earliest known philosopher in Western traditions to speculate on the possibility of species extinction based on findings of fossilised bones that bore no relation to known animals in classical Greece.[4] But most scientists in the Christian Era until the eighteenth century thought that extinctions of species were impossible. Medieval theologians held that all life – though not ensouled in the Greek sense of possessing *rational* souls – was animated by soul-like powers, and interconnected in a 'Great Chain of Being' which stretched from earth to heaven.[5] This chain of being was said to have been created by God at the beginning of time and preserved with all its parts intact in the state in which they appear lately to humans through the divine providential upholding of 'nature's economy.'[6] This theology of creation, and theological accounts of nature's economy, prevented early moderns from recognising the long eras in which Earth and life had formed before humans arrived; that there had been extinctions in the past; and that human activities, if too domineering and invasive, could also cause species extinctions.

The trigger for modern scientific recognition of past extinctions, and their temporal implications, was the discovery in North America of buried bones of very large mammals, and particularly of Mammoth bones in what is now the State of Kentucky. Large bones had been regularly discovered from Antiquity and had given rise to the belief that humans had in an earlier era lived alongside a race of giants or cyclops, an idea whose influence is discernible in Biblical and Greek mythology.[7] But the large find of bones on the Ohio River in 1739, including tusks, giant skeletal bones, and teeth, were clearly of animal origin and unlike those of the then-known largest mammal, the elephant. Discovered by a Frenchman, they were brought back to France where the belief gradually emerged that they represented the bones of an animal that had become extinct.

Georges Cuvier worked in the Paris Museum of Natural History where the bones ended up. He spent much time researching the bones from Kentucky, and comparing them with elephant bones and other finds, and in a lecture in 1796 on his findings he controversially suggested that the finds thus far indicated that they belonged to two different lost species.[8] In the following years he tracked down bones, and studied drawings of others, and eventually documented 49 extinct species, including a flying reptile – the pterodactyl.[9] Cuvier also argued that the Earth had gone through numerous ages over time long before humans existed, and that extinct animals in buried rock strata indicated that the Earth had been disturbed by catastrophes – which he called 'revolutions' – at a number of points in its long history and that these had wiped out whole worlds of lost species that stratigraphic finds were uncovering.[10] The use of the word revolution is striking, revealing the role of culture in shaping the human history of nature. The French Revolution had proven a signal break with France's past. This included her Catholic past, and the 'extinguishing' of the Mass in France for three years.

Together with the execution of many clergy and monks, and the destruction of art, and monastic and religious buildings, this must have felt like a cultural catastrophe to many of the French people: it provides therefore a striking metaphor for Earth System catastrophes and the extinctions they presaged.

While Cuvier resisted the theory of evolution as an explanation for the extinction of species, he proved remarkably prescient in his catastrophe theory since it is now widely believed that catastrophes, including giant volcanic eruptions and the impacts of large extra-terrestrial objects, played pivotal roles in the five biggest extinctions since the first Cambrian explosion of species 500 million years ago. Cuvier's senior colleague at the Natural History Museum in Paris was Lamark, who originated the theory of transformation of species over time which Wallace and Darwin later called the theory of evolution. When the young Darwin encountered fossils of extinct species in Patagonia it led him to ask about the possible causes of their extinction. But it was a visit to New Zealand, where he observed the range of introduced species and their effects on native flora and fauna, which led him to speculate that in due course 'a multitude of British forms would become thoroughly naturalised there and would exterminate many of the natives.'[11]

The Cambrian geological era is notable because the forms of life it birthed were so unlike those that had gone before. The large number of novel body structures and parts, particularly in animal species, which arose included placentally nourished embryos, hard skeletal structures protecting internal organs, and articulated knee and elbow joints. Biologists still have no satisfactory explanation for the extent of innovation that the Cambrian species explosion generated in the relatively short period, in geological terms, of just 20 million years. Some speculate that genetic adaptation was the principal driver, but as Simon Conway Morris points out, the genes themselves were already present in Precambrian life forms.[12] The Cambrian explosion was also preceded by the extinction of the *majority* of Precambrian life forms. Thousands of species which preceded the Cambrian were extinguished, and scientists only know of their existence from the vast number of fossils so well preserved in the Burgess Shale in the sedimentary foothills of the Rocky Mountains of British Columbia.

As we have seen, the geological record indicates there have been five major waves of extinction, and three less major, before the present, the last being at the end of the age of the dinosaurs 50 million years Before Present. The causes of these successive waves of extinction were long speculated upon but the principal driver seems to have been climate change. Sudden climate changes are detectable in fossilised tree-rings and studies of these indicate a contiguity between such changes and extinction waves. The likely causes of these relatively rapid climate shifts are still hotly debated but the most favoured candidates are sudden cooling events caused by particles in the atmosphere from huge volcanic eruptions, and from the impact of large extra-terrestrial objects such as asteroids; and small changes in the tilt of the Earth in relation to the sun.[13] Sudden climate changes cause species extinction in part because some creatures are unable to cope with rapid changes in temperature, in part because plant foods are reduced in quantity, and in part because rapid cooling leads to increased glaciation and ice cover.

This reduces the available landmass for those species that survive temperature changes. Species tend to develop in relation to their capacity to find 'ecological niches.' When land area is reduced, the number of ecological niches is reduced, and competition ensures that some are pushed out.

Evolutionary Convergence towards Maximal Species Diversity and *Homo Sapiens*

There is an unexplained trend in the punctuated history of mass extinction and speciation events that converged on the evolution of mammals and ultimately *Homos* and there is no known biological or chemical reason for this trend.[14] This trend combines with another less often noted convergence of the deep time record of life on earth towards maximal biodiversity. Each mass extinction event has been followed by an appearance of far more varieties of species than existed before. The result is that there has never been a time in Earth history when as many different species of micro-organism, plant, insect, reptile, and animal existed as now. Morris, who is an expert on the numerous fossils found in the Burgess Shale of British Columbia, argues that this convergence indicates that science absent metaphysics is insufficient to explain the evolution of life on Earth and that a larger cosmic and divinely originated force is at work, hidden from scientific instruments and the data models derived from historic and deep time proxies such as subterranean fossils and ice cores.[15] Morris is in accord with adherents of most of the world's traditional religions in ascribing life's trajectory towards greater abundance, beauty, and diversity to a divine principle set into creation and known among other names by those of the Tao, Vishnu, and the *logos*.

That there is a driving force behind evolution is strenuously resisted by ultra-Darwinians such as Richard Dawkins and Stephen Jay Gould. But they have not provided theoretical reasons for the convergence of Gaia towards the current state of maximal species diversity, combined with an atmosphere ideal for conscious life. Morris amasses a considerable body of fossil data from the Burgess Shale which demonstrates that evolutionary processes are convergent towards the emergence of conscious humans, an argument Tippler named the 'anthropic principle.'[16] But though Gaia may have converged on the evolution of conscious humans many argue that this lately evolved primate is the first terrestrial species to cause a wave of species extinctions.

A Sixth Extinction Wave?

In the late 1960s, biologist Paul Ehrlich, adopting Henry George's metaphor of the Earth as a spaceship, argued that modern humans had failed to recognise the role of biodiversity in their own evolution, and were causing a species extinction crisis which was analogous to astronauts pulling out the rivets from a spaceship as it hurtles through the vacuum of space.[17] Ehrlich is also known for his apocalyptic claim that the Earth is suffering from over-population by humans and that by their sheer numbers they are provoking a present-day wave of species extinctions.[18] Richard Leakey, the Kenyan-based primatologist, argued that present-day human extinctions amount to a wave of despeciation analogous to earlier extinction waves.[19] Stuart Pimm argues that while the number of humanly caused extinctions in the last 100 years is not analogous to previous extinction waves it is occurring at many times the speed at which any previous wave of extinction took place.[20] Moreover, he says the present wave is unique because it is the first extinction event provoked by the behaviour of only one species, and the first major extinction event in the history of the human species.

Against Ehrlich, Leakey, and Pimm, others argue that if huge waves of extinction have occurred in the past, then the contemporary extinction wave is not a grave cause for concern. Gould suggests that humans are such a lately arrived species in evolutionary history that their possible impact on the history of life is insignificant. While humans have a legitimate interest in their own survival, they need not over-concern themselves as to how their activities affect other species. Evolution will reassert itself in the end since, according to Gould, humans are 'stewards of nothing,' 'nature does not exist for us,' and it is arrogant for humans to imagine they can dramatically change the history of life.[21] This argument is given greater weight by the recognition that there have been species extinctions before in human history without 'spaceship Earth' collapsing, including the extinctions of mammoths in the last 8,000 years in Eurasia and North America, and other large mammalian extinctions such as rhinoceros, steppe bison, aurochs, cave lions, and bears. While these extinctions coincided with Northward migrations of humans, some attribute them to climate change not humans. And in evidence for this latter view it is notable that megafauna in Africa were not extinguished by humans in the same period. Extinctions in Africa are of much more recent origin and coincide with the colonial move of Europeans into Africa, and the subsequent industrialisation of agriculture.

There is, however, no doubt that European farmers began eradicating species in significant numbers in the late Middle Ages. Britain led the way in this respect and the killing was legalised by State laws and enacted through payments for dead animals and birds by each parish.[22] To this day, Britain is the most nature depleted country in Europe, if not the world. Huge areas of upland Britain are maintained in a degraded condition for the State-subsidised activities of bird shooting and sheep farming. The safest refuges for the few wild animals that remain are urban waste ground, riverbanks, and parks and so the urban fox population now exceeds the rural while the governments of England, Wales, and Scotland continue to promote, or at least refuse to effectively prevent, the slaughter of badgers, beavers, foxes, hares, and other alleged threats to farm animals. Raptors such as marsh harriers, buzzards, and even eagles and kites, are similarly trapped and killed by gamekeepers who work for the large landowners. None of these animals or birds are, however, endemic to Britain and hence their eradication there, though it depletes the British landscape of biodiversity and deprives the British people of encountering landscapes that have significant numbers of non-domesticated creatures in them, does not result in their extinction from the

Eurasian land mass to which Britain was once joined by a land bridge and is again by the undersea Channel tunnel.

Instead of modern declines in *numbers* of each species, which began on a significant scale in Britain, and later spread to Europe and the colonies, it is extinctions of species per se which scientists argue is now a cause for concern since the extent of contemporary habitat modification – principally for farming and plantations – is so large, with over 70 per cent of the global land area under human management.[23] Contemporary concern with species extinctions has even led to the popular view that humans are putting their own existence at risk, in a reprise of Ehrlich's spaceship earth argument. But is this the case? The problem is that scientists do not agree on this, as on many things.

The best way to assess a matter of scientific disagreement is to obtain reviews drawing on hundreds, or better thousands, of different data points. A review essay by a group of biologists that includes Ehrlich concludes that since 1500, 76 species of mammal, 134 of birds, 159 of amphibians, 91 of fishes, and 21 of reptiles have become extinct. Conversely it finds that the known numbers of different species still alive on Earth are: mammals, 5,488; birds, 9,990; reptiles, 8,734; amphibians, 6,347; and fishes, 30,700. This survey indicates that, in percentage terms, known extinctions of animal creatures in the last 500 years at 381 are less than 1 per cent of extant animal species. If extant species, and not numbers of each species, is the measure of the alleged extinction crisis, this does not yet look like a global catastrophe.[24]

Another survey, the UN-backed *Global Assessment Report on Biodiversity and Ecosystems 2019*, puts the percentages of extinct species in the last 500 years a little higher, at amphibians, 2.4%; land mammals, 1.9%; birds, 1.7%; reptiles, 1.1%; fishes, 0.95%.[25] But nonetheless, according to this report, which includes newly discovered species in the last 30 years, the total number of known-to-science animal species still alive in the present era is 76,457: of these only 608 are extinct in the wild. Again this does not sound like a catastrophe. Furthermore, many species that are extinct in the wild are still conserved in captive breeding programmes in zoos, botanical gardens, aquariums, and laboratories, from where the potential for reintroductions in restored habitats remains. On a field trip in 2008 I visited a number of sites in which this process had been conducted by the New Zealand Department of Conservation and it has proven remarkably successful in a number of small islands off the North and South Islands from which introduced mammals had been eradicated and onto which flightless and other birds that had once been common, and were still preserved in captive breeding programs, were reintroduced and are now re-established as viable populations.[26]

The global number of extinctions of invertebrates – principally insects – is far higher than for animals but it is impossible to estimate a reliable percentage of those that have been extinct since 1500 as there are still millions, possibly billions, of insect species which are yet to be recorded. Plant extinctions are also a cause for concern but again an accurate global figure is hard to come by. Much about extinction is therefore still unknown. However, what scientists are reasonably sure of is that the current rate of extinction is somewhere between 10 and 100 times the background rate over the last ten million years. This is not yet a catastrophic event, and predictions of a 'sixth extinction' may therefore be unnecessarily alarmist.

The largest number of species extinctions since 1500 occurred on islands, like the islands of New Zealand, or Pacific Islands such as the Hawaii group or the Marshall Islands, which, by being islands far from land and out of the reach of other land-dwelling birds and mammals, have significant numbers of endemic – unique to the habitat – species. The peak period since 1500 for island extinctions was the end of the nineteenth century which was the first era of fully global connection and trade, and the increasing size and speed of European ships, which introduced rats to many islands, resulting in significant numbers of bird extinctions.[27] The present-day extinction rate for birds is actually *lower* than the late nineteenth century as conservation programs including habitat restoration, eradication of non-native species, and reintroductions of endangered species have reversed the colonial trend. But the contemporary concern with extinctions is that the combination of increases in habitat modification, domestic animal breeding, the huge intercontinental wildlife trade, and growing encroachments on remaining old-growth forests for bushmeat and timber now threaten endangered species on continents as well as islands. The lead nation in all these endangerment activities is China which is the global centre of the illegal wildlife trade. China also deploys the largest number of industrial trawlers of any nation, and industrial trawling is the principal cause of endangerment of species in the oceans.

Despite alarmist predictions of mass extinctions going forward, it is important to note that the proven numbers of extinctions of animals – the species category which includes birds, animals, humans, fish, and reptiles – are far lower than alarmists indicate. If we take the oceans which cover two-thirds of the planet, most *species* diversity is still present in the oceans. The core of the problem of contemporary extinctions is that the *numbers* of extant species are being vastly reduced on land by industrial agriculture, and in the oceans by industrial fishing. The rate of this reduction degrades and destabilises the viability of terrestrial and marine ecosystems which have evolved over millions of years as the cradles of species diversity, and in which the numbers of different species are a critical adaptive feature of ecosystems, enabling them to respond resiliently to external shocks such as climate change. The decline in numbers of creatures on land and in the oceans is also the principal cause of the reduced resilience of marine and terrestrial ecosystems in the face of natural disasters, and of their reduced capacities to regulate the Earth's climate, capacities which were first properly theorised by James Lovelock in his Gaia theory, of which more below.

European Expansion as the Root of the Extinction Crisis

The historical roots of major reductions on land and in the oceans of the numbers of living species are in the late Middle Ages of Western Europe at the end of the era of the Holy Roman Empire and during the rise of Renaissance mercantilism. The key instruments of increased human control and resource harvesting of the nonhuman were new technologies, including the deep plough, the telescope, the chronograph, the sextant, and the compass, which were invented in the late Middle Ages and during and after the Renaissance. These new devices gave Europeans capacities to travel across and map oceans, and to control and instrumentalise ecological processes, which no predecessor cultures are believed to have possessed. These powers also enabled new kinds of agriculture and forestry which led to the large-scale eradication of nonhuman species on the European mainland and its islands, including Britain.

With the aid of advances in marine technologies, Europeans from the fifteenth century expanded out of Europe into the Americas, Asia, and Africa. Initially travelling to explore, to find 'treasure,' and to trade, Europeans from the sixteenth century establish settlements, initially on the coasts, and then inland, as settlers followed traders. Agricultural colonies in shore from coastal settlements were initially established in the temperate and semi-temperate zones of the Americas and Africa which were more favourable to the European constitution and enabled the rapid deployment of European plants and animals without significant modification. The tropics were more resistant to this deployment as many early European settlers died young from malaria, and Asian lands and coastlines were in any case already more densely settled by Mongoloid peoples.[28]

The biological effects of European colonialism, which occurred on a scale far beyond that of previous empires because of the developed maritime capacities of Europeans in circumnavigating the Earth, are well characterised by Alfred Crosby as 'ecological imperialism' and the 'Columban exchange.' Foremost among the reasons for the global biological influence which European expansion occasioned was European success in depopulating indigenous peoples – often through overt genocide but also by introduced diseases to which indigenous peoples had no resistance – while at the same time growing their own numbers in the new colonies.[29] Upwards of 500 million people of European descent now live in the temperate colonies set up in the Americas, Africa, and Australasia. And their ancestors introduced and vastly expanded the numbers of European species – plants and animals mainly – as they practised European agriculture in their newly established domains.

The second driving force of ecological imperialism was the introduction of European fauna and flora, some of it inadvertent, such as rats off ships, and much of it intentional, as European settlers sought to replace native plants and ecosystems with European agricultural crops and livestock, decorative plants, and of course weeds such as dandelion and gorse which ran wild and often overwhelmed native plants.[30] It is therefore possible to see large parts of the Big Island of Hawaii aflame with gorse in yellow bloom for much of the year, while rats eradicated huge numbers of native birds on the same island group, as on islands throughout the Pacific.

The principal agents of these introductions from 1600 to 1900 were the Dutch and English East India Companies. Many plants also went from the New World, and the 'East Indies,' to the old, either as exported crops or as plants to be grown in Europe, including cane sugar, coffee, corn, potatoes, spices, tobacco, and tomatoes. But ecosystems and cultures were far more dramatically transformed in the Americas, Asia, and Africa by European introductions, and by European agriculture, than the other way around. Charles Darwin wrote in his journal when he travelled in New Zealand that he anticipated that the extent of non-native introductions would lead to the extinction of native species, and indeed this was the case in New Zealand and in Africa, Eurasia, and the New World.

The third driver of extinctions was the European imperial extractive project. It was the potential for treasure which persuaded investors to put up the huge sums for the first explorations of the New World, and East Asia, and which led to the setting up of the Indies Companies, the Hudson Bay Company, and so on. The ultimate purpose, beyond mere exploration, was to export plant, animal, and mineral wealth from what were to become colonies back to the European bankers and royal households which funded and chartered the boats and the Companies.[31] Among the key devices for extraction of wealth were the supplanting of diverse ecosystems with mostly human slave sustained monocultures of 'cash crops' for European consumption including coffee, sugar, and tobacco in the 'West' Indies and North America; rubber, oil palm, and teak in tropical Asia; and opium, cotton, and tea in temperate Asia.[32] Temperate latitudes of South and North America, together with temperate zones of Africa and Austronesia were in large swathes converted into European-owned and -run farms for sheep and cows together with European staples including wheat and barley and the newly adopted corn from South America.

Ecological Imperialism as the Root of the Ecological Crisis

The connection between the contemporary extinction crisis and the colonial spread of European monocultures is less discussed than it should be in the burgeoning extinction literature. This is partly because of the failure of colonial records, initially controlled by the companies that ran the colonies, to observe the contrast between the species rich and verdant nature of colonised lands when the Europeans arrived and their denuded but allegedly more 'productive' state after a century or more of European agricultural conversion and economic 'development' that from an indigenous perspective was often reverse development. A classic example of this was the East India Company's purposeful wreckage of the Indian cotton industry, which was far superior to Britain's, with scandalous tariffs against 'home spun' that Mahatma Gandhi resisted by himself spinning cotton, and which were designed to force Indians to buy inferior Lancashire cloth. Another and even more egregious set of examples is the genocidal destruction of so many thousands of independent, pre-colonial, socially, and ecologically resilient indigenous cultures and communities in both the West and East Indies.[33]

Alongside the general tendency of Company settlers and traders to destroy indigenous ecosystems and cultures, and supplant them with European plants and agriculture and slave-powered settler-run plantations, warning reports were sent back to London and Amsterdam about the dangers of ecological breakdown from representatives of the Companies from the early eighteenth century, and these reports represent the first documentary evidence of the rise of modern environmental concern.[34] These reports from Company-run islands including Barbados, Madeira, and Mauritius were combined with a scientific drive to collect and investigate non-European species and landforms, and together these gave impetus to the formation of some of the first scientific societies.[35] But nonetheless, the dominant trend of Company and Imperial policy remained an extractive one and was based on the idea – going back to John Locke who developed his theory from observations of encounters between settlers and natives in the Colony of Virginia – that the European colonial and settler project would make 'nature' more 'productive' than it had been under indigenous or 'native' customary practices and governance institutions. Productivity was not though measured in terms of the health and longevity of indigenous populations, or the diversity and numbers of species that their indigenous ways of life enabled to flourish, but rather in terms of monetary and resource extraction to Europe.

European-introduced diseases – some inadvertent and in some cases intentional – resulted in the decimation of local populations who had no resistance to them. Those who survived introduced diseases and violent land expropriation were rendered far less free and noble by their increasing servitude to colonial agricultural and trading systems, either through direct slavery or through pernicious and arbitrary forms of expropriation of native crops and products as 'taxation.' For the Companies, and the British, Dutch, and other colonial offices that took over colonial terrains in the nineteenth century, the 'success' of colonial economies was measured not in the improvement or 'development' of indigenous peoples, and their ecosystems, but in the quantity of loot exported back to the home countries – either in the form of treasure such as gold, silver, diamonds, and copper, or in the form of valuable export crops such as coffee, tea, and spices. Oil was a later export product which proved to be the most valuable of all, but only began to shape colonial relationships from the late nineteenth century with the gradual shift from coal and steam power to oil. Nonetheless, in the twentieth century it was oil companies, including Royal Dutch Shell, the Anglo-Persian Oil Company (later BP), Exxon, and the French ELF which took over the commanding extractive roles, and accompanying influence over colonial policy, once occupied by the East India Companies, together with the Hudson Bay Company, the Company of Virginia and their French, Spanish, Portuguese, and Italian counterparts.

The ecological destruction wrought by colonial deforestation, introduced species, mining, monocultures, and settler expansion was on a scale and took place at a speed that had not occurred before in history. There had been previous imperial extractive projects which had similar results, as for example, the denudation of trees and wild animals of parts of North Africa which were made to serve as the breadbasket of the Roman Empire. These led to the northward expansion of the Sahara as extractive methods, together with denudation of indigenous vegetation and wild animals, ultimately destroyed the hydrology and native ecosystems of much of the region.[36] But the European colonisation projects from the eighteenth to the twentieth centuries were the first to be global in their effects in reducing both biological *and* cultural diversity. And the two are inextricably connected, as is now acknowledged by ethnobiologists and other scholars.

To acknowledge that there is an inextricable connection between imperialism and ecological destruction requires resisting the misanthropic environmentalist assumption that it is humans *per se* who degrade ecosystem resilience and biological diversity.[37] This assumption gives an antihuman and eugenicist gloss to the catastrophic ecological effects of the spread of industrial 'development' from Europe to the colonies. It also reflects the tendency in European encounters with alien lands and ecosystems to see them as 'virgin' territory in the sense of being 'unmodified' by humans, as, for example, in the journal of John Muir's journey to the Amazon.[38] This idea of humans against nature, as we observed above, is also evident in Darwin's and Wallace's theories of evolution which excluded *Homos* as a force in the evolution of the distribution of species.

Biocultural Theory and Symbiosis in Cultural and Species Diversity

Studies in ethnobotany and linguistics have converged since the 1980s on 'biocultural theory' according to which indigenous humans were shapers of ecosystems long before moderns arrived on the scene, and that their shaping sustained far greater numbers and kinds of species than the shaping of moderns. These studies also reveal that human cultures were in turn shaped by the variety of ecosystems, including forest, montane, marine, riverine, and savannah habitats in which humans dwell.[39] Luisa Malfi defines bioculturalism as follows:

> Biocultural diversity comprises the diversity of life in all of its manifestations: biological, cultural, and linguistic, which are interrelated (and possibly coevolved) within a complex socio-ecological adaptive system.[40]

Biocultural theory goes against 200 years of evolutionary science in acknowledging the proactive roles of groups of humans in the distribution and evolution of species. It also suggests that the core assumption of much modern environmentalism – that humans are intrinsically 'bad' for 'nature' or non-humans – is wrong. If bioculturalism holds as a theory – and it is relatively novel and case studies and data which underwrite it are still emerging – then much received opinion about the human role in the evolution of Gaia needs revising. Furthermore, if human cultures have shaped the distribution of species and the organisation of ecosystems for thousands of years, then the claim of the uniqueness of the modern 'Anthropocene' is more about Eurocentric and industrial 'development,' and the need to end it, than it is a novel geological epoch.

Bioculturalism significantly revises human evolutionary theory since the contiguity of endemic species diversity and endemic languages indicates that human cultural diversity played a role in sustaining species diversity; and that there is also a contiguity between the erosion, and extinctions, of living languages, and major declines in species numbers. Among the places on earth where these interlinked processes can be seen most clearly is Papua New Guinea, which has more extant languages per head of population than any other region of the earth. As Malfi explains, the reason linguistic and species diversity tend to go together is that highly linguistically and species-diverse regions are characterised by 'ease of access to abundant resources found in place' and such abundance 'seems to favour localized boundary formation and diversification of larger numbers of small human societies (and languages).'[41] So, in tropical forests, hundreds of thousands of small groups of humans lived isolated from other human groups – or exchanged only briefly with them from time to time – for much of the history of humans before the Columban beginning of European colonialism.

The existence of groups of humans over tens of thousands, in some cases even hundreds of thousands of years, in separated and diverse ecosystems inhabited by diverse assemblies of species meant that each group developed different customs, languages, practices, and rituals. There is therefore a significant overlap between highly biodiverse countries containing many endemic species, and human linguistic diversity. Among the linguistically most diverse regions are the Amazon, the Congo, Java, Sulawesi, the Philippines, New Guinea and Borneo, the Congo, India, China, the United States, and Australia. But in many of these countries and regions forest clearance, settler cultural diffusion, and industrialisation – including the introduction of chemical agriculture, and the construction of roads and riverine and sea-ports – have dramatically reduced human cultural and linguistic diversity, and species diversity in these countries and regions is declining in tandem.

The role of human cultural and linguistic diversity in sustaining species diversity, and the implication of their loss in declines in species numbers are insufficiently understood by conservation scientists. The lead scientific explanation for species loss in Africa, Latin America, and Asia is human population increase. This claim is widespread in conservation and environmental literature, and I reprised it already with reference to Ehrlich above. Conservation scientists far less often reference the introduction of European colonial agriculture, non-native species, and indigenous cultural destruction as root causes of biodiversity loss. Even less do they connect biodiversity loss to the ongoing post-colonial trajectory of globalisation, as led by globalist agencies which claim to be addressing the biodiversity crisis, and especially the United Nations, the World Bank, and international conservation and 'development' agencies. These agencies have taken over many of the roles of the colonial companies of the eighteenth and nineteenth centuries. In partnership with these agencies are multinational corporations founded in the twentieth century as shareholding vehicles for the collective pursuit of energy, grains, minerals, and seeds in non-European lands.

Doctrines of Development

The key ideological device for advancing the legitimacy of expropriating other peoples' lands, minerals, and species, and their labour via slavery or punitive taxes, was the European claim of 'trusteeship' over less 'developed' or 'civilised' non-Europeans, and the associated nineteenth-century 'doctrine of development.' This doctrine is the principal ideological cover for ongoing destruction of species and ecosystem diversity and human cultural diversity. The globalising and homogenising trajectory of human–nonhuman relations which this doctrine promotes is merely restated and reaffirmed in the revised doctrine of 'sustainable development' as expressed in the United Nations' 'Sustainable Development Goals,' Agenda 21, and Agenda 2030 of the United Nations Environment Programme (UNEP).

The doctrine of development emerged in the nineteenth century and has ideological roots in the Scots and German Enlightenments, especially the writings of Adam Smith and Immanuel Kant, and in evolutionary and theological ideas of 'progress.'[42] The doctrine encapsulates Smith's argument that there is an innate progress of 'stages' in human cultures from hunting and fishing through pastoralism and agriculture to capitalist trade and industrial manufacture.[43] Smith suggested such development was inevitable, but it was a convenient way of justifying the forceful movement of the British people off the land, via the Enclosures and the Clearances, and the creation of a mobile work force, detached from the land, and forcibly put to work to provide labour – in workhouses or via poverty wages – in the factories of the industrial revolution. For Smith it was the extent to which nations were prepared to follow seventeenth-century Britain in Enclosing or 'clearing' the commons and converting 'land' and 'labour' into rent and wages which would advance the 'progress' of trade and manufacturing which were the ultimate sources of the wealth, or conversely the poverty, of nations.

As Cowen and Shenton argue, Smith's influential account elided the horrific scale of immiseration imposed on the English, and later the Scots, to force them off the land and into the wage slave industrial system over more than two centuries. It also neglected the many attempts of citizens in England and Scotland to resist their own 'development' in the 'rough and tumble of industrialisation.'[44] Development was the Victorian aristocratic justification for the subjugation of working people in their own land in miserable conditions. The Victorian claim of 'trusteeship' of the colonies which led to the immiseration and denudation of the lives and ecosystems of millions of non-Europeans was an extension of the original aristocratic and commercial claim that forcible dispossession by the rich of the lands and means of production of the poor was justified as means to achieve 'agricultural improvement' and technological and cultural 'progress.' It is a little-known part of British history, but there were numerous attempts by groups of peasants, sailors and workers to resist the dispossession and horrific working and living conditions to which they were subjected from the seventeenth century. History is written by the winners, and the documentation of peasant and worker revolts in Britain is meagre but among the better known incidents are the Liverpool Sailors Revolt of 1775 and the Glasgow Clydeside Revolt of 1919.[45]

The Luddites, and other radical and socialist movements which attempted to resist de-peasantisation and proletarianisation in Britain, were the predecessors in the home country of the indigenes who resisted colonisation overseas.[46] Resistance to colonisation was much greater and led to far greater loss of life than resistance in the home country. And it was in part to justify the immensely coercive nature of the colonial enterprise, and the scale of both ecological destruction and human immiseration involved, that the idea of development was deployed more extensively in the colonies than in the imperial homelands. It was in the European colonies, or what are still called 'less developed' countries, that those who lived by hunter gathering supplemented by small-scale agriculture such as forest gardens were described as barbarians, and even subhuman savages, in need of development, and such descriptions became central to justifications of colonial enterprises or the 'white man's burden.'

The idea of development as 'progress' is a second and equally influential strand to the doctrine. As Cowan and Shenton demonstrate, it is rooted in such texts as Kant's *Idea for a Universal History from a Cosmopolitan Perspective*, John Henry Newman's *An Essay on the Development of Christian Doctrine*, and the positivist philosophy of Saint Simone and Auguste Comte.[47] For these theorists of the inevitability of 'progress' there is as an intrinsic, even 'objective,' process which runs through human history and at its heart are two transformations. The first is the human cultural transformation of what Smith called barter, and what Marcell Mauss called gift exchange, into the 'rational' processes of commerce and trade. John Stuart Mill, among others, considered this as the transformation of society into a newly constructed set of mechanical processes. The second is the ecological transformation of organic and sun-fuelled human production processes – in agriculture, and in building and craft procurement – into industrial processes characterised by plant monocultures, and the use of fossil fuelled energy and mechanistic production techniques and systems.

Although Mill, like Smith, Kant, and Newman, represented the process of development as inevitable and even a 'natural' process through which those with the 'greatest capacity' to advance the 'material interests' of society come to the fore, he also, and inconsistently, argued that development is about subjective choices by those 'doing' development and in particular those men of capability, power, and wealth who, as the 'trustees' of the process, make choices about its direction and form.[48] But if development is not an inevitable process but instead depends upon *choices* made by the agencies and actor-networks of individuals that direct development, then it is not an evolutionary or inevitable process but rather the outcome of choices by those who take it upon themselves to subject lands and peoples to economic, extractive, and technological priorities which emanate from European culture and are not choices that would have been made by 'Third World' or 'less developed' countries in the absence of what Mill acknowledges as the necessary despotic powers of the colonisers.[49] Mill, like his father, worked for the East India Company for most of his career and argued that when Britain conquered India the 'Hindu' were 'in a state of degradation.'

It is notable that against this utilitarian and liberal progressive view of empire, the conservative Edmund Burke argued that it was British rule or 'protection' which destroyed India: 'Every rupee of profit made by an Englishman is lost forever to India. With us there are no retributory superstitions, by which a foundation of charity compensates, through ages, to the poor, for the rapine and injustice of a day.'[50] Burke contrasts the linear and secular idea of colonial development as a voluntarist process of forceful economic extraction and wealth accumulation with the Indian sacred cosmology of karma and the circle of life. And as Burke points out, Indians had a constitution long before the English or the Mughals conquered them and it was 'a constitution made by what is ten thousand times better than choice, it is made by the peculiar circumstances, occasions, tempers, dispositions, and moral, civil, and social habitudes of the people which disclose themselves only in a long space in time.'[51]

Burke's conservative argument for recognising the superior value of constitutions and customs that evolved in diverse ways in different cultural settings over many generations is borne out by the recognition that modern and often revolutionary 'development' almost invariably leads to the erosion of biological and cultural diversity. And it is notable that this trajectory of reduced diversity is against the tendency of Gaian history to produce ever greater species diversity in each geological wave of speciation, and the evolved human trajectory towards a diverse range of customary practices and cultural forms fitted to the biological, climatic, and ecosystemic character of the places in which myriad human cultures evolved.

Prior to the Columban exchange there was *fittingness* between the diversity of human communities, constitutions and customs, the diversity of habitats in the places in which humans dwelled, and the biodiversity of species which had evolved ever more maximally in each successive phase of Earth history.[52] In Burkean terms, a conservative approach to diverse indigenous non-Western cultures, which liberal utilitarians like Mill regarded as 'backward' and systematically destroyed, would have produced a conservationist approach to species diversity. This suggests that instead of continuing with Western-style development after independence, former colonised nations would have done better to reshape their economic, political, and social priorities around recovering and reviving the cultural, customary, and constitutional diversity they enjoyed before the European conquest. But most post-colonial development has trended in the same direction as pre-colonial development and accelerated both the pace of the destruction of indigenous cultures, customary and commons governance systems and languages, and the destruction of resilient native ecosystems and associated species diversity.

The current direction of 'development' towards homogenisation of biological life and human cultures was set in train by European colonial expansion from the fifteenth century. The regions that were most heavily homogenised in terms of human culture and biology were initially the temperate regions which European settlers preferred as places to transfer European agriculture, plants, and domesticated animals. Smaller islands also suffered considerable losses from introduced animals and plants and sometimes cultural losses from European settlement as well. Contemporarily, the biggest declines in biodiversity are in the tropical regions and in the oceans. Deforestation, as discussed above, remains the principal cause of biodiversity and cultural destruction in the tropics and it is still being driven in the main by conversion of tropical forests into monocultural production of biofuels, cooking oil, and agricultural crops – and especially genetically modified soya and corn. The principal agents of the destruction remain the large industrial food, fishing, and mining corporations, such as Unilever, whose historic roots are colonial but which retained and often grew their powers to determine what grows in postcolonial tropical regions, and what animals and humans eat there.

Industrial Fishing, Forestry, and Agriculture and Biodiversity Decline

The greatest loss of numbers of species from the industrial harvesting and homogenisation of Earth in the twentieth century has occurred not on land but in the oceans. Long-range estimates of depletion of the quantity of life in the ocean by industrial fishing since 1945, drawing on the journals and logs of fishers before scientific monitoring began, put it at 90 per cent, which is an astonishing decline.[53] Those primarily responsible are not indigenous fishers who constitute over 95 per cent of those employed in fishing and who operate with small coastal fishing boats and without sonar.[54] The culprits are giant sonar-equipped industrial freezer trawlers deployed by companies principally based in China, Japan, the Netherlands, Russia, South Africa, Spain, and the United States. Just one Dutch company – Parlevliet & Van der Plas – runs two of the biggest trawlers at sea with a combined freezing capacity of 13,000 tons of caught fish. Their trawlers are so huge they are capable in just one calendar year of depleting whole fishing grounds, as did the *Helen Mary* off the South Pacific coast of Chile, and later off the coast of West Africa.[55]

A majority of the seven largest 'super trawlers' identified by Greenpeace are based in and licensed by the European Union. Wealthy developed nations also dominate inshore as well as deep-ocean catch in many former colonised nations. The European Union, Japan, and China all deploy huge ships in developing country waters, in the main with the agreement of the developing countries. But the terms of such agreements rarely favour the home nation, and the rate of catch by foreign boats leads to declining catches by the far larger number of small but technologically unsophisticated boats deployed by developing nation coastal fishers. The result is that in a recent survey developed nations were responsible for 97 per cent of tracked industrial fishing on the high seas, and they also obtained 78 per cent of all the fish in national waters of lower-income nations.[56]

A growing proportion of fish hoovered up by these vast ships is fed to aquaculture farms whose outputs of fish have overtaken consumption of wild-caught fish in Asia.[57] The rearing of so many domesticated fish is having growing impacts on riverine and ocean fish as microbial resistance spreads from the farms into water catchments and the oceans because of the use of antibiotics as growth promoters.[58] Wild salmon are in major decline in Scotland, Norway, and British Columbia also because of infertile escapees from fish farms, and the extensive pollution of wild salmon runs by lice and toxins from the farms.[59]

On land, the conversion of old-growth forest and other species-rich ecosystems to monocultures to produce fast-growing timber, animal feeds, biofuels, and palm oil, as discussed above, is the biggest cause of ongoing reductions in numbers of non-domesticated species on land (Figure 3.2). The demand for timber globally is the main driver of deforestation. But although carefully managed old-growth forests produce more lumber, both for board and wood chips and

FIGURE 3.2 Old growth, species-rich forest, Blue Tier, Northern Tasmania, 2008. Author photo.

FIGURE 3.3 Clear-cut and burnt forest, Upper Florentine, Tasmania. Author photo.

pulp, than industrial monocultures, foresters prefer monocultures of fast-growing trees such as Eucalyptus and Sitka spruce because they are more amenable to mechanical harvesting and more predictable in terms of Returns on Investment. Having harvested the best wood from old-growth forests, the rest is burnt, and the biodiverse forest is replaced with fast-growing trees or non-timber monocultures such as oil palm and soya (Figure 3.3).

Soya grown on much formerly forested land in the Amazon is mainly used in the production of industrial animal feed. The demand for animal feed arises principally from industrial animal husbandry in modern intensive 'factory farm' facilities.[60] The global numbers of large domesticated animals at around 4.3 billion exceeds the weight and numbers of all non-domesticated large animals by over 50 times.[61] The principal keepers of this vast number of domesticated animals, and the principal converters of ecosystems from wild habitat to animal-feed crops, are not small farmers but large industrial food and agricultural corporations and government agencies. Domesticated animals when managed ecologically can prevent desertification, encourage plant growth, and improve soil fertility as I discuss more fully in Chapter 8. But when corralled in industrial sheds and feed lots, and fed cereal, waste from domesticated animals, far from fertilising the Earth, results in significant pollution to the air and water sources.

Industrially reared meat is harmful to human health as well as the environment. Marketing of cheap grain fed and intensively reared meat as a 'cheap' luxury has led to significant increases in meat consumption as a proportion of human diets in Europe, the Americas, and Asia relative to fruit and vegetable consumption. Given the toxicity of much industrial meat, from antibiotics used as growth promoters and pesticides used on animal feed, there is an evidential association between rising levels of industrial meat consumption and chronic diseases such as cancers, heart disease, high blood pressure, obesity, and strokes.[62]

The problem with industrial meat production is not that animals are bad for the environment. On the contrary, as I discuss more fully in Chapter 8, there are vastly fewer animals now living in almost all ecosystems than there were before the colonial and industrial eras. The problems for human health and the environment arise from *industrial* animal husbandry. Traditional animal husbandry of chickens, cows, ducks, goats, horses, pigs, turkeys, and yaks incorporates the animals into arable, montane and rice farms, and forest gardens, where the presence of animals *increases* biodiversity, and especially soil microbiota. The contribution of animals living mainly off the by-products of agricultural production to protein in the diet in developing countries such as Indonesia is also considerable. Similarly in highland areas, such as Nepal, Mongolia and Bhutan, the European Alps, and the Andes of Latin America, animals increase food productivity from marginal food-growing lands as they are raised in, and help to sustain, biodiverse forests and high mountain alpine plants, mosses, and so on.

Corporations as Agents of Biodiversity Decline

The principal drivers of deforestation, industrial trawling, the destructive spread of factory farms, and the related rise of animal-feed monocultures, are not peasant farmers or city dwellers, nor even the global increase in human numbers. The principal drivers are large State- and privately owned corporate agencies which market resource-hungry lifestyles, including a meat-intense diet, as the way of life to which 'under-developed' people progress when their lands and waters are subjected to 'economic development' which, as we have seen, is another name for the ongoing process of Enclosure and corporate capture of the economy of nature that commenced in seventeenth-century Europe. It is private corporations, acting in concert with governments and globalist agencies and especially Multilateral Development Banks, including the World Bank and the Asian and African Development Banks, which drive fishers, farmers, and forest dwellers from their customary lands and waterways, and destroy traditional and more ecologically sustainable practices of food gathering and growing. Similarly, it is corporations in the urban hinterlands to which former forest dwellers, fishers, and subsistence farmers are forced to migrate, which construct rural–urban food procurement and consumption patterns which conform to industrial development, including its inbuilt drive to simplify and reduce the complex interrelations between myriad species and myriad human customs and traditions which before the industrial age had co-evolved between humans and their co-inhabitants, and their shared habitats. In this way, long-established dietary patterns and modes of farming and fishing, and associated cultural and traditional beliefs and practices, are still being replaced with more 'Western-style' diets and ways of living, which are less healthy for both people and planet.

The large corporations – public and private - that continue to 'advance' industrial development are a relatively novel feature in human history. They originated in Europe in the formation of jointly owned deep coal mines in Germany in the sixteenth century, and in the same century, the formation of overseas trading organisations, and most notably the Dutch and English East India Companies, the Hudson Bay Company, and Virginian plantation companies. Corporate personhood which legally separated private corporations from the public domain originated in a legal case between Dartmouth College and the State of New Hampshire.[63] Their modern equivalents are the energy, chemical, pharmaceutical, agricultural, and technological behemoths, with incomes the size of many national Gross Domestic Products, such as Royal Dutch Shell, Exxon, Cargill, Monsanto-Bayer, and tech companies such as Amazon, Alphabet, and Apple.

The decisions and ownership of the top 200 global corporations, together with investment bankers, and heads of international agencies such as the World Bank, the International Monetary Fund, and the European Commission, represent the actor-network – to adopt Latour's helpful metaphor - by which a small number of principally North American, European, Japanese, and Chinese individuals turn so much of the biomass of the oceans and land into cereals, industrial meat, and fish. The resultant and increasingly interconnected global culture is displacing human cultural diversity with an increasingly monocultural global matrix dominated by North American, European, and, latterly, Chinese industrial corporations so also threatening extinctions of species.

The clearest indicator of the link between the homogenisation of human cultural practices and the homogenisation of ecosystems and species is the narrowing range of species from which more and more humans derive their diet. Humans have access to a greater range of species than ever in their history: 4,000 plant species are regularly consumed by humans around the world.[64] But just 30 agricultural species now dominate human diets across much of the planet and this trend towards reliance on a reduced number of species is growing.[65] Fifteen crop plants now supply 90 per cent of human calorific intake, with rice, maize, and wheat being the most dominant, and as the growing and marketing of these crops increases, traditional, more diverse, and ecologically more benign food cultures decline.[66] This has implications not only for biodiversity because of the continued growth of monocultures but also for human food security since the small number of plant species on which humans rely reduces the resilience of the human food supply during external shocks such as financial crises, natural disasters, plagues, and wars.

Sustainable Development and the Rise of Global Technocracy

In recognition of the growing and deleterious ecological impacts of industrial development, development agencies and scholars began to talk about 'alternative development' and 'another development' from the 1980s. André Gunder Frank influentially argued that what the West calls development is really wealth extraction, and that the process, especially in its early stages, often produces immiseration, reduces human life expectancy, wrecks environments, and ultimately results in 'under-development.'[67] The principal reason is that development was colonial ideological cover for the raiding of other peoples' lands and waterways for treasure to take home to Europe. Post-colonial development is no different and often accelerates the processes of primitive accumulation and economic 'integration' by which formerly independent and sovereign communities, regions, and nations become *dependent* culturally, and for key resources, on Western nations, and lose their capacity to provide the essentials of life for their own people while tending their own ecosystems.

In 1987 international concerns at the questionable trajectory of development and global economic integration led to the setting up of a United Nations World Commission on Environment and Development. The resultant report, *Our Common Future*, defined what it claimed was a more beneficent form of economic development as follows: 'Sustainable development seeks to meet the needs and aspirations of the present without compromising the ability to meet those of the future.'[68] But 'sustainable development' is primarily associated with efforts to rebrand globalist and corporate-led industrial development as a benign process which improves the wellbeing of peoples subjected to it and enhances their environments. The main globalist agency promoting this rebranding of development is the UN Environment and Development Program whose first major program was the organisation of the 'Rio Earth Summit' in 1992. The summit promulgated what came to be known as 'Agenda 21' which was a new 'blueprint' for global governance of human and planetary 'resources.' The UNEP also developed the 'Millennium Development Goals' which were later reframed as the 'Sustainable Development Goals' and then Agenda 21 was updated as 'Agenda 2030' at the UN General Assembly in 2015. The principal sponsoring agency of this new revised agenda, in addition to the UNEP, is a new and portentously named body called the UN 'Trust Fund for Human Security.'[69]

It is notable that the key concept of Sustainable Development remains 'development' as 'progress' through civilisational stages to a 'transformed' state of life, as first promulgated in the nineteenth century. This 'needed' development and 'transformation' will not be spontaneous or organic as it was before the colonial era: and nor will it be left to the self-organising capabilities of indigenous, organic, and sovereign communities. It requires, instead, external direction, engineering, funding, and management by globalist agencies which remove the sovereignty and economic independence of local communities. Instead of the East India Company, or colonial officers, the managers of 'sustainable development' are 'experts,' engineers, economists, agronomy, energy, hydrology and mining consultants, social engineers and behavioural psychologists, public health managers, deployed by private companies, intergovernmental agencies, and public–private partnerships. I discuss in Chapter 8 a case study of the attempts of such 'experts' to destroy

Bali's 1000-year-old Temple-governed rice-growing and water-sharing culture and replace it with fast-growing seeds and chemical fertilisers.

The shift from the metaphor of 'trusteeship,' which characterised nineteenth- and most twentieth-century development discourse, to the discourse of 'partnership' fits neatly with the neoliberal conception of public–private partnerships which is the preferred language of governments and corporations for the agencies which pursue the technocratic post-colonial project of 'sustainable development.' In these partnerships States license private companies to take over land, waterways, and public services such as education and health care, electricity, and water supplies. The companies turn publicly owned property into shareholder-owned property and at the same time are licensed to charge fees to the public while being promised State subsidies where user fees do not match expected shareholder Returns on Investment.

Sustainable development based on the neoliberal public–private partnership model is said to be designed to empower private and non-governmental agencies to transform the lives and environments of non-Western peoples in line with the strategic priorities of Agenda 2030 and the SDGs. But while these priorities are focused on 'the needs of the poorest and most vulnerable,' the 'Global Partnership for Sustainable Development' will nonetheless require that the lives and habitats of the poorest are transformed into private property belonging to globalist and corporate agencies who know best how to 'transform' them and who will then charge the people for products from lands and water courses they or their ancestors formerly owned as sovereign communities.[70]

By comparison with Agenda 2030, the aspirations of nineteenth-century colonial officers to control the lives of those in their areas were modest. Agenda 2030 proposes that the United Nations and its public and private partners will press forward a range of goals and practices in every country on earth that will cover the whole gamut of human life from the womb to the grave by 2030. Agenda 2030 is a technocratic and managerial blueprint for the industrial, scientistic, and technological transformation of every nation, community, culture, and form of life on Earth. The intended reach of the plan is so vast, so invasive, that hubris is too weak a word to describe it. The framers of the plan claim that it will foster 'biodiversity' and reduce species extinctions, when it involves the greatest project of human cultural integration, *homogenisation* of biological and cultural forms of life, and top-down control of human actions and ways of living ever envisaged in human history.

The agenda is also more ethnocentric and Eurocentric than the nineteenth-century doctrine of development. To give just one illustration, Sustainable Development Goal 4, concerning education, indicates that by the age of seven, children in classrooms everywhere in the world should be able 'to unfold a degendered understanding of professions, sports and family roles' and distinguish between biological functions – such as the ability to bear and nurse children – and the socially constructed idea of 'gender.'[71] The idea of 'gender identity' as distinct from biological form, emanated from 'gender theory' which was advanced in the second half of the twentieth century by scholars and writers primarily located in the United States and Britain, and most influentially in an article entitled 'The meaning of difference' by two US psychologists as recently as 1987.[72] The ideology of gender identity as distinct from biological sex is a modern Caucasian cultural construct and has no given biological locus across cultures equivalent, for example, to the universal recognition of the biological ability to give birth, and to nurse infants, of those called 'mothers' in English. The proposed imposition of this novel ideology globally on young children under the aegis of 'sustainable development' indicates the extent to which sustainable development is a Trojan horse of an ongoing post-colonial and even post-human agenda.

This post-human agenda came to the fore in the context of the 'COVID-19 pandemic' in the World Economic Forum project of a 'Great Reset' of human society and of human beings in which it is proposed by Klaus Schwab, the WEF founder, that COVID-19 represented 'a great opportunity to reset our world.' The character of this great reset is laid out in remarkable depth and detail in the reports and website of the World Economic Forum which has been preparing for this project of global transformation in its annual meetings of a globalist elite at Davos, and in its programme of mentoring of 'young global leaders' since 1994. In *COVID-19: The Great Reset*, Schwab, with Malleret, argues that after COVID-19 'there will be a gradual merging of biological, digital and psychological processes in human beings who will increasingly interact with each other via machines, screens, chips and artificial intelligence rather than through face to face and skin to skin bodily encounters.'[73] But this post-human agenda has nothing to do with human health or the health of the environment. It is a post-human dystopia advanced under the dubious heading of 'biosecurity' even as bioweapons laboratories in the United States, China, the United Kingdom, and Europe continue to genetically engineer ever more dangerous human infective agents from wild-caught species, as I discuss more fully in Chapter 5.

The project of 'sustainable development,' and the related offshoot of Schwab's World Economic Forum–sponsored 'Great Reset,' has at its heart the deep-set belief, stretching back to the European Enlightenment and the colonial development project, that Western, which is to say North American and European, forms of cultural, economic, and technological development are the ideal towards which all cultures and communities ought to be directed by global coordinated interventions of empowered 'agencies,' and that the 'sustainable industrial development' these agencies will promote in every domain and culture and ecosystem on earth will be the means to bring about human 'transformation.'[74]

Until 2020, the chances of bringing about such a vast globally coordinated and cultural transformation of human life from birth to death in every community, from land-locked Himalayan provinces to small island Pacific states, would have seemed very slim. But then, a chimeric virus was unleashed on the world from a bioweapons laboratory in Wuhan, the development of which was funded by a globalist partnership between US development and scientific research agencies and Chinese Communist research funding agencies.[75] The globally interlinked and highly invasive restrictions on normal human life and social interaction which followed in almost every nation revealed a novel level of global political and technological coordination, or 'lockstep,' of the actions of governments worldwide which was first proposed in a strangely prescient report on technology and international development by the Rockefeller Foundation in 2010. This proposed 'lockstep' of global governance was achieved using electronic media of communication and the agency of the World Health Organisation based in Geneva, and in partnership with national health ministries, mainstream media and social media, public relations, pharmaceutical and technological corporations the majority of which are headquartered in the United States.[76] Before 2020 the technocratic top-down globally invasive reach of Agenda 2030 looked impossible to achieve. But, as Schwab imperiously puts it, the globally coordinated response to COVID-19, though a 'mild disease' compared to global pandemics such as the Spanish Flu, represented a 'great opportunity' to 'transform and reset our world.'[77]

Central to the global emergence of a novel coronavirus in 2019 with certain unique features is the new technology of genetic engineering. Both the development of the coronavirus SARS-CoV-2, and the novel mDNA and mRNA drugs which were issued on an experimental basis as vaccines for the virus, involved genetic engineering techniques first developed in the field of human agriculture. These techniques represent a significant heightening of the scientistic and technocratic control of the elements of life on Earth, both nonhuman and human. In the next two chapters I consider the ecological and human pathologies which this agenda of deepened scientistic control is spawning and the anti-Gaian re-engineering of ecosystems and human communities which it advances.

Notes

1 Ella Vazquez-Dommguez and Hector T. Arita, 'The Yucatan Peninsula: Biogeographical History 65 Million Years in the Making.' *Ecography* 33 (2010) 212–19.

2 See further Michael S. Northcott, 'Buddhist Rituals, Mosque Sermons and Marine Turtles: Religion, Ecology and the Conservation of a Dinosaur in West Malaysia.' *Journal for the Study of Religion, Nature and Culture* 6 (2012) 196–214.

3 Richard Leakey, *The Sixth Extinction: Patterns of Life and the Future of Humankind*. New York, NY, Anchor Books, 1996.

4 Nikos Solounias and Adrieene Mayor, 'Ancient References to the Fossils from the Land of Pythagoras.' *Earth Sciences History* 23 (2004) 283–96.

5 Arthur O. Lovejoy, *The Great Chain of Being: A Study of the History of an Idea*. Cambridge, MA, Harvard University Press, 1936.

6 On medieval accounts of the economy of nature see Joel Kaye, *Economy and Nature in the Fourteenth Century: Money, Market Exchange, and the Emergence of Scientific Thought*. Cambridge, Cambridge University Press, 1998, 163–9.

7 M.J.S. Rudwick, *Bursting the Limits of Time: The Reconstruction of Geohistory in the Age of Revolution*. Chicago, IL, University of Chicago Press, 2005, 76.

8 Rudwick, *Bursting the Limits of Time*, 76–7.

9 Elisabeth Kolbert, *The Sixth Extinction: An Unnatural History*. New York, NY, Henry Holt, 2014.

10 Georges Baron Cuvier with Robert Kerr, *Essay on the Theory of the Earth*. Edinburgh, William Blackwood, 1815.

11 Charles Darwin, *On the Origin of Species by Means of Natural Selection*. New York, NY, D. Appleton & Co., 1860, 294.
12 Simon Conway Morris, 'The Cambrian "Explosion": Slow-Fuse or Megatonnage?' *Proceedings of the National Academy of Sciences* 97 (2000) 4426–9.
13 L.W. Alvarez, F. Asaro and H.V. Michel, 'Extraterrestrial Cause for the Cretaceous-Tertiary Extinction.' *Science* 208 (1980) 1095–108; A. Berger, 'Milankovitch Theory and Climate.' *Reviews of Geophysics* 26(4) (1988) 624–57; M. Milankovitch, *Kanon der Erdebestrahlung und seine anwendung auf das eiszeitenproblem*. Belgrade, Königlich Serbische Akademie, 1941.
14 Simon Conway Morris, *Life's Solution: Inevitable Humans in a Lonely Universe*. Cambridge, Cambridge University Press, 2004.
15 Morris, *Life's Solution*, 324.
16 Morris, *Life's Solution*, 146–7.
17 Paul R. Ehrlich, *Extinction: The Causes and Consequences of the Disappearance of Species*. New York, NY, Random House, 1981.
18 Paul R. Ehrlich, *The Population Bomb*. New York, NY, Ballantine Books, 1968.
19 Leakey, *Sixth Extinction*.
20 Stuart Pimm, Gareth J. Russell, and John L. Gittleman, 'The Future of Biodiversity.' *Science* 269 (1995) 347–50.
21 Stephen Jay Gould, 'The Golden Rule: A Proper Scale for Our Environmental Crisis.' *Natural History* 9 (1990) 24–9.
22 For fuller discussion and references see Michael S. Northcott, *Place, Ecology and the Sacred: The Moral Geography of Sustainable Communities*. London, Bloomsbury, 2015, 60–2.
23 Stuart Pimm, *The World According to Pimm: A Scientist Audits the Earth*. New York, NY, McGraw Hill, 2001.
24 Gerardo Ceballos, Andrés García, and Paul R. Ehrlich, 'The Sixth Extinction Crisis Loss of Animal Populations and Species.' *Journal of Cosmology* 8 (2010) 1821–31.
25 S. Díaz, J. Settele, E.S. Brondízio, et al., eds., *Summary for Policymakers of the Global Assessment Report on Biodiversity and Ecosystem Services of the Intergovernmental Science-Policy Platform on Biodiversity and Ecosystem Services*. Bonn, Germany, IPBES Secretariat, 2019, https://zenodo.org/record/3553579#.X45dqS2B3OQ accessed October 20, 2020.
26 Doug P. Armstrong, R. Scott Davidson, Wendy J. Dimond, et al., 'Population Dynamics of Reintroduced Forest Birds on New Zealand Islands.' *Journal of Biogeography* 29 (2002) 609–21.
27 S.H.M. Butchart, A.J. Stattersfield and T.M. Brooks, 'Going or Gone: Defining 'Possibly Extinct' Species to Give a Truer Picture of Recent Extinctions.' *British Ornithologists Club Bulletin* 126A (2006) 7–24.
28 Alfred W. Crosby, 'Ecological Imperialism: The Overseas Migration of Western Europeans as a Biological Phenomenon.' In Donald Worster ed. *To the Ends of the Earth: Perspectives on Environmental History*. Cambridge, Cambridge University Press, 1988, 103–17.
29 Crosby, 'Ecological Imperialism,' 105.
30 Alfred Crosby, *The Columbian Exchange: Biological and Cultural Consequence of 1492*. Westport, CT, Praeger, 1972.
31 John Bellamy Foster and Brett Clark, 'Ecological Imperialism: The Curse of Capitalism.' *Socialist Register* 40 (2004) 186–202.
32 Foster and Clark, 'Ecological Imperialism.' See also Eduardo Galeano, *Open Veins of Latin America: Five Centuries of the Pillage of a Continent*. New York, NY, 1973, NYU Press, 72–3.
33 On the wreckage of Indian culture and ecosystems by British imperialists see William Dalrymple, *The Anarchy: The Relentless Rise of the East India Company*. London, Bloomsbury, 2019; on the destruction of indigenous cultures, the best multi-domain overview is the collected multi-author essays in Martin Daunton and Rick Halpern,

eds., *Empire and Others: British Encounters with Indigenous Peoples, 1600–1850.* London, UCL Press, 1999.

34 Richard H. Grove, *Green Imperialism: Colonial Expansion, Tropical Island Edens and the Origins of Environmentalism 1600–1860.* Cambridge, Cambridge University Press, 1995, 6–10.

35 For an engaging account of the imperial quest for plants and its connection to the establishment of Kew Gardens and other botanical gardens, see Jordan Goodman, *Planting the World: Joseph Banks and His Collections: An Adventurous History of Botany.* London, Harper Collins, 2020.

36 Clive Ponting, *A Green History of the World: The Environment and the Collapse of Great Civilizations.* New York, NY, St Martin's Press, 1991.

37 Luisa Malfi, 'Linking Language and Environment: A Co-evolutionary Perspective.' In C.L. Crumley ed. *New Directions in Anthropology and Environment: Intersections.* Walnut Creek, CA, Altamira Press, 2001, 24–48.

38 John Muir and Michael P. Branch, *John Muir's Last Journey: South to the Amazon and East to Africa: Unpublished Journals and Selected Correspondence.* Washington, DC, Island Press, 2001.

39 Luisa Malfi, 'Biocultural Diversity and Sustainability.' In Jules Pretty, et al. eds. *Sage Handbook of Environment and Society.* Thousand Oaks, CA, Sage Publishing, 2007, 267–77.

40 Malfi, 'Biocultural Diversity and Sustainability,' 269.

41 Malfi, 'Biocultural Diversity and Sustainability,' 270.

42 M.P. Cowen and R.W. Shenton, *Doctrines of Development.* London, Routledge, 1996, ix–xi.

43 Cowen and Shenton, *Doctrines of Development,* 11.

44 Cowen and Shenton, *Doctrines of Development,* 11.

45 Gomer Williams, *History of the Liverpool privateers and letters of marque : with an account of the Liverpool slave trade, 1744-1812.* Liverpool, Liverpool University Press, 2004; Gordon Barclay, ' "Duties in Aid of a Civil Power:" The Deployment of the Army to Glasgow, 31 January to 17 February 1919.*Journal of Scottish Historical Studies* 38 (2018) 261-92.

46 On the Luddites, Levellers and other worker resistance movements in England see Frank Peel, *The Risings of the Luddites, Chartists & Plug-Drawers.* Brighouse, Yorks, J. Hartley, 1895; and Tony Benn, *Levellers and the English Democratic Socialist Tradition.* Nottingham, Russell Press, 1976.

47 Cowen and Shenton, *Doctrines of Development,* 19–30.

48 Cowen and Shenton, *Doctrines of Development,* 33–9.

49 Cowen and Shenton, *Doctrines of Development,* 40.

50 Edmund Burke, 'Speech on Mr Fox's India Bill.' In Peter J. Stanlis ed. *Edmund Burke: Selected Writings and Speeches.* Abingdon, Routledge, 2017, 321–9.

51 Burke, 'Speech on Mr Fox's India Bill,' 326.

52 On the relevance of the concept of fittingness to environmental concerns see Michael S. Northcott and Sreven van den Heuvel, 'Introduction.' In Michael S. Northcott and Steven van den Heuvel eds. *Fittingness and Environmental Ethics: Philosophical, Theological and Applied Perspectives.* London, Routledge, 2023.

53 Loren McClenachan, Francesco Ferretti, and Julia K. Baum, 'From Archives to Conservation: Why Historical Data Are Needed to Set Baselines for Marine Animals and Ecosystems.' *Conservation Letters* 5 (2012) 349–5.

54 D.V. Pauly, S. Christensen, S. Guenette, T.J. Pitcher, U.R. Sumaila, C.J. Walters, R. Watson, and D. Zeller, 'Towards Sustainability in World Fisheries.' *Nature* 418 (2002) 689–95.

55 'Monsters of the Oceans: 7 Criminal Super Trawlers That Threaten Our Waters.' *Greenpeace,* November 19, 2014, https://www.greenpeace.org.au/blog/monsters-oceans-7-criminal-super-trawlers-threaten-waters/ accessed October 22, 2020.

56 Douglas J. McCauley, Caroline Jablonicky, Edward H. Allison, et al., 'Wealthy Countries Dominate Industrial Fishing.' *Science Advances* 4 (2018) eaau2161.
57 Taryn Garlock, Frank Asche, James Anderson, et al., 'Global Blue Revolution: Aquaculture Growth Across Regions, Species, and Countries.' *Reviews in Fisheries Science and Aquaculture* 28 (2020) 107–16.
58 Joy E.M. Watts, Harold J. Schreier, Lauma Lansk, et al., 'The Rising Tide of Antimicrobial Resistance in Aquaculture: Sources, Sinks and Solutions.' *Marine Drugs* 15 (2017) 158.
59 On the toxicity of Scottish farmed salmon, and the environmental destruction caused by fish farms, see Michael S. Northcott, 'Farmed Salmon and the Sacramental Feast: How Christian Worship Resists Global Capitalism.' In William F. Storrar and Andrew Morton eds. *Public Theology for the 21st Century*. London, T&T Clark, 2004, 213–30.
60 Stefan Wirsenius, Christian Azar, and Göran Berndes, 'How Much Land Is Needed for Global Food Production under Scenarios of Dietary Changes and Livestock Productivity Increases in 2030?' *Agricultural Systems* 10 (2010) 621–8.
61 Vaclav Smil, *Harvesting the Biosphere: What We Have Taken from Nature*. Cambridge, MA, Massachusetts Institute of Technology Press, 2013, 228–9.
62 H. Charles J. Godfray, Paul Aveyard, Tara Garnett, et al., 'Meat Consumption, Health, and the Environment.' *Science* 361 (2018) eaam532400026199.
63 See further Northcott, *Place, Ecology and the Sacred*, 69–70.
64 Procheş Şerban, John R.U. Wilson, Jana C. Vamosi, et al., 'Plant Diversity in the Human Diet: Weak Phylogenetic Signal Indicates Breadth.' *Bioscience* 58 (2008) 151–9.
65 J. Janick, 'New Crops and the Search for New Food Resources.' In J. Janick ed. *Perspectives on New Crops and New Uses*. Alexandria, VA, ASHS Press, 1999, 104–10.
66 Tony Loftas, Geoffrey Lean, Don Hinrichsen, et al., eds., *Dimensions of Need: An Atlas of Food and Agriculture*. Rome, Food and Agriculture Organization of the United Nations, 1995, 21–2.
67 Andre Gunder Frank, *Capitalism and Underdevelopment in Latin America: Historical Studies of Chile and Brazil*. New York, NY, Monthly Review Press, 1987.
68 *Report of the World Commission on Environment and Development: Our Common Future*. Geneva, United Nations, 1987, Chapter 2.1. At https://www.are.admin.ch/are/en/home/sustainable-development/international-cooperation/2030agenda/un-_-milestones-in-sustainable-development/1987--brundtland-report.html accessed March 9, 2021.
69 *Resolution Adopted by the General Assembly on 25 September 2015: 70/1: Transforming Our World: The 2030 Agenda for Sustainable Development*. At https://www.un.org/en/development/desa/population/migration/generalassembly/docs/globalcompact/A_RES_70_1_E.pdf accessed March 10, 2021.
70 *Transforming Our World*, 'Introduction,' 20.
71 *SDG Resources for Educators – 5 Gender Equality*. Paris, UNESCO, 2019, https://en.unesco.org/themes/education/sdgs/material/05.
72 Rachel T. Hare-Mustin and Jeanne Marecek, 'The Meaning of Difference: Gender Theory, Postmodernism, and Psychology.' *American Psychologist* 43 (1988) 455–64.
73 Klaus Schwab and Thierry Malleret, *Covid-19: The Great Reset*. New York, NY, World Economic Forum Publ., 2020.
74 *Transforming Our World*, 'Introduction,' 21.
75 See further Sharri Markson, *What Really Happened in Wuhan*. Sydney, Harper Collins, 2021.
76 Rockefeller Foundation and Global Business Network, *Scenarios for the Future of Technology and International Development*. New York, NY, Rockefeller Foundation, 2010 and at http://www.nommeraadio.ee/meedia/pdf/RRS/Rockefeller%20Foundation.pdf accessed May 21 2021.
77 Schwab and Malleret, *Great Reset*, 142, 169.

4

REVERSE ENGINEERING LIFE

Abstract

Modern industrial agriculture and aquaculture have an inbuilt trajectory towards homogenising ecosystems. In the twentieth century, scientists developed new techniques for plant and animal hybridisation which grew de-speciation impacts. The global spread of genetically modified organisms from the 1980s further increased the de-speciation tendencies of industrial agriculture while at the same time inflicting increasing harms on human health as GMOs are novel foods for which the human microbiome is not well adapted. In tandem with increased scientific control and financialisation of food systems, nature conservation strategies are also increasingly being financialised with the aid of technologies such as blockchain and satellite-linked biosensors.

Agriculture and Religion

Industrial civilisation and the Earth System are on opposite trajectories. Gaia, over 3.5 billion years, as we have seen, has an inbuilt trajectory towards greater species diversity. Modern industrial civilisation, particularly in combination with modern agricultural monocultures, and industrial fishing, as facilitated by science and technology, has an inbuilt trajectory towards homogenising Gaia, reducing species diversity, and simplifying ecosystems so as to render them more controllable and, in the short term, profitable sources of wealth extraction in the form of monetary accumulation. In important respects this project has its roots in the original human move into seed-crop agriculture which occurred simultaneously in the Levant in the Middle East, and in Mesoamerica and Peru, around 10,000 years BC.[1] Whereas seed-cropping arose in drier areas, root-based cultivation was gradually adopted as a supplement to forest harvesting in tropical Amazon and Southeast Asia.

DOI: 10.4324/9781003110750-5

The cultivation of wild forest grasses and especially emmer wheat, and barley, was the basis of Levantine agriculture which developed as the climate warmed after the last interglacial or 'Younger Dryas.' The productivity of primitive Galilean agriculture was high as a single hectare of wild emmer and barley 'could provide more than two million calories – or about the number required by a family of three during the course of a year.'[2] But history shows that seed-based agriculture is more prone to centralised and ultimately imperial control, such as arose in ancient Egypt and Babylon. James C. Scott argues this is because seed crops are visible and the quantity of harvest predictable. Together with their above-ground storage, this renders them more easily taxable.[3] Root crops are hidden in the ground and can remain there after they reach full size, and therefore in tropical regions plant domestication was less disruptive to low-hierarchy hunter gatherer social structures.

From its inception agriculture, and especially grain-based agriculture, had a big impact on human cultures, as evidenced in significant changes in symbol systems and religious beliefs and practices contiguous with the development of agriculture. The Neolithic period in the Levant around 10,000 BC is associated with increased cultural tendencies towards symbolic thinking, abstract concepts, and 'worldviews.' Along with this development arose the idea of a female divinity often represented symbolically in Neolithic stone carvings, and a male divinity represented as a bull, also painted or sculpted. Jacques Cauvin identifies with this development new psychological capacities to think 'up' and 'down,' 'high' and 'low,' to distinguish between a transcendent or supernatural realm and the human realm of existence, and to identify sovereign power with the abstract higher realm first imagined in Anatolian and Levantine mythologies as the creator God goddess.[4]

A reasonable explanation for why agricultural modes of dwelling and production fostered this mentality is the observation that when humans lived in forests they experienced a realm of agencies or beings which influenced their existence including other animals, tree canopies, trunks, leaves and root systems, the soil, myriad insects, birds and plants that lived in and around trees, and the rain, rivers, the sun, and the sky. In a forest, no one being is completely dominant. When humans moved from forests onto grasslands and began to clear trees to grow more grasses for milling cereals, and for grazing animals, the beings which dominated their landscape and which they did not control were far fewer. There would have been large mammals such as deer, wild ox, and wild cats, and images of these animals are depicted in the oldest preserved cave art in late Neolithic agricultural areas such as in Sulawesi and Southwest France. The other newly highly visible beings were the sun and moon, rain, rain clouds, and wind which dominated the new experience of Being which Neolithic agrarians had in the Levant. This is why it was in this region that humans first developed conceptions of supernatural beings who dwell above and beyond the Earth, as do the sun and the moon whose phases so significantly influence the growth of crops.[5] These newly conceived beings included the Neolithic goddess cult, and the cults of Pharaoh, Baal, and Yahweh which gradually supplanted the goddess cult in the Levant and later, in the form of Abrahamic monotheism, did the same in the Mediterranean region and much of Southwest Asia, North Africa, mainland Europe and beyond.

Agriculture favoured the evolution of what Ara Norenzayan calls 'big gods' because agriculture, and agricultural developments, and especially food storage and irrigation, made it possible for the product of the land to be stored, and so reduce dependence directly on the seasonality of food.[6] Large stores of surplus grain changed the relationship between humans and the Earth and they made possible the evolution of large-scale societies and large-scale infrastructure projects such as the pyramids, and the irrigation schemes of ancient Egypt and Babylon. Large granaries gave power to those who controlled them over other humans, as the Genesis story of Joseph, an Egyptian prince, mythically indicates: Joseph as vice-regent of the Pharaoh used granaries to store up the product of the land in the 'seven plentiful years' and in the 'seven lean years' that followed, the grain was doled out to the people in exchange for their ownership of the land.[7] The story is a neat mythic description of the historic roots of feudalism, slavery, and sovereign rule in agricultural technology and of the power of food technologies to shape destinies, imaginaries, lands, and societies.

Hunter gatherer societies, and root-crop based agriculture, are historically characterised by low hierarchies, polymorphous sacred beings, and weak States, as Scott shows consummately in his study of the highland regions of Southeast Asia, *The Art of Not Being Governed*. Tribes in this extensive region, such as the Karen and Kacin, have long resisted the cult of the modern nation state but are even now being militarily subjugated and taxed by the China-backed military junta which rules Myanmar.[8]

The Fall into Agriculture

In the Gilgamesh Epic of ancient Babylon, in the Hindu Vedas, and in the Book of Genesis, forest dwelling and nomadism are depicted as ideal forms of society from which humans descended, or 'fell,' into settlement, seed selection, and slavery. Genesis neatly encapsulates this contrast mythology. After the acts of Creation, Paradise is said to be a state enjoyed by the original human hunter gatherer couple who are not beholden to an earthly lord and are made in the divine image, which in ancient Egypt was the ascription given only to human rulers – the Pharaoh god-kings. Being fated to live by plant selection – agriculture – is depicted as the outcome of the original eviction of early humans from the forest garden of Paradise in the Genesis story. Eve and Adam are said to have been forced out of a life of relative leisure and ease because they selected the wrong plant to eat – the 'forbidden' tree of the knowledge of good and evil. I read this as a metaphor for the knowledge Neolithic humans acquired to control plants and animals – choosing which plants and animals to domesticate and which to suppress. The outcome of their eviction from a life of relative leisure in which they plucked fruits and seeds from trees and wild grasses was agriculture: a life dependent on sweat and toil in which they had to grow their food in competition with 'weeds.' The saga in Genesis ends, after many plot twists and turns, including the Great Flood, with the descendants of Adam and Eve caught up in the feudal power of the first large-scale agricultural civilisation from which, in Exodus, an Israelite born Egyptian Prince – Moses – ultimately is said to liberate them.

The rest of the Hebrew *Tanakh* can be read as a contrast story to the agricultural civilisation and enslaving cult of the Pharaohs of ancient Egypt.[9] The ancient Hebrews, as I have explained more fully elsewhere, are depicted by their historians as being called by their God to a new monotheistic society in which only one Being is truly Lord, and all other persons and creatures, including judges, chief priests, merchants, kings, lands, rain, sun, and wind are subjects of the one Lord.[10] The effect of the desacralisation of earthly powers was intended to be democratic: each household was to have a share of the land under the Israelite covenant, and no ruler or merchant could legally claim an everlasting ownership of land divinely gifted for the *nagalah* – the inherited property – of each true Israelite. But as the story proceeds, the Israelites demanded kings to rule over them who gradually turn into a reflection of the enslaving idolatrous cult of Egypt whence their ancestors had escaped. Solomon enslaves his own people to build a vast temple, and the decline of Israel's kings thereafter, which ultimately issues in the division of the Kingdom's imperial invasions, and exile from the land, is a foregone conclusion.

Other religious traditions also encapsulate this contrast story between the centralising and imperial possibilities agricultural technologies realised in the evolution of human societies and shifting cultivator or peasant agriculture in which people gather and plant their own food on a mix of shared and household-owned land, build their own houses, and make their own clothes and household artefacts. The Hindu Vedas have an equivalent, and older, contrast story, though as Hinduism becomes routinised this contrast gets encapsulated in the Hindu caste system in which food growers and forest dwellers are 'low' caste and subjugated by the priestly and urban-dwelling 'high' castes in a social system which discriminates against those who grow food and in favour of those who live from the labour of others, a system which lives on to this day in contemporary India. The contrast story recurs in Greek mythology though, as in Hinduism, it is again reversed as compared to the romantic depiction of hunter gathering in the Tanakh. Classical Greece was a culture primarily of City States in which manual labour and food growing were activities performed by slaves and not free men. The same reversal occurred in ancient China where the contrast is sustained by two different but interrelated religious traditions – those of Taoism and Confucianism. Taoism in ancient China was the religion of forest dwellers and rice growers while Confucianism was an urban and courtly cult. To this day Confucianism is the preferred cult of rulers and city dwellers, while Taoism is still the cult more often followed by rural peasants.

One of my aims in this book is to explore the interactions between human relations with the Earth System, or Gaia, and the evolution of religions.

Cultures of plants, ecosystems such as forests, savannah, montane, or maritime areas, and humanly domesticated ecosystems, such as grain fields, root crops, and rice padi, exercise powerful shaping forces on human symbolic representation, and on religions. The emergence of monotheism in association with the Levantine origin of agriculture is strong evidence for this association. So too are the influences of cultures of plants, landscapes, and climate regions on religions in the polytheistic or pantheistic tendencies of indigenous religions; in peasant Hinduism, Buddhism, and Taoism; and in the myths of classical Greece, including the Gaia myth.

Industrial Agriculture against the Grain of Gaia

Contemporary industrial agriculture, and especially in its latest forms of the Green Revolution and Genetic Engineering, represents an emergent techno-mythology of a supreme all-powerful deity which we might simply call 'Science.' The goal of the modern cult of science – 'scientism' – is to make visible, countable, and controllable all the subjects of the food-growing domain including the crops, the machines, the chemicals, the field workers, and the owners. This quasi-religious quest has no empirical association with improved living standards and wellbeing for it *dispossesses* peasants and hunter gatherers from their ancestral lands. But in its quasi-religious character the cult of scientific agriculture is a kind of perverse monotheism in which there is only one god – science itself – and the priesthood are scientists and agronomists. The overarching goal of agronomy is to maximise production of individual species in engineered ecosystems in which as many elements as possible are brought under the control of the scientist, and the farmers who labour under, and follow, 'science.' Scientistic agronomy is a supreme example of what C.S. Lewis presciently called 'government in the name of science' which 'is how tyrannies come in.'[11] For Lewis scientism is above all a spiritual error because it displaces the God of love, which for Christians is the transcendent meaning of the 'God who is in Jesus,' with the idol of control.[12] The one who truly loves desires the freedom and fulfilment of the other, but the one who idolises control desires power over the other.

The founder of the modern cult of scientism was Francis Bacon, who believed that through the control, manipulation, and re-engineering of all life, including ultimately the human body, Paradise would be re-established on Earth and humans might even develop the capacity to overcome death.[13] The Western scientistic myth of Baconian control over nature, far more than the Genesis myth of the Garden of Eden and agrarian humanity's eviction from Paradise, is the root of the contemporary Promethean overreach of humanity. For thousands of years after the discovery of plant selection and agriculture, agrarian, as well as fishing practices, had little impact on species diversity. As already discussed, premoderns do seem to have extinguished a small number of mega-fauna including mammoths in Siberia and Northern Canada, and moa – giant flightless birds – in New Zealand, though the extinction of mammoths is at least as likely to have been caused by climate change as by human over-hunting.[14] But since the scientific and industrial revolutions, humans have begun to eradicate species at an increasing rate and in ways that far exceed earlier agrarian and aquacultural tendencies to selectively domesticate species and reduce species diversity in ecosystems in the course of human food production. The key reason for this tendency is not the numbers of people, and domesticated animals on the land, as Western environmentalists such as David Attenborough argue, but industrial agriculture, which is an enterprise still largely in the control of the West because of the prominent role of corporations, and Western agricultural science in directing the agricultural agencies of formerly colonised nations towards policies and programs which favour the conversion of traditional mixed and subsistence agriculture into monocultures of cash crops, cereals, and industrialised animal factories.

Central to this scientistic turn of modern agriculture are scientific techniques for selecting genetic traits in agricultural plants using methods such as radiation to speed up the 'natural' growth cycles of plants and so precipitate faster heritable genetic selection.[15] With these methods from the 1960s crop scientists developed new fast-growing hybrids of wheat, corn, and rice which, in concert with increasing use of herbicides and pesticides, produced dramatic rises in yields in the early years of these new technologies. At the same time scientists adopted new synthetic chemical technologies which interfere in electrical and hormonal signalling systems of multi-cellular beings to suppress competitor plants and insects.

In the developing world, the significant increases in production per hectare these technologies facilitated was dubbed the 'Green Revolution' and in countries such as China, India, and Indonesia staple crops – and particularly rice – are still largely grown using these methods.[16] But as ecosystems and competitor plants and insects began to adapt to the new levels of manipulation to which they were subjected, greater chemical inputs were required as the hybrid seeds were weaker than traditional seeds in resisting pests and competitor plants. This required annual increases in crop chemicals which drove many farmers into debt and landlessness and led to high suicide rates among farmers in the developed and developing worlds. The political economy of chemical farming eventually removed millions of peasant and subsistence farmers from the land and resulted in large Chinese, European, and North American food corporations such as Cargill, Bayer-Monsanto, and the Bill and Melinda Gates Foundation, becoming the principal owners of agricultural land, and the principal growers of the new hybrids.

The constant dousing of plants with life-killing chemicals also impacts the crucial ecosystem of the soil rendering it virtually lifeless in many farming areas. In Indonesia, as in much of Europe and North America, agrochemicals have killed most farmland worms and other biota in industrially tilled soils, thus reducing available food for wild birds and other species. Killing life in the soil stops the millions-of-years-old process by which Gaia produced life from soil. It takes 3,000 years for Gaia to generate six inches of topsoil sufficient to grow plants. Agrochemicals, tillage, irrigation, and wind erosion also result in soil erosion on a huge scale and a global average of six kilograms of soil lost per kilogram of food produced. The Mississippi River, which drains the heavily industrialised prairie fields of the United States carries 300 million tons of soil a year into the Gulf of Mexico, causing a huge dead zone visible on Google Earth.[17]

The outcomes of 70 years of agricultural technologies along these lines has rendered large areas of the Earth System's single most generative life force other than the ocean – soils which took billions of years to evolve by rock weathering, plant growth, microbial growth, and fungal decomposition – increasingly lifeless. The lack of microbial and organic matter – humus – in soils renders them more liable to erosion and flooding and destroys the capacities of soils to soak up humanly produced carbon and turn it into oxygen, just as pelagic fishing is wrecking the oceanic photosynthetic carbon pump.[18] Ploughed soil erodes at least twice as fast as natural slopes and this counteracts evolved soil production and erosion processes which is a key agent of Gaian climate homeostasis or stabilisation.[19]

Despite the problematic, anti-evolutionary direction of industrial agriculture and aquaculture in the twentieth century, considerable resources are still invested by governments, development agencies, and private corporations in ever more invasive forms of re-engineering of life down to the cellular level in an increasingly centralised and technologically controlled global food project. The most striking feature of the increasingly globally interconnected industrial production of food is that it replaces the 500-million-year-old and highly biodiverse, complex, and interconnected nutrient cascades which constitute pre-anthropic food webs with human food chains in which most nutrients are synthetically produced from oil and other harvested or mined minerals. Often represented by globalist organisations such as the UN Food and Agriculture Organisation and the World Bank as an endeavour to 'feed the world,' the real objective is not to support resilient, community-based and locally owned farms, or the food sovereignty of indigenous, subsistence, and small farmers who still produce the majority of human foods. The real objective is to maximise the opportunities of bankers and investors to increase Returns on Investment (ROI) in agricultural and food products and land purchases. In the pursuit of the idol of ROI, increased crop production per acre regardless of ecosystem impacts remains the goal of agricultural scientists and agencies, and of corporate and government-regulated farm management systems.[20]

Industrially produced food stuffs are bought and sold and speculated upon as economic commodities such as the influential Chicago futures market. This process of speculation combined with increasing corporate control of food production has resulted in a situation in which 90 per cent of humans presently derive most of their nutrients from just 85 crop species, an estimate obtained from a valuable study of national food supply chains.[21] Even this number of species in the human diet is declining from the continuing corporate narrowing of human plant dependence by agricultural agencies, biotechnologies, and multinational corporations. Globally, these trends have resulted in countries depending for 69 per cent of their food production on non-indigenous or 'foreign' seeds and plants and a declining number of seed commodities dominated by corn, rice, soya, vegetable oils, including palm oil, and wheat.[22]

The process of de-indigenisation of food production began with the Columban exchange. But despite the formal end of colonialism, post-colonial regimes continue to reduce global/local biocultural diversity in former colonies. The dominant response of national governments, agronomists, and globalist agencies to the ongoing decline in indigenous and traditional seeds, and the threat to long-term human food security from over-reliance on a small number of seeds, has not been an effort to re-diversify peasant and indigenous farming, and to reduce farm size, since smaller farms have a historically much better record than large farms in conserving soils and sustaining more species. Instead, food globalists are attempting to patent and secure genetic diversity in 'gene' and 'seed' banks such as the Svalbard Seed Bank in sub-arctic Norway.[23] This banking approach to genes and seeds underwrites the top-down monocultural and oligarchic tendencies of a food system which continues to reduce local food sovereignty, and resilience to shocks such as climate change and violent conflicts. At the same time, it fails to address the continuing die-off of soils and species as a result of continued intensification of industrial chemical agriculture and its increasing spread into relatively unmodified ecosystems in both tropical and temperate zones.[24]

The Centrality of Agriculture

As we have seen, agriculture – the selection and domestication of plants and animals – originated at scale approximately 10,000 years ago. It represents a set of practices which together constitute the means by which humans produce food and fibre for nutrition, clothing, shelter, and warmth – for life. In pre-modern societies, food growing and gathering are central activities to society in which every person plays a part. In hunter gatherer cultures, which also practice agriculture on a small scale, such as Swiden in Borneo, each member, from child to adult, has a particular set of tasks, while in settled agrarian societies most participate in agricultural production at certain points in the year, especially the harvest. Viewed in this light, agriculture has cultural and ecological significance which goes far beyond the production of food, for through participation in the practices of food production and animal husbandry people are socialised into a cooperative relationship with the land and with nature which is expressed in symbols and rituals which in their turn give shape to human life itself.[25]

Agriculture is more than human survival or food production. In some senses agriculture *is* civilisation as most of humanity has traditionally understood the meaning of this word. In pre-modern society, agricultural rituals are not solely focused on the production of food but are also concerned with the integration of land and life, humanity and nature, with cosmic harmony and peace with the divine. This understanding of the cultural and ecological significance of agriculture is not absent from industrial societies but it is more frequently found in the imagined farms of the artist or the photographer, or in the quest for authentic wholeness in the ex-urbanite's quest for rural self-sufficiency on a small holding. E.F. Schumacher believed agriculture had three purposes, only one of which was the production of food: the others are to keep people in touch with living nature, and the humanisation and ennoblement of the whole habitat of humanity. Schumacher also saw the spiritual significance of agriculture which is ecologically as well as ethically sound:

> Although ignorance and greed have again and again destroyed the fertility of the soil to such an extent that whole civilisations foundered, there have been no traditional teachings which failed to recognise the meta-economic value and significance of 'the generous earth'. And where these teachings were heeded not only agriculture but also all other factors of civilisation achieved health and wholeness. Conversely, where people imagined that they could not 'afford' to care for the soil and work with nature, instead of against it the resultant sickness of the soil has invariably imparted sickness to all the other factors of civilisation.[26]

The American ecologist Aldo Leopold was pointing to the same interconnections between agriculture and civilisation, agriculture and human relations, and agriculture and religion when he said there are two spiritual dangers in not owning a farm: 'one is the danger of supposing that breakfast comes from the grocery, the other that heat comes from the furnace.'[27]

Agriculture in pre-industrial societies is essentially the cultivation and nurture of life in all its diversity. It represents a set of practices and of inherited wisdom by which human life and nonhuman life are brought into a community of interest for a range of goals including the nurture of the landscape, the preservation of biodiverse soils, ecosystems, and water catchments for future generations, the celebration of the cyclical nature of life, the quest for harmony between human and divine purposes, and the nurture of the soul. This sense for the spiritual significance of food and farming is central to a huge range of pre-industrial religious practices including the religion of ancient Israel. The ancient Israelites were in the main agriculturalists, and they treated the land as part of their ethical, covenantal community. The Torah, the law of Moses, charged them to respect the land, and to recognise moral constraints on its use, because they understood that it was given to them as gift:

> When you enter the land which I am giving you, the land must keep Sabbaths to the Lord. For six years you may sow your fields and prune your vineyards and gather the harvest, but in the seventh year the land is to have a sacred (sabbatical) rest, a Sabbath to the lord.[28]

Agriculture was practised by the Egyptians, the Babylonians, and the Persians as well as the ancient Israelites but according to Torah, Israelite agriculture was distinguished from other nations by the tradition that the land was not Israel's possession but rather the gift of God received after the Exodus. This tradition is reflected in Jubilee and Sabbath laws which limited mercantilist concentration of land ownership and prevented members of the tribes of Israel from becoming again serfs or slaves on the lands of others: 'No land may be sold outright, because the land is mine, and you came to it as aliens and tenants of mine.'[29]

Land in ancient Israel was understood primarily in terms of covenant: in Israel's ancestral memory, the land was said to have been gifted to Israel to rescue their forefathers and mothers from slavery in Egypt, Egypt being the first large-scale agricultural civilisation in human history. In Israel, God was said to be the original owner and hence the Israelites were stewards rather than owners and justly distributed patterns of land use were supposed to be the means by which all Israelites continued not to be slaves.[30] This covenantal concept of divine ownership of the land had implications not only for patterns of land tenure but also for land use: the Israelites were not to abuse the land by harshly ploughing and forcibly irrigating it as the Egyptians did. Nor were they to use up so much of the land for farming that no space was left for wild animals.[31] Such practices would have helped to conserve soil, in contrast to the irrigation-dependent practices of Egyptian agriculture. Preservation of the fragile soil of this dry hilly region of the Eastern Mediterranean was therefore enshrined in the sacred texts of the children of Israel for guidance in their use of the land. The Israelites were also warned in Torah that however successful their farming, they should not forget that the fertility of the land arose from its divine givenness as the means of their liberation from slavery in Egypt and not from human manipulation.[32]

In the cultic traditions of Israel, and in particular in certain of the Kingly Psalms, the fertility of the land is rhetorically linked to divine justice as reflected in the society of God's people:

> O God, endow the king with thy own justice,
> and give thy righteousness to a king's son,
> that he may judge thy people rightly
> and deal out justice to the poor and suffering.
> May hills and mountains afford thy people
> peace and prosperity in righteousness.[33]

The fruits of the land, its original redemptive abundance, in this poem are said to be dependent upon Israel keeping the terms of the covenant. Where God is worshipped in truth, where every household in the covenant community has access to the means of livelihood, where neighbour does not oppress neighbour with usury or debt slavery, where the poor, the widows, and the fatherless are offered hospitality, and where the animals are not cruelly abused, then there will the land remain fertile and meet the needs of the people with abundance. Correlatively, the neglect of the worship of God, and neglect of the respect for land and social justice which the covenant required, would bring about real threats to livelihood and security.

According to the Hebrew prophet Jeremiah, the divinely originated created order of the cosmos was subverted when the rulers of the people failed to worship God and respect the covenant law:

> But this people has a rebellious and defiant heart,
> they have rebelled and gone their own way.
> They did not say to themselves,
> 'Let us fear the Lord our God,
> who gives us the rains of autumn
> and spring showers in their turn,
> who brings us unfailingly
> fixed seasons of harvest.'
> But your wrongdoing has upset nature's order,
> and your sins have kept from you her kindly gifts.[34]

The prophet linked ecological devastation with the abandonment of the worship and commands of the Lord. Because the people of Israel had turned from the Lord, their land, its mountains and streams, animals and crops, was being laid waste, polluted, and destroyed. The prophetic reading of ecological breakdown points to a conflict between the unequal society spawned by the late Israelite monarchy and the fertility of the land, the welfare of the created order. The pride of kings had denuded valleys of tree cover for their massive building projects, wealthy landowners were abusing the earth from greed, generating soil erosion, while the new landless poor went hungry, and so the land suffered and shared in the alienation that human corruption produced:

> Woe to those who add house to house and join field to field until everywhere belongs to them and they are the sole inhabitants of the land.
> Yahweh Sabaoth has sworn this in my hearing,
> 'Many houses shall be brought to ruin, great and fine,
> but left untenanted;
> ten acres of vineyard will yield only one barrel,
> ten bushel of seed will yield only one bushel.'[35]

Jeremiah and Isaiah argued that such ecological effects as soil erosion, desertification, and declining rainfall, as well as the Exile of the people of Israel from their once fertile land, were connected consequences of the abandonment of the terms of the covenant under which the Israelites had become resident in the land. They were the consequence of the practices of idolatry, mercantilism, and the dispossession by the rich of the sacred inheritance of many in Israel.

The contemporary devastation of rural communities by industrial agriculture in Europe, North America, and much of the Southern hemisphere, and the enforced exile of rural communities to the cities, is a consequence of modern farming methods which also destroy the health of the soil, decimate resident species, and reduce the nutritional quality of human foods. North American farmer and writer Wendell Berry describes the cultural condition to which chemical and industrial agriculture had reduced his own rural farming community in rural Kentucky:

> Counting the post office, the town has five enterprises, one of which does not serve the local community. There is now no market for farm produce in the town or within forty miles. We no longer have a garage or repair shop of any kind. We have had no doctor for forty years and no school for thirty. Now, as a local economy and therefore as a community, Port Royal is dying.
>
> My part of rural America is, in short, a colony, like every other part of rural America. Almost the whole landscape of this country – from the exhausted cotton fields of the plantation South to the eroding wheatlands of the Palouse, from the strip mines of Appalachia to the clear cuts of the Pacific slope – is in the power of an absentee economy, once national and now increasingly international, that is without limit in its greed and without mercy in its exploitation of land and people.[36]

Graham Harvey describes the ecological as well as human impacts of modern industrialised farming and the heavy public subsidies for output quotas which sustain it. In most of Britain chemicalised and mechanised farming reduced employment on the land to the point that many rural communities are now little more than commuter belts for the urban economy. Meanwhile, the communities of species of all kinds which once lived on farmland are rapidly disappearing because of the rain of pesticides and herbicides which government policies have directed onto Britain's industrialised and increasingly silent fields:

> Scores of arable land species are currently under threat. Among the flowering plants corn gromwell, pheasant's eye, shepherd's needle and corn buttercup, while endangered bird species include the corncrake, stone curlew, Montagu's harrier, the curl bunting and the grey partridge. Half a century of public funding has given farmers unprecedented power to re-design rural Britain. Their legacy to a new generation is a sterile, lifeless landscape.[37]

Whereas traditional farmers regarded as their first duty the stewarding of the land, and in particular their duty to pass on soils on their land in better condition than they found them to the next generation, modern industrial farmers, and the scientists and agronomists who advise and subsidise them, push their farms to the edge of infertility by their failure to conserve soils and species:

> Topsoils are disappearing from susceptible land throughout the length and breadth of Britain. The rush to intensive farming is robbing a future generation of its capacity to grow food. We are stripping away the nation's true wealth and dumping it in our rivers and streams. Not that soil erosion is new.

> The signs have been there for decades. But farm support policies that make intensive arable cropping profitable on marginal land have added enormously to the scale of loss. Everywhere old grassland has been ploughed up, particularly on the marginal land of hill and chalk down. The shift to arable crops leaves the land bare or thinly covered in the autumn and winter, when most of the damaging rains are likely to fall.[38]

Vandana Shiva charts a similar process in farming in India. Government subsidies for the chemicalisation and industrialisation of farming, and in particular the move to the monocrop planting of high-yield varieties of rice and other cereal crops associated with the Green Revolution, have had major impacts on plant and animal biodiversity while damaging the sustainability of formerly self-sufficient rural farmers. The new technologically enhanced high-yield seeds only offer a high yield with the expensive application of herbicides and pesticides whose high cost has created much indebtedness in rural communities and an epidemic of farmer suicides. The widespread planting of single strains of foreign seeds have created major problems of pest control and disease which makes the new crops much more chemically dependent, and more susceptible to climatic and other environmental threats, than traditional plantings of different seed varieties. The abandonment of traditional systems of rotation planting leads to a reliance on expensive chemical inputs which make the new agriculture too expensive for many small farmers, driving them into debt, even though small farmers remain the principal source of food in agrarian communities in the South:

> The Green Revolution created the perception that soil fertility is produced in chemical factories, and agricultural yields are measured only through marketed commodities. Nitrogen fixing crops like pulses were therefore displaced. Millets which have high yields from the perspective of returning organic matter to the soil, were rejected as 'marginal' crops.[39]

By treating essential organic inputs that maintain the integrity of nature as 'waste,' the Green Revolution strategy ensured that fertile and productive soils are laid waste. The 'land-augmenting' technology has proved to be a land-degrading and land-destroying technology.[40]

'Famine at the Heart of the Feast'[41]

In addition to deleterious effects on rural communities, on the economics of small and subsistence farms, on biodiversity, and on the quality of the soil, the global embrace of intensive agriculture by science-informed official agricultural agencies in partnership with corporate lenders and seed and chemical companies is accompanied by changes in the quality of diet and the availability of fresh. The United Kingdom is a prime example of this tendency. Not only does the United Kingdom have one of the most intensively farmed land areas in Europe but it also supports one of the most heavily processed and centralised food supply systems in the world. From the Enclosures of the eighteenth century to the subsidised silent fields of the twenty-first century, British agriculture is characterised by a growing alienation between the food consumer and the food producer. Giant supermarket chains control two-thirds of British food supply and, along with the largest farmers, they are the principal beneficiaries of the subsidy system which has laid waste rural communities and the quality of the soil.

Supermarkets are purveyors of a vast range of processed and junk foods of which the British eat far more than any other nation in Europe. The British also eat much less fresh and locally produced food than their continental counterparts and suffer correlatively higher rates of dietary disease of the kind that consumption of fresh fruit and vegetables are known to protect against. Pesticide residues in fresh foods are also high in British supermarkets: one-third of foods showed residues above recommended levels in a 1995 government survey. If the quality of diet is generally poorer in fresh foods in Britain than on the continent, the diet of the poor in Britain is particularly damaging to the health of themselves and their children. Residents of many inner-city estates often have no local access to purveyors of fresh foods, and the limitations of the minimum wage or government social security payments mean that they are forced to resort to the cheapest available calories which are to be found in frozen and processed foods such as meat burgers and pasties.

What Harvey calls the 'famine at the heart of the feast' is also a consequence of the declining quality of chemically farmed foods. Industrially produced cereals, fruits, and vegetables have fewer minerals and nutrients naturally occurring in them than traditionally grown produce and this means that dietary deficiencies in certain key minerals such as zinc and selenium are increasing in Western countries. Food manufacturers sometimes put back certain essential dietary elements into foods, for example into factory produced bread and breakfast cereals. Consumers in the West may also take food supplements as awareness has grown that the vitamin and nutrient content of industrial foods is lower than traditional foods. Consumers in the South are unable to supplement their diets in this way and malnutrition is increasing in many rural communities where farmers have been encouraged by governments and seed and chemical companies to adopt the new seeds and growing methods.

Even more serious are the many millions of people whose ability to grow their own food has completely collapsed because of changes in food production systems. These changes are mostly visited on developing countries by agronomists deployed by such agencies as the World Bank, the International Monetary Fund, and the Bill and Melinda Gates Foundation. Somalia is a particularly tragic example of Western-directed agricultural 'improvement.' Until the 1970s Somalia was self-sufficient in food, but policies imposed by the IMF and World Bank forced a shift in land use from nomadic pastoralism and subsistence agriculture to food production for export which resulted in growing dependence on imported grain. Cheap surplus wheat and rice had the effect of undermining local producers of traditional crops such as maize and sorghum. In addition, privatisation of veterinary services and water supplies enforced by the World Bank decimated the livestock herds of pastoralists. Somalia is now wrecked by civil war, widespread hunger, piracy, and terrorism and is a classic example of a 'failed state.' But the root cause of the failure was not 'Islamic terrorism' but globalist interference in Somalia's successful subsistence farming village society.[42]

Similar globalist policies imposed upon Ghana greatly increased the amount of land devoted to the cultivation of cocoa in the 1980s. At the same time IMF-enforced devaluation of Ghana's currency reduced the value of export earnings from cocoa, while trade-liberalisation measures imposed by the IMF and the Word Bank encouraged the importation of cheap surplus cereals and vegetable oils from Europe and North America. Local producers were hit very hard, and malnutrition in poor communities has greatly increased in Ghana, as in many countries in sub-Saharan Africa in recent years after the external imposition of 'structural adjustment,' 'free market,' 'deregulated' agricultural practices by the World Bank and other globalist agencies.[43]

The relentless global shift towards unsustainable industrial agriculture is closely related to cultural processes amongst both farmers and agricultural scientists which have generated a focus on the goals of increasing productivity per planting and per hectare above other agricultural goals such as preserving rural communities, conserving the soil, and improving human health through dietary improvement. Berry argues that this productionist approach to agriculture originates in the scientific laboratories of the public land-grant universities of the American mid-West.[44] Paul Thompson finds the same approach replicated in agricultural laboratories in Europe and the developing world:

> Despite the positivist ethic of value neutrality and the doctrine of academic freedom, the work ethic of farmers has been carried over into agricultural science with little revision. The scientist's products are a sign of worth as much as the farmer's bales and bushels. Agricultural scientists regard their work as successful when it is widely adopted, and the surest path toward adoption is to increase the productivity of a farming operation. Thus, scientists have laboured hard to increase productivity as a sign of their own virtue.[45]

The principal justification for biotechnological research is the enhancement of farming productivity. Scientists and agronomists claim that biotechnology is the only way, with current levels of soil erosion, that it will be possible to feed the growing population of the world into the next century. Such assertions are made with no reference to the social scientific literature on hunger and malnutrition and its causes even though we know from this literature that the principal cause of global hunger is not low farm productivity but lack of access to available food among landless rural and urban poor. In Britain, to take just one example, malnutrition is reappearing among children of the poor while farmers complain that the industrial food system wastes up to 40 per cent of the food they grow.

The Rise of Genetically Modified Organisms

The next stage in the industrial race to increase monocultural production per hectare arrived in the 1990s in the form of genetically modified organisms which are developed via the laboratory manipulation of genes in the novel technologies of genetic editing and plant and animal engineering. This new science allows humans not only to select plants and animals and progressively breed them for traits such as high gluten content in wheat or increased breast size in chickens, but to introduce or 'recombine' DNA from one plant or animal and insert it into the genetic structure of another plant or animal so as to confer a desired genetic characteristic on it. So, for example, in the early days of plant engineering scientists introduced an anti-freeze gene from a deep-sea fish into a strawberry plant to improve frost resistance. Insightfully described as 'a form of molecular agriculture,' genetic engineers use microbiology in combination with greenhouses and livestock laboratories to create new hybrid life forms that did not and would not evolve in 'nature' without human interventions.[46] The most widely dispersed of these novel life forms are genetically modified soya and rape/canola seed plants. Both plants have been modified in their DNA not to express a hybrid trait from another plant but to resist the widely used herbicide glyphosate. This trait allowed farmers to dowse their fields liberally with a chemical concoction that disrupts the electronic hormonal signalling system in all plants except GM crops and so kills them.

My own introduction to the science fiction world of genetic engineering came in a research project on the ethics of genetic engineering set up under the auspices of the Science, Religion and Technology Project of the Church of Scotland in Edinburgh. At the invitation of Donald Bruce, a British nuclear physicist, I joined a group of scientists in a collaborative research project on the subject of genetically modified foods whose collective deliberations were published in 1999 in the book *Engineering Genesis*. One of the scientists who joined our group was a man called Ian Wilmut from the Rosslyn Institute. Wilmut was catapulted from relative obscurity onto the front cover of *Time* magazine a few months after joining the group because of his laboratory's success in perfecting and using the technique of nucleic cell transfer in the reproduction of a cloned animal. Wilmut had succeeded for the first time in human history in artificially cloning a living mammal from an adult animal.

As part of our research for *Engineering Genesis* we visited the Rosslyn Institute and were introduced to Dolly, who enjoyed a pristine enclosure in the outbuildings of the Rosslyn biotechnology campus. She was visually no different from thousands of other sheep though her face became familiar to millions. Dolly was marked not so much by her appearance as by her location, not on a farm or in an abattoir, but in a pen at the heart of a large pharmaceutical research institute. She had to be kept in an enclosed pen because scientists feared for her safety, and for her health in the Scottish elements. Dolly was conceived not in the uterus of a sheep but in a laboratory, not through the intercourse of a female with a male sheep, but in a Petri dish by the insertion through a needle thinner than a human hair of DNA extracted from the udder of a six-year-old sheep which was then inserted into an embryonic sack obtained from another sheep and from which all the genetic material had been removed.[47] Five months after cloning Dolly, Wilmut's team successfully cloned another sheep, Polly, who was genetically modified with *human* genes, which was the ultimate end in view. Wilmut's team performed nuclear transfer 277 times – and this statistic implies a large number of surgical procedures on sheep – before their success with Dolly. But as the techniques became more refined more predictable results could be achieved.[48] In the production of subsequent cloned animals only 50 procedures were typically involved.[49] Next, Wilmut's team succeeded in producing a clone, Tracy, genetically modified with human genes, whose milk contained the human protein *alpha-1-antitrypsin* used in the treatment of the lung disease emphysema and cystic fibrosis[50] (Figure 4.1).

The birth of Dolly produced a worldwide controversy. Wilmut was alternately hailed and pilloried, as the scientist who had taken humanity dramatically closer to the Brave New World imagined by Aldous Huxley where humans clone their young in artificial wombs and engineer them into social groups or classes by genetic selection of intellectual and physical attributes of each clone before birth. Both the Clinton administration in the United States and Jacques Santer of the European Commission directed panels of experts to deliberate on the legal and ethical issues associated with cloning technology. Later President Clinton wrote a memorandum prohibiting the use of Federal Government funding for the cloning of human beings.[51] A US bioethics Commission called for this ban to be extended to all private companies, clinicians, researchers, and professional and scientific societies because 'to create a child by somatic cell nuclear transfer into a mother's womb would at this time be an irresponsible, unethical, and unprofessional act.'[52] Pope John Paul II also established a panel to deliberate on cloning at the Pontifical Academy of Life which concluded that human cloning would be against the will and purposes of the Creator.[53]

Principled religious opposition to cloning of humans, *and* animals, arises from the sense that with this technique humans have begun to play around with the constitution of life itself, and that this takes human creativity beyond the boundaries of morality and of wisdom. Leon Kass, a biochemist and a philosopher, in testimony to the American National Bioethics Advisory Commission quoted the Christian ethicist Paul Ramsey: 'a man of serious conscience means to say in raising urgent ethical questions that there may be some things that men should never do. The good things that men do can be made complete only by the things they refuse to do.'[54]

The accusation that scientists who want to clone life are abrogating to themselves quasi-divine powers – that they are 'playing God' – is given added weight by the observation that the desire to clone life originated in the same romantic spirit which infused the scientists and artists of the eighteenth and nineteenth centuries. They recorded the minute details of life itself, including life in embryonic, microscopic mode, in all its artistry and infinitesimal complexity because

FIGURE 4.1 Dolly the Sheep at a photo shoot on the cover of Donald and Anne Bruce, eds., *Engineering Genesis*. 1st Edition, London, Earthscan, 1998. © Taylor & Francis.

they believed that seeing, understanding, and creatively mimicking the embryonic seeds of life gave access to, purchase over, the meaning of the universe.[55] The act of cloning life enables humans not only to mimic the process of embryonic reproduction but to come to understand more deeply the structure of life itself, and so to grasp more securely their own identity as makers and shapers of a biosphere without God. As Stanley Hauerwas put it, in the advent of cloning we can see 'a kind of drive behind this for us to be our own creators.'[56]

Engineering Genesis

In *Engineering Genesis* we found that one of the principal objections to Dolly the Sheep, and the cloning of further animals, was that it was contrary to the fundamentals of created life since cloned animals are virtually identical to the adults from whose DNA they are artificially reproduced.[57] The production of biological replicas, we suggested, represents a rejection of the divinely originated diversity which characterises life on earth. Some have contested this point, partly on descriptive grounds. They point out that cloned sheep are merely exemplars of naturally occurring biological twins – the phenomenon in itself is not new to nature and cloning is not therefore evidently opposed to evolved, and/or divinely given, creative diversity.[58] They further point out that traditional animal breeding has selected for particular animal traits which after centuries of refinement means that animals on an average farm are much less biologically diverse than they would be under conditions of 'nature.' But the issue of copying, and the denial of diversity, has real force and is a fundamental part of the problematic that Dolly, that cloning, and especially the possibility of cloning humans, represent. Cloning is a form of creativity which seems to be a denial of the possibility of true creation. This is scientific creativity as replication and not as a bringing forth of the true possibilities of created order. It is in a way a denial of life, not its affirmation: a stultifying mechanistic copying of the life God and Gaia confer on humans and other kinds.

Objections to Dolly, and to cloning more generally, derive their force from the sense that there is a technological advance in her creation or invention which goes way beyond previous animal breeding technologies, and previous technological interventions in reproduction. The technique of cloning new individuals from the cells of living adult beings is said to represent an overstepping of the legitimate boundaries of human creativity: the creature is usurping the place of the creator or 'playing God.' This odd language involves the claim that this kind of invention represents a degree of technological mastery over nature which is arrogant, morally wrong, and of which the consequences are incalculable.

In 'The Question Concerning Technology' Martin Heidegger described modern technology – *techne* – as a kind of 'unconcealment' of the real within nature: 'Technology is a mode of revealing. Technology comes to a presence in the realm where revealing and unconcealment takes place, where *aletheia*, where truth, happens.'[59] Heidegger drew a distinction between older forms of technology and modern technology. From the Greeks to the medieval era, Heidegger argued that technology involved a 'bringing forth,' a *poesis*, of what was already within the givenness of the

FIGURE 4.2 Electricity-generating turbine hall, Lake Manapouri, South Island, New Zealand, January 2008. Author photo.

natural order. But modern technology involves a challenge, an interrogation, even a torturing of nature which does not so much call forth its essence as wrest from nature a 'standing reserve' of energy, a force which humanity turns towards the controlling ends of technological production. The form of standing reserve that Heidegger describes is the modern hydroelectric dam in which the energetic flow of a river is turned into a standing body of water for the generation of electricity (Figure 4.2).

For Heidegger, modern technology is indicative of an underlying spiritual disorder which is that the only revealing, the only truth, which can be brought forth in modernity is mediated through the human, and that therefore the real cannot be revealed within nature apart from human projects. The answer to this sickness he argues is artistic *poesis*. As for Friedrich Nietzsche, the new metaphysics, the new morality Heidegger proposes, is aesthetics. Where technology and science have imposed an alien order on nature as standing reserve or resource bank, art, poetry, music can bring forth the truth of being, including other-than-human being. Poetic dwelling is the only truly authentic form of the revealing and unfolding of the human vocation to be a 'guardian of being.' But in the birth of the clone, even the claimed distinction between aesthetic creativity and technological invention is swept away. In the future it will be possible to engineer strawberries that glow in the dark; to engineer genetically modified sheep with wool which is pink or brown or whatever the fashion industry desires in a particular year.

The birth of the first clone of an adult sheep, even though it contained no human genes, marked a significant heightening of the modern agricultural project not only to select and domesticate nonhuman beings but to redesign them, re-engineer them, in such a way as to render them more deeply fitted to human intents and desires. Dolly was therefore

> a symbolic site for the exploration of identity, heredity, destiny, and the social meaning of science, she is a spectacular beast. She is one more step in a series of dreams about science – evoking for some, euphoric fantasies; for others, horrible nightmares and the fear of science out of control. She offers up the possibility of hyper-rationality in the management of bodies and of complete genetic control of cows, sheep, and humans as well. She offers the spectre of technical decisions that will turn all bodies (human and animal) into intentional products, manufactured and designed 'on purpose'. She evokes a way of thinking about bodies as little more than efficient mechanisms for the production of 'value' – be it milk, or meat, or creative imagination.[60]

The defenders of cloning claim that it is simply a technological advance in a gradual line of progress from other previous advances, including the conventional hybridisation of animals to express more milk or grow more muscle for less feed inputs, the discovery of DNA, the identification and mapping of the genome of humans and other animals, artificial insemination, and in vitro fertilisation. But Dolly symbolised at least as dramatically as the first in vitro conception of a human foetus a new stage in human stewardship of the planet, where humans are in a position literally to *invent* life after human purposes and designs, and in particular after the commodity form: it was the economic *efficiency* of a potential new therapeutics production method, not the creation of new therapeutics in itself, which was the driver behind the invention of nucleic cell transfer technology.

In *The Technological Society*, Jacques Ellul argues that the ethical problem when efficiency becomes the overarching goal of the technological society is that it shapes human life towards the reductive end of technical and administrative efficiency. Societies so shaped set aside all other values, including goodness, beauty, and truth, in the pursuit of efficiency. This pursuit also results in 'automatism'; the automatic choice to use new technologies regardless of the possible consequences because the primary managerial goal of efficiency directs the choice.[61] On this account, the moral question concerning cloning technology is not that this technology enables humans to do things unnaturally – humans do lots of things which are unnatural all the time – but the ends towards which the technological reproduction of life directs human life. For Elul technologies are social-shaping and this can be discerned in influential interactions between technological inventions and the larger institutional and social settings in which they are adopted. Technologies also mediate the scientific and socio-economic forces in which they arise: they do not spring unmediated from the genius of the inventor.[62] In technological mediations human ends tend to be adapted towards technical means. Examples are legion: the levels of security and control required by the creation and recycling of plutonium, the centralised managerial and administrative systems needed to run railways or chemical plants, the patterns of human dwelling fostered by motor cars and freeways. As Langdon Winner suggests, artefacts have politics, and the history of technology includes many 'instances in which the invention, design, or arrangement of a specific technical device or system becomes a way of settling an issue in a particular community.'[63]

During our collaborative research for *Engineering Genesis*, the question of ends was one of which most natural scientists in our group were dismissive. They did not see its relevance. Whether their science was funded by a pharmaceutical corporation or a public university, whether a technical invention such as a genetically modified seed relying on a particular patented herbicide advances a corporate form of agriculture – these questions were not seen as relevant to their role as inventors.[64] But if Ellul and Winner are right, the view of invention as the product of the lone genius, and of technics as means and not ends, are misapprehensions. The scientist dwells in a community and his research experiments and inventions manifest the larger ends and accounts of reality which are authoritative for that community.[65]

The outcomes of the Church of Scotland project on GMOs were reported in the form of a collaboratively written book which was published just as the controversy over GMOs spilt over into a major public campaign against them.[66] The very fraught and public debate led to the widespread banning of GMOs in human foods in Europe. GMOs were viewed by European regulators as novel organisms because the mix of genes the new methods made possible could not occur under natural selection, or in conventional plant breeding techniques. But in the United States, GM foods were widely introduced into the human food chain with little public debate. This was because the US Food and Drug Administration adopted the claim of some of those involved in patenting GMOs that food crops derived from these patents were 'substantially equivalent' to non-GM foods and therefore required no special licensing, labelling, or regulation. The claim rested upon laboratory investigation by crop scientists, who might be said to have an interest in the outcome of the chemical composition of GMOs.[67] But no investigations of the effects of these new substances on mammals beyond 30 days were undertaken before the claim was made, although long-term mammalian tests are mandated by the US FDA for novel substances created in pharmaceutical laboratories using analogous biochemical technologies to those used in the creation of GMOs. Subsequent efforts by public scientists to investigate the long-term effects of GMOs on mammals were resisted by public and private agencies, and there remains a paucity of such studies.

Genetic Engineering and Food Futures

Two French peer-reviewed studies of the effects of GMOs on the mammalian gut beyond 30 days of dietary introduction found significant toxic effects, including the production of cancerous cells in rats fed Roundup Ready maize for 90 days.[68] However, the findings were contested by biotech companies, and one of the two articles was retracted. In Scotland, Hungarian scientist Arpad Pusztai, who described the toxic effects of GMOs on animals in a TV documentary, was sacked by the Rowett Laboratory, his computers and hard copies of data taken from him, and efforts made to prevent him publishing his results as an independent scientist.[69] A fourth scientist, Ignacio Chapela, was similarly subjected to an exceptional campaign of intimidation and reputation destruction on his research into the contamination of traditional Mexican maize varieties by GM corn.[70]

The considerable efforts of biotech companies and laboratories to suppress research into potential toxicity of GM foods, and so protect the profits of biotech companies such as Monsanto-Bayer and DuPont, mean that there are no long-term studies of the effects of GM foods on mammalian or human guts. But despite this, there is still considerable scientific evidence that genetic engineering techniques in both animals and plants promote genetic mutations including in gene-edited animals and in animals and humans which consume GMOs. The best attested of these mutations are cancers, including leukaemia, in transgenic mice, and other GM animals. In humans such cancerous mutations may also be provoked by the consumption of gene-edited foods because these are broken down in the gut where they may cross over the gut–blood barrier and promote changes in DNA.[71] The absence of a sufficient number of rigorous studies of the health effects of the consumption of GMOs, because of the suppression of scientific investigation by the biotech industry, makes definitive scientific judgement impossible: nonetheless anecdotal evidence from farmers and consumers indicates increasing concern about the effects of GMOs on human health and the environment.[72]

There is more extensive scientific evidence of environmental and reproductive health problems in humans associated with glyphosate, which is the most widely used herbicide on GM crops engineered for herbicide tolerance. One study found that exposure to glyphosate at levels lower than those found in glyphosate sprayed crops and in the water catchments which drain them is injurious to human placental cells and so to unborn children.[73] Other studies show associations between exposure for just two days a year to glyphosate and the cancerous condition non-Hodgkin lymphoma.[74] Glyphosate is now so prevalent in GM corn and in water catchments in the United States that exposure to it on a daily basis is routine for hundreds of millions of people. And yet when Monsanto applied for a licence to market GM corn in the United States, which it claimed was 'substantially equivalent' to non-GM corn, it submitted studies of GM corn to US food regulators which had *not* been exposed to glyphosate even though GM corn is designed to respond genetically to the application of the herbicide and the herbicide therefore impacts on the genetic structure of the plant.[75] No safety studies were performed on GM corn treated with glyphosate.

Given the growing toxification of food and water promoted by GMOs, a few State legislatures in the United States are considering State laws requiring the labelling of GMOs in foods in their states to allow consumers to avoid these products. The first to pass such a law was Vermont. But Vermont State was sued by a coalition of American food processors in an attempt to overturn the law.[76]

Against the background of growing controversy over GM crops and foods in the developed world, I proposed to colleagues at Durham University that we put together a research project on the outcomes of two decades of genetically modified foods which the John Templeton Foundation funded. Our aim was to investigate whether the widespread adoption of GM crops in developing nations provided any lessons for understanding (and potentially overcoming) the controversy in North America and Europe. The project research outcomes reveal precisely the same disconnect between crop scientists and the lay food-eating public that characterise the GM crop controversy in Europe and in North America. They also indicate that the disconnect between agricultural science and ecologically sustainable forms of agriculture identified by Berry and others is not unique to GM crops but a broader feature of the science–food relationship.[77]

The 'GMFuturos' research data show how percipient was Berry's analysis of the ecologically destructive trajectory of science-informed industrial farming in the unfolding saga of GM foods in Mexico. Mexico is the origin of the largest agrarian social movement in human history – La Via Campesina.[78] This movement originated in the 1980s among peasant farmers and urban food growers and consumers who saw that the mechanisation, chemicalisation, and hybridisation of corn, and other staple foods, threatened both the ability of Mexicans to feed themselves in the future, and put at risk the enduring cultural pattern of food growing on small plots which for many Mexicans (those not living in big cities) remains central to a good life.[79] Anthony Montoya's research on the symbolism of food in Mexico reveals that cultural meanings around maize-growing and meal preparation are ontological, sociolinguistic, moral, politico-economic, and cultural. They are situated in embodied relationships between peasant farmers, cooks, and householders, and between people and land. When an agricultural system that neglects these is imposed on a people, it inevitably courts political controversy, including contestation over GM foods. The laboratory interviews, as well as the interviews with farmers, in the GMFuturos project reveal that this relationality between people and land, culture and agriculture, is missed by crop scientists whose primary training is focused on maximising production of an individual crop in a laboratory or on a university or crop institute plot.[80] In other words, when agricultural science is primarily about what is done in a food laboratory, absent of considerations of what impacts what is done in the laboratory will have on farmers, food consumers, or other species, then the likely outcome will be a growing disconnect between science-informed farming and the health of farming communities, of farming ecosystems, and of food consumers. Science is not neutral for those interviewed for our project in Brazil. Instead they perceive GM crops, and the advocacy of them by multinational corporations, as a value-laden and politically and economically portentous project which was locking farmers into dependence on expensive hybrid seeds, and into an expensive and polluting chemical arms race against increasingly herbicide resistant weeds.

Heisenberg and Poesis

The reflexive quality of the nature–culture relationship in agriculture is at odds with the perception of science–society relationships held by many in the scientific community, and among the corporate funders of much scientific research. That laboratory science and physical reality are reflexive was first proposed by Walter Heisenberg who argued that laboratory instruments such as electron microscopes are capable of modifying the behaviour of the physical subjects their users investigate, and this finding is now called the 'uncertainty principle' or the Heisenberg principle.[81] The Heisenberg principle was brought into popular culture by an influential US TV series called *Breaking Bad*. The main protagonist of the series, Walter White, is a chemistry teacher in Albuquerque, New Mexico, who develops lung cancer, treatment for which his health insurance does not cover. To meet the costs of treatment, and to provide financial security for his family on his death, he sets up a mobile methamphetamine laboratory with a junkie who has contacts with drug sellers. He makes exceptionally pure crystal meth and the product acquires a mythic reputation. He moves his operation into an underground laboratory hidden beneath a Chinese laundry and as the product grows in quantity he has to get into higher gear in terms of selling, which brings him into competition with Mexican-based drug barons. In recognition of the exceptional purity of the product, White acquires the market name 'Heisenberg.' But as the name becomes better established, and the quality and quantity of his production methods continue to increase, so the bodies start piling up. Despite the purity of his product, and the underground location of his lab, White is ultimately drawn into the violence associated with drug dealing. His lab, his cover day-job, his domestic suburban lifestyle, and his police family in-laws, are no longer sufficient to insulate him from the death-dealing drugs market, and the consequences of what he is making in his laboratory start to visit him in real life: he becomes caught up in a downward spiral of successively more bloody acts of violence among those he chooses as partners in crime.[82]

Breaking Bad works well as a metaphor for GM crops. GM crops are more 'pure' from a scientific perspective than a Mexican farmer's inherited seeds. They are designed for a specific purpose and with laboratory instruments: each gene has been supposedly charted, counted, and inserted at a level of microbiological precision that is unavailable outside of a highly insulated and 'clean' laboratory. But this 'clean' lab underwrites the belief that a laboratory-made crop will not influence the environment, the farmers, the eaters, and other species who interact with it. In the field, however, it is inevitable that GM characteristics edited into plants in the laboratory spread to other plants. In Brazil, GM Futoros researchers took photographs of triffid-like weeds up to 4 metres high growing in soya crop which were immune to glyphosate and other agrochemicals: the only way of removing them without wrecking the soya crop was by hand with machetes and spades.[83]

GMOs manifest even more deeply than non-GM crop science the anti-holistic, mechanistic, and reductionist frame of scientistic epistemology. Biotech scientists refuse that their considerable expertise and 'advances' in creating novel genetic mutations in animals and plants can have problematic effects outside the laboratory. And in so refusing, they are in denial of one of the greatest discoveries of twentieth-century natural science, which is the principle of reflexive uncertainty. Walter Heisenberg who pioneered the splitting of the atom in Germany and the United States in the 1930s, discovered that electron microscopy influences the behaviour of subatomic particles. The act of isolating an electrically charged particle with the use of electric scanning techniques *influences* the behaviour of the particles. In other words, insulation does not work. And hence the naming of the leader character in *Breaking Bad* as Heisenberg: his brilliance as a chemistry teacher in purifying crystal meth ultimately could not isolate him from the toxic effects of the drug among drug users, and from the violence of those who deal in drugs.[84]

For Vandana Shiva, the most influential Indian campaigner against GM crops in the GMFuturos study area, GMOs are products that are deeply intertwined with neoliberal economics. This is because GM crops offer private corporations such as Monsanto the ability to privatise gene races and the fundamentals of human food cultures on every continent, and Monsanto, now Bayer, in particular have pursued that corporate goal with considerable success, and despite extensive resistance from civil society. [85]

Poesis is the word given by classical Greek philosophers to the unique capacity of humans to generate and sustain aesthetic and ethical ideals through their creative powers. For Plato, makers such as craftsmen and cooks, when they combine knowledge of the ideals with their practices of making, become exponents of the 'liberal arts' just as do philosophers or sculptors.[86] This same conception of *poesis* is found in other ancient stories, including those of Jews and Christians, about the cultural powers involved in making temples or furniture, and including the making of food.[87] In the origin story of the Jewish people, their ability to grow food on their own plots of heritable land is seen as the source of their freedom from the slavery of collectivised agriculture in Egypt. Similarly in Christian history, cultures of farming, and of crafts and workshops, become sources of freedom when they are mediated by communitarian craft Guilds and by land distribution patterns that enabled yeomen farmers to flourish, and which restrained the powers of nobles and princes.

The gradual neoliberal collectivisation of agriculture by globalist agencies and private corporations repeats earlier collectivising projects from ancient Egypt to Maoist China, and it threatens the same losses of freedom and sovereignty that these earlier collectivisation projects imposed on the peoples subjected to them. The contemporary neoliberal collectivisation project in the spheres of food-growing and making is underwritten by scientific agriculture, and GM crops. The latest technological turn of industrial food producers is to the large-scale construction of artificial growing environments such as multi-storey buildings in cities in which foods are grown hydroponically with light-emitting diodes to replace sunlight, and nutrients in water to replace soils.[88] The expert knowledge which creates these new science-based food systems, and the high-cost inputs required, are in the vanguard of this new collectivisation project in the developing, and developed, world which transfers power and deliberation over the growing and making of food from farmers and householders to private corporations and venture capitalists.[89]

Resistance to GM crops in the developed, and developing, world reflects not only uncertainty about the risks of GM foods to the environment or to human health. It also resonates with the reflexive relationships between culture and agriculture, and between political freedom and distributed powers over land and making. Neither the walls of the science laboratory, nor the biota-free environment of the petri dish, can isolate the products created therein from contestation over these relationships. And yet the drive to *control* Life, and even hack the software of Life by engineering the information system of genetic code, remains the Holy Grail of biotech scientists. This domineering and Promethean quest for control is presented by biotech advocates as the means to achieve 'biosecurity' not only in the human supply of food but also in nature protection, and, as we will see in Chapter 5, in human health.

Control, Exclusion, and Surveillance as 'Nature Protection'

Techniques of biological control and engineering associated with the Statist and corporate projects of biotechnology and 'biosecurity' are not confined to food-growing: they are also colonising nature conservation. Scientists and public and private conservation agencies increasingly seek to secure the future of threatened species by excluding evolved human and nonhuman agency in protected ecosystems or landscapes designated as 'protected areas' so that these areas can be managed scientifically for the protection of specific endangered species. This approach is combined with techniques of satellite surveillance and the use of digital biosensors, which are small digital tracking devices on animals, trees, and other biological objects for remote data collection from conservation sites by satellite and for recording in computer systems.

One such project, which was rolled out in 2021 during the COVID-19 pandemic by a Brazilian company called Moss, involves the establishment of a new asset class of crypto-backed Non Fungible Tokens (NFTs) linked to 735 million trees that the company proposes to plant in the Amazon.[90] Similar schemes in many countries are connected with the 'carbon credit' market set up by the United Nations Framework Convention on Climate Change (UNFCCC), and are associated with claims concerning 'carbon emissions savings' which are fictional. Such schemes are often characterised by large-scale timber extraction and destruction of old growth forests belonging to indigenous communities, and their replacement with fast-growing monocultures such as eucalyptus. Two examples of this in Southeast Asia are a 'nature conservation agreement' granting monopoly rights over 2 million hectares of forest in Sabah, North Borneo, to a Singapore-based shell company against which indigenous peoples and rainforest conservation organisations have protested to the United Nations.[91] Another such corrupt scheme grants to an Indian company, Kanaka Management Services, monopoly rights to 397,000 hectares of forest in Oro Province of Papua New Guinea for 100 years, and was approved by the United Nations as meeting its Verified Carbon Standard despite its clear violation of the rights of the indigenous population, and its corrupt attribution of 60 per cent of income from the UN 'carbon credits' scheme to the company.[92] Norwegian scholar Bram Buscher leads the academic analysis of this new financialisation of nature project.[93] The ultimate aim of this approach is to turn the project of nature conservation into a digitised blockchain information system in which conserved plants, animals, and even insects perform as units in information systems and are rendered amenable to financialisaton and marketing as biodiversity credits and other novel tradeable financial assets.[94]

The protected area approach to conservation of certain environments in the industrial era goes back, as discussed above, to the romantic protest against industrialism in the eighteenth century which was also where Modern environmentalism originated. This approach, as I argue more fully elsewhere, had religious and spiritual roots in the Protestant adulation of unaltered or 'wild' nature as revelatory of the divine Spirit, or the 'sublime.' Places such as the Lake District in England, the Trossachs in Scotland, the European Alps, and the Californian Sierra Mountains were set aside as places for human physical and spiritual renewal through such activities as hiking and mountain climbing. But as we have seen, this setting aside involved a problematic conception of a human–nature divide, and the assumption that humans are invariably degraders of natural environments when human activities are uncontrolled by conservation agencies and park authorities. While it is the case that humans for centuries have tended to remove undomesticated mammals from the land through hunting and the use of fire, the degradation of land and ocean ecosystems to the point of desertification, biodiversity loss, and local climatological change is principally a feature of the industrial-era. Far from reversing these trends, protected areas, as a conservation strategy, are signally failing as a nature restoration strategy in both temperate and tropical regions, and in the oceans. And they are not only failing but, in most cases, as discussed in Chapter 8, they are *advancing* desertification and other kinds of ecological breakdown because they are based on false assumptions about the causes of these.

A classic example of the problems that ensue when humans are evicted from ancestral guardianship of place concerns the national parks in the Western United States. Scotsman John Muir, who is honoured as the 'founding father' of the US National Park Service, began his engagement with the American west as a shepherd, and later as a recreational walker and campaigner for the conservation of the mountainous landscapes of the Californian Sierra from industrial developments and in particular hydroelectric dams.[95] Muir also campaigned for the extension of the park areas, and before he died, much of what he worked for had come to pass. In Yosemite, the extension of the national park to include its current large area from high montane to the lower plateau, and rocky valleys through which flows the mighty Yosemite River, led to the eviction of the native Americans whom Muir himself in his diaries had regarded as superfluous to the 'wilderness' he so much admired, and even as a kind of pollutant or desecrating presence to the new sacred of 'wilderness.'[96] Within ten years of the indigenous inhabitants being evicted, it soon became apparent that they had for centuries performed certain essential functions in managing the landscape, such as setting controlled fires, which were also beneficial to other resident species. And without their presence, the balance of species present when the park was set up, began to change.

In recent years, the park authorities have determined that one crucial species that the native Americans had permitted to live on their lands alongside them was the wolf. And it turns out the wolf provides crucial land management characteristics that keep other grazing and browsing species – especially deer – away from certain areas., The presence of wolves enables a richer flora and fauna to grow than when these grazing species over-dominate. The reintroduction of wolves into the American West was therefore sponsored by the National Park Service, beginning in Yellowstone National Park, and wolves have now spread through the national parks system right down to the southwestern United States. Beneficial ecosystem results are seen in consequence, and have been charted in a number of scientific papers, as well as in other media such as televisual documentaries.[97]

The recovery of the agency of other creatures in the management of ecosystems reflects a realisation about the asymmetric relationships that industrial humans – even when science-informed and involving 'conservation projects' such as national parks – have typically developed with other creatures and their habitats. I think it is reasonable to say that Jean Calvin anticipated something along these lines when he wrote in *Institutes of the Christian Religion* that the problem with the city – the most humanly made landscape of his time – is that it is most representative of the work of humans, and since they are fallen, the glory of God is less manifest in cities than it is in those parts of the world that have been less modified since the origin of creation.[98] In effect, industrial technologies, developed in and managed by residents of industrial cities, have allowed the humanly dominated urbanscape – in which human agency, human acts, and intentions overbalance the agency, acts, and intents of other creatures – to spread from cities across almost the whole surface of the Earth, including even protected areas.

Calvin's belief that the excess of human influence in cities reduces their capacity to reflect the divine glory of the creator, as compared to forests and 'wild' landscapes, is a theological imaginary of 'nature' that, as I have argued elsewhere, is a significant root of the development, first among Protestants in Western Europe and North America, of the modern idea of 'wilderness' as a place where human influence is little in evidence, and where humans are overawed by the magnificence, scale, and 'sublime' qualities of the landscape.[99] But there is a paradox in this imaginary which, as it has become increasingly detached from its theological roots represents wilderness as intrinsically nonhuman, and humanity as an alienated, overpowering and ultimately destructive presence in the landscape, and even on Earth. In this way of thinking, James Lovelock, of whom in a more positive vein I write below, argued in one book that at some point in the future, the Earth, or Gaia will finally slough off this presence, so allowing life to return to its former path of speciation, making the Earth fertile and benign for all creatures, not just for humans. Weirdly then, as its theological underpinnings fell away, a Protestant theology of fall and redemption has produced misanthropy, and even despair, about human influence on the rest of creation that was not part of Calvin's theology, and which is even more alien to the theologies, both Catholic and Orthodox, that preceded the Reformation.

This misanthropy is most evident in various kinds of ecological apocalyptic, or dystopias, such as Lovelock's metaphor of humanity as a virus, or the essay, and later book, of Alan Wiseman entitled *The World Without Us*.[100] Such ecological apocalyptic is marked by an absence of hope. Some even suggest that hope is the enemy of action to restrain and reverse the current trajectory of human influence and that a seasoned pessimism is a more 'useful' attitude than hope.[101] But ecological restoration, as in the case studies which I describe in Chapter 8, indicates that what works best is not the total absence of humans, but intentional restraint by humans in the management of ecosystems, and a preparedness to give space to other creatures so they recover agency and formative influence on habitats and places. But ecological hope does not arise from the scientistic belief that humans are the only agential beings on Earth and that only by excluding their agency can species be 'biosecured' for posterity. Instead it arises from recognition of the co-evolved agency of indigenous humans and other animals and plants as described in in biocultural theory as reprised in Chapter 3.

Despite their misanthropic character, 'protected areas' where normal human activities are excluded have spread to many domains around the world where they usually result in environmental exclusion, migration, poverty, and even genocide as the cultures of indigenous peoples in these areas die out after they are forced to migrate to urban areas. Ironically, the first such protected area, the English Lake District, is still filled with non-native grazing animals, shopping malls, and other commercial activities. But protected areas in Africa, Asia, and the Americas follow the exclusionary model first commended by Muir. Scientists such as E.O. Wilson, along with global conservation agencies such as the International Union for the Conservation of Nature, the UN Environment and Program, and the World Wildlife Fund, propose that one-third to one-half of the land area of the Earth should now be taken from local people and put in the hands of government and globalist agencies and scientists, in order to address the alleged global extinction crisis.[102] This is despite the well-evidenced fact that this approach has genocidal implications for indigenous peoples as it leads to the forcible, and often armed, exclusion of local people from their ancestral forests, grazing areas, and water catchments, and despite the fact that they and their animals have a better record than scientific conservation agencies at preserving forests, soil quality, species diversity, and unpolluted water catchments.

At the core of the protected area approach are not only misanthropy – and especially towards non-Europeans – and the claim that there is an intrinsic conflict between humans and 'nature,' but also the scientistic dream of total control of every element in ecosystems, a dystopic approach that may only be realised when the spontaneous agency of humans and even of other creatures is controlled or excluded. This quest for full-spectrum control of ecosystems, and even the Earth System, was advanced at increased pace and scale by the measures to control human activities allegedly to prevent the spread of the SARS-CoV-2 virus from 2020. During lockdowns, citizen journalists in countries as distant from each other as Germany and Brazil posted videos and photographic evidence of increased deforestation activities in the absence of people, and especially people deriving livelihoods, in forests. In other words, once the eyes of citizens were removed from forests because of forcible home detention, agencies, both governmental and private, moved into forests to extract more lumber, and lay them waste for conversion into yet more monocultures. There were also increases during the pandemic in forcible exclusions of indigenous peoples from forests and other areas, including in the Congo and in the Amazon. In tandem with this increased rate of exclusion of humans, new nature financialisation projects were rolled out in which biosensors and blockchain were linked via satellite and GPS surveillance systems.

In Chapter 8 I explore more hopeful and alternative approaches to the revival of ecosystems in the shape of forms of agriculture involving the reintroduction of large mammals at scale in food-growing areas as agents of conservation, and holistic ecosystem management. Unlike scientistic conservation, these approaches utilise long-evolved symbiotic relationships between diverse species and people, including domesticated animals, and between species, soils, and water catchments, in way that regenerate soils and plant growth, produce food, increase ecosystem resilience, and sustain human health and livelihoods. It is in association with such forms of agriculture that indigenous and peasant communities are able, when free from predatorial corporate development and the priorities of globalist 'sustainable development,' to recover their sovereignty as the most conscious beings on the planet to work with their fellow creatures in creating dwelling places and deriving nutrients, and building and cultural materials, for living creatively and imaginatively without destroying their fellow creatures' habitats and continued existence

In conclusion, and to anticipate the chapter which follows, industrial agriculture, as shaped by agrochemicals, productionist agronomy, and biotechnological gene editing, creates the illusion that humans are in control of life on Earth, and that life on Earth is best advanced, in terms of both available nutrition to humans and of species biodiversity, by *increased* scientistic control of all elements in the nonhuman environment, including both agricultural production and nature conservation. In the next chapter I explore the role of gene editing in the evolution of the 'COVID-19' pandemic of 2020–22 and the related quest of medics, bioweapons experts, and the pharmaceutical industry to control not only the movements and meetings of people – as in 'public health' measures widely adopted in the pandemic – but even the nutrients in foods they are permitted to grow and eat, and the microbiota in their stomachs.

Notes

1 Kent V. Flannery, 'The Origins of Agriculture.' *Annual Review of Anthropology* 2 (1973) 271–310.
2 Flannery, 'Origins of Agriculture.' 278.
3 James C. Scott, *Against the Grain: A Deep History of the Earliest States*. New Haven, CT, Yale University Press, 2017.
4 Jacques Cauvin, *The Birth of the Gods and the Origins of Modern Agriculture*. Trans. Trevor Watkins, Cambridge, Cambridge University Press, 2000, 71–2.
5 Ara Norenzayan, *Big Gods: How Religion Transformed Cooperation and Conflict*. Princeton, NJ, Princeton University Press, 2013.
6 Norenzayan, *Big Gods*, 23–47.
7 Genesis 37-50.
8 James C. Scott, *The Art of Not Being Governed: An Anarchist History of Southeast Asia*, New Haven, CT, Yale University Press, 2009.
9 See further Ellen Davis, *Scripture, Culture and Agriculture: An Agrarian Reading of the Bible*. Cambridge, Cambridge University Press, 2008.
10 Michael S. Northcott, 'Food Sovereignty from Joshua to La Via Campesina.' In *Place, Ecology and the Sacred: The Moral Geography of Sustainable Communities*. London, Bloomsbury, 2015, 121–36.
11 C.S. Lewis, 'Willing Slaves of the Welfare State.' *The Observer*, 1958.
12 This phrase is from my PhD supervisor, and late Bishop of Durham, David E. Jenkins in his *The Glory of Man*, Oxford, Clarendon Press, 1968, 43.
13 Francis Bacon, *Novum Organum*. Ed. Thomas Fowler, Oxford, Clarendon Press, 1889; for an influential consideration of the ecological impacts of Bacon's ideas see Carolyn Merchant, 'The Scientific Revolution and the Death of Nature.' *Isis* 97 (2006) 513–33.
14 Paul S. Martin, *Twilight of the Mammoths: Ice Age Extinctions and the Rewilding of America*. Berkeley ÇA, University of California Press, 2007.
15 See further Paul S. Thompson, *The Spirit of the Soil: Agriculture and Environmental Ethics*. New York, Routledge, 2005.
16 For a fine account of the Green Revolution with particular focus on the impacts in South Asia see Vandana Shiva, *The Violence of the Green Revolution: Third World Agriculture, Ecology and Politics*. Lexington, KY, University Press of Kentucky, 2016.
17 Statistics on soil erosion from a United States Department of Agriculture presentation 'Soil Erosion,' at https://www.nrcs.usda.gov/Internet/FSE_DOCUMENTS/nrcs142p2_010152.pdf accessed September 15, 2021.
18 For a highly cited piece on the science of soil respiration and its impacts on the Earth System see William H. Schlesinger and Jeffrey A. Andrews, 'Soil Respiration and the Global Carbon Cycle.' *Biogeochemistry* 48 (2000) 7–20.
19 David R. Montgomery, 'Soil Erosion and Agricultural Sustainability.' *PNAS* 33 (2007) 13268–72.
20 The reductionist focus of agricultural science on seed productivity to the exclusion of the health of soils, water courses, or farmers, is brilliantly documented in Thompson, *The Spirit of the Soil*.
21 Robert Prescott-Allen and Cristine Prescott-Allen, 'How Many Plants Feed the World?' *Conservation Biology* 4 (1999) 365–74.

22 Colin K. Khoury, Harold A. Achicanoy, and Anne D. Bjorkman, 'Estimation of Countries' Interdependence in Plant Genetic Resources Provisioning National Food Supplies and Production Systems.' *The International Treaty on Plant Genetic Resources for Food and Agriculture*, Research Study 8 (2014) Food and Agriculture Organisation and at https://www.academia.edu/16329978/Estimation_of_Countries_Interdependence_in_Plant_Genetic_Resources_Provisioning_National_Food_Supplies_and_Production_Systems accessed June 25, 2021.

23 Beatrijs Bossuyt and Olivier Honnay, 'Can the Seed Bank Be Used for Ecological Restoration? An Overview of Seed Bank Characteristics in European Communities.' *Journal of Vegetation Science* 19 (2008) 875–84; 'The Role of the Svalbaard Seed Vault in Preserving Crop Genetic Diversity.' In M. Ehsan Dalloo, ed., *Plant Genetic Resources: A Review of Current Events and Future Needs.* Cambridge, Burleigh Dodds Science Publishing, 2021, 217–28.

24 For an excellent survey of the continuing threat to biodiversity represented by current trends in global agriculture, including a full range of reference, see A.M.J. Henk ten Have, *Wounded Planet: How Declining Biodiversity Endangers Health and How Bioethics Can Help.* Baltimore, MD, Johns Hopkins University Press, 2019.

25 Colin A.M. Duncan, *The Centrality of Agriculture: Between Humankind and the Rest of Nature.* Montreal, McGill Queen's University Press, 1996.

26 E.F. Schumacher, *Small Is Beautiful: Economics as if People Mattered.* London, Abacus, 1974, 113.

27 Aldo Leopold, *A Sand County Almanac and Sketches Here and There.* Oxford, Oxford University Press, 1949, 6.

28 Leviticus 25: 2-3.

29 Leviticus 25: 23.

30 Northcott, 'Food Sovereignty from Joshua to La Via Campesina.'

31 Davis, *Scripture, Culture and Agriculture.*

32 Deuteronomy 8: 11–18.

33 Psalm 72: 1–3.

34 Jeremiah 5: 23–25.

35 Isaiah 5: 8–10.

36 Wendell Berry, *Sex, Economy, Freedom and Community.* New York, NY, Pantheon Books, 1993, 6–8.

37 Graham Harvey, *The Killing of the Countryside.* London, Random House, 1998, 21.

38 Harvey, *Killing of the Countryside*, 142.

39 Vandana Shiva, *Monocultures of the Mind: Biodiversity, Biotechnology and the Third World.* Penang, Third World Network, 1993, 74.

40 Shiva, *Monocultures of the Mind*, 58.

41 Harvey, *Killing of the Countryside*, 126.

42 Julius Ihonvbere, 'The World Bank and the Somali Crisis.' Paper prepared for the symposium on "Towards Conflict Resolution in the Horn of Africa," organised by the African Studies Program, Central Connecticut State University, New Britain, CT, November 19, 1994, and published at Julius Ihonvbere.com: Where Theory Meets Praxis, https://juliusihonvbere.com/the-world-bank-and-the-somali-crisis/ accessed February 2, 2022.

43 These and other impacts of World Bank and IMF structural adjustment programs were well documented in Susan George, *A Fate Worse than Debt.* London, Penguin, 1990.

44 Wendell Berry, *The Unsettling of America,* San Francisco, CA, Sierra Club, 1979.

45 Thompson, *Spirit of the Soil*, 68.

46 Desmond S.T. Nichol, *An Introduction to Genetic Engineering,* 3rd Edition, Cambridge, Cambridge University Press, 2008, 4.

47 The process is described with greater precision in I. Wilmut, A.E. Schnieke, J. McWhir, et al., 'Viable Offspring Derived from Fetal and Adult Mammalian Cells.' *Nature* 385 (1997) 810–13.

48 Gina Kolata, *Clone: The Road to Dolly, and the Path Ahead.* New York, NY, William Morrow & Co., 1998.
49 Arlene Judith Klotzko, 'The Debate about Dolly.' *Bioethics* 11 (1997) 427–38.
50 Ian Wilmut, 'Case Study 8: Pharmaceuticals from Milk: Producing Pharmaceuticals from Sheep Milk.' In Donald Bruce and Anne Bruce eds. *Engineering Genesis: The Ethics of Genetic Engineering in Non-Human Species.* London, Earthscan, 1998, 60–2.
51 W.J. Clinton, 'Prohibition on Federal Funding for Cloning of Human Beings: Memorandum for the Heads of Executive Departments and Agencies.' The White House, Office of the Press Secretary (March 4, 1997).
52 National Bioethics Advisory Commission Report, *Cloning Human Beings.* Washington, DC, 1997, cited in Richard A. Epstein, 'A Rush to Caution: Cloning Human Beings.' In Martha C. Nussbaum and Cass R. Sunstein eds. *Clones and Clones: Facts and Fantasies about Human Cloning.* New York, NY, W. W. Norton & Co., 1998, 263.
53 The Pontifical Academy of Life cited by Wendy Donniger in 'Sex and the Mythological Clone.' In Nussbaum and Sunstein eds. *Clones and Clones,* 114.
54 Leon Kass, Testimony before the National Bioethics Advisory Commission, March 13, 1997, cited in Kolata, *Clone,* 15.
55 Kolata, *Clone,* 36.
56 Stanley Hauerwas cited in Gina Kolata, 'With Cloning of a Sheep, Ethical Ground Shifts.' *New York Times,* February 24, 1997 archived at https://web.stanford.edu/dept/HPS/NYTimes/WithCloningofaSheepEthics.html accessed February 2, 2022.
57 Bruce and Bruce, eds., *Engineering Genesis,* 140.
58 Ted Peters, 'Genes, Theology and Social Ethics.' In Ted Peters ed. *Genetics: Issues of Social Justice.* Cleveland, OH, Pilgrim Press, 1998, 11–35.
59 Martin Heidegger, 'The Question Concerning Technology.' In Martin Heidegger. *The Question Concerning Technology and Other Essays.* Ed. and Trans. William Lovitt, New York, NY, Harper & Row, 1977, 3–35.
60 Dorothy Nelkin and M. Susan Lindee, 'Cloning in the Popular Imagination.' *Cambridge Quarterly of Healthcare Ethics* 7 (1998) 145–9.
61 Jacques Ellul, *The Technological Society.* Trans. John Wilkinson, New York, NY, Vintage Books, 1964.
62 Langdon Winner, 'Do Artifacts Have Politics?' *Daedalus* 109 (1980) 122.
63 Winner, 'Do Artifacts Have Politics?,' 123.
64 With the increasingly business-oriented research plans of university funding councils the distinction between public and private funding of scientific research is of declining significance: see further Wilson Poon and Tom McLeish, 'Real Presences: Two Scientists' Response to George Steiner.' *Theology* 807 (1999) 171.
65 T.S. Kuhn, *The Structure of Scientific Revolutions.* Chicago, IL, University of Chicago Press, 1970.
66 Bruce and Bruce, eds., *Engineering Genesis.*
67 W.K. Novak and A.G. Haslberger, 'Substantial Equivalence of Antinutrients and Inherent Plant Toxins in Genetically Modified Novel Foods.' *Food and Chemical Toxicology* 38 (2000) 473–83.
68 G.-E. Séralini, D. Cellier, and J. de Vendemois, 'New Analysis of a Rat Feeding Study with a Genetically Modified Maize Reveals Signs of Hepatorenal Toxicity.' *Archives of Environmental Contamination and Toxicology* 52 (2007) 596–602; and G.-E. Séralini, E. Chair, R. Mesnage, et al., 'Long Term Toxicity of a Roundup Herbicide and a Roundup-Tolerant Genetically Modified Maize.' *Food and Chemical Toxicology* 50 (2012) 4221–31.
69 Andrew Rowell, *Don't Worry (It's Safe to Eat): The True Story of GM Food, BSE and Foot and Mouth.* London, Routledge, 2015.
70 Rowell, *Don't Worry,* 152–7.
71 Rowell, *Don't Worry,* 95–7.

72 John Smith, *Seeds of Deception: Exposing Industry and Government Lies About the Safety of Genetically Engineered Foods You're Eating*. White River Junction, VT, Chelsea Green Publishing, 2005.
73 Sophie Richard, Safa Moslemi, Herbert Sipahutar, et al., 'Differential Effects of Glyphosate and Roundup on Human Placental Cells and Aromatase.' *Environmental Health Perspectives* 113 (2005) 716–20.
74 Manisha Pahwa, Laura E. Beane Freeman, and John J. Spinelli, 'Glyphosate Use and Associations with Non-Hodgkin Lymphoma Major Histological Sub-Types: Findings from the North American Pooled Project.' *Scandinavian Journal of Work, Environment and Health* 45 (2019) 600–10.
75 Rowel, *Don't Worry*, 142–3.
76 Thomas Hallenbeck, 'Vermont Defends GMO Labelling Law.' *Burlington Free Press*, August 8 2014.
77 See further Michael S. Northcott, 'Crop Science, the Heisenberg Principle and Resistance to Genetically Modified Organisms.' In Phil McNaughten and Susanne Ricarro eds. *Governing Agricultural Sustainability: Global Lessons from GM Crops*. Abingdon, Routledge, 2015, Ch. 13.
78 Mark Redclift, 'Agrarian Populism in Mexico - The "via campesina".' *The Journal of Peasant Studies* 7 (1980) 492–502.
79 Anthony Montoya, *The Theology of Food: Eating and the Eucharist*. Oxford, Wiley-Blackwell, 2010.
80 Northcott, 'Crop Science, the Heisenberg Principle and Resistance to Genetically Modified Organisms.'
81 Walter Heisenberg, *Physics and Philosophy*. New York, NY, Harper, 1958.
82 *Breaking Bad* (TV Series 2009–12) Dir. Colin Bucksey, AMC Productions Inc.
83 Northcott, 'Crop Science, the Heisenberg Principle and Resistance to Genetically Modified Organisms.'
84 David P. Pierson, 'Breaking Neoliberal? Contemporary Neoliberal Discourses and Policies in AMC's *Breaking Bad*.' In David P. Pierson ed. *Breaking Bad: Critical Essays on the Contexts, Politics, Style, and Reception of the Television Series*. Plymouth, Lexington Books, 2013, 13–42.
85 Vandana Shiva, *Earth Democracy: Justice, Sustainability and Peace*. London, Zed Books, 2006.
86 Plato, *Statesman*. Trans. Waterfield R, Cambridge, Cambridge University Press, 2005.
87 Michael S. Northcott, 'Concept Art, Clones and Co-creators: The Theology of Making.' *Modern Theology* 21 (2005) 219–36.
88 On the turn from conventional farming to hydroponics and 'vertical farming' by venture capitalists in the United States see Corey Morningstar, 'New Controlled Food System is Now in Place and They Will Stop at Nothing to Accelerate Their Control.' *Corey's Digs*, April 27, 2022, published online at https://www.coreysdigs.com/global/new-controlled-food-system-is-now-in-place-and-they-will-stop-at-nothing-to-accelerate-their-control/ accessed May 10, 2022.
89 Michael S. Northcott, 'Behold I Have Set the Land before You (Deut 1. 8): Christian Ethics, GM Foods and the Culture of Modern Farming.' In Celia Deane-Drummond, Bron Szerszynski with Robin Grove-White eds. *Re-Ordering Nature: Theology, Society and the New Genetics*. London, T&T Clark, 2003, 85–106.
90 'The Amazon: NFTs Minted to Save the Forest Sell Out in One Hour.' *CoinDailyNews*, February 11, 2022, https://www.coindailynews.io/2022/02/11/the-amazon-nfts-minted-to-save-the-forest-sell-out-in-one-hour/ accessed February 16, 2022.
91 Cynthia Ong, et al., 'Communication Regarding a Nature Conservation Agreement Granting Monopoly Rights on 2 Million Hectares of Sabah's Forested Lands to a Private Company – Extremely Urgent.' Addressed to UN Special Rapporteur on the Rights of Indigenous Peoples, UN Office of the High Commissioner for Human Rights, Geneva, March 8, 2022, and published online at https://sarawakreportdocs

.s3.eu-west-2.amazonaws.com/UN+Special+Rapporteur+Letter+-+Final_Revised.pdf accessed March 27, 2022.

92 'Carbon Credit Rush Hits Papua New Guinea.' *Sarawak Report*, March 25, 2022, and published online at https://www.sarawakreport.org/2022/03/carbon-credit-rush-hits-papua-new-guinea/ accessed March 27, 2022.

93 See further Brian Büscher, Sian Sullivan, Kath Neves, et al., 'Towards a Consolidated Critique of Neoliberal Conservation.' *Capitalism, Nature, Socialism* 23 (2012) 4–30.

94 For forensic analyses of the commoditisation of nature conservation, including the use of biosensors and blockchain, and related exclusions of indigenous peoples, see the collected essays in the special issue of *Political Ecology* 28 (2022) 696–888 entitled *Political Ecologies of Extinction,* edited by Brian Buscher, and available at https://brambuscher.files.wordpress.com/2022/02/jpe-pe-of-extinction-si-2021-1.pdf accessed February 4, 2022.

95 John Muir recoded his early sojourn in the Californian Sierras in a book that has become a classic of North American nature writing, *My First Summer in the Sierra.* Fairbanks, AK, Houghton Mifflin, 1911.

96 Mark Spence, 'Dispossessing the Wilderness: Yosemite Indians and the National Park Ideal, 1864–1930.' *Pacific Historical Review* 65 (1996) 27–59.

97 W.J. Ripple and Robert L. Beschta, 'Restoring Yellowstone's Aspen with Wolves.' *Conservation Biology* 138 (2007) 514–19; John W. Laundré, Lucina Hernández, and Kelly B. Altendorf, 'Wolves, Elk, and Bison: Reestablishing the "Landscape of Fear" in Yellowstone National Park, U.S.A.' *Canadian Journal of Zoology* 79 (2001) 1401–9; for a popular account of 'wolf restored' habitats see Nate Bakerslee, *American Wolf: A True Story of Survival and Obsession in the American West.* New York, NY, Crown Publishing, 2017.

98 For full citations of Calvin, see Michael Northcott, 'Reformed Protestantism and the Origins of Modern Environmentalism.' *Philosophia Reformata* 83 (2018) 19–33.

99 Northcott, 'Reformed Protestantism and the Origins of Modern Environmentalism.'

100 The metaphor of humans as a 'virus' was first mobilised by James Lovelock in his *Gaia: A New Look at Life on Earth.* Oxford, Oxford University Press, 1995; see also Alan Weisman, *The World Without Us.* London, Virgin Books, 2008.

101 As Bruno Latour puts it: 'we are all climate quietists when we hope, while doing nothing about it, that "everything will be all right in the end".' Latour, *Down to Earth: Politics in the New Climate Regime.* Trans. Catherine Porter, Cambridge, Polity Press, 2018.

102 Edward O Wilson, *Half Earth: Our Planet's Fight for Life.* New York, Liveright, 2017.

5

BIOSECURITY, COVID-19, AND HUMAN–EARTH HEALING

Abstract

The emergence of an allegedly novel coronavirus – SARS-CoV-2 – in 2020, though it turned out to have a mortality rate equivalent to influenza, was used to impose a novel global regime of social and technological controls on human life that parallel the biosensor-connected and digital-surveillance character of the neoliberal financialised regime of nature conservation discussed above. Lockdowns and other pandemic regulations had numerous and now well-evidenced harmful effects, including cutting people off from each other, from nature, and from sunlight, all three of which are known sources of health. Fear propaganda was widely used in the media, and by governments, to promote conformity to the new controls. Some suggest increased social control over human activities and even over human DNA, and isolation of humans from nature, are also needed to address the ecological crisis. But contemporary discoveries concerning the human biome show that the health of the microbiome – which is also the immune system – is closely related to the variety of bacteria it sustains and that this in turn is related to highly biodiverse diets and foods and water grown in healthy soils and water catchments.

Against Lockdowns for Health *and* Climate Change

During the COVID-19 pandemic in 2020–2 there was an eerie convergence in public discourse between novel 'public health' policies which isolated human beings from one another and the natural environment, and led to a huge rise globally in hunger, mental ill health, poverty and suicides, and discourses concerning the scientific and technological policies, including social controls said to be needed for responding to the ecological crisis. Bruno Latour suggested

DOI: 10.4324/9781003110750-6

that 'the climate crisis resembles a huge planetary lockdown trapping humanity within an ever-deteriorating environment' and that

> It would be a mistake to believe that the pandemic is a crisis that will end, instead of the perfect warning for what is coming, what I call the new climatic regime. It appears that all the resources of science, humanities and the arts will have to be mobilised once again to shift attention to our shared terrestrial condition.[1]

The term 'lockdown' originates in the US prison system.[2] The United States keeps more prisoners in jail per head of population than any other domain in the world: in 2021 there were 1.77 million prisoners, and the majority are black Americans, many of whom are in jail for drug use.[3] The prison system is mainly privatised and corrupt links between judges sending prisoners down and the for-profit companies that run US prisons are well attested.[4] The term lockdown refers to the increasingly common practice in the mass incarceration gulag of the United States under which prisoners are locked 24 hours a day in their cells by means of centrally controlled bolts on the bars of prisoner cells. The conventional US jail cell is barred at one end and open to the corridor, reducing privacy and increasing ease of prisoner surveillance. In 'supermax' jails, cells are fully enclosed and lockdown in such cells also means 24/7 social isolation of the prisoner, often without even a window for daylight, which produces major and often irreparable psychological and physiological harms.[5]

The established scientific consensus for dealing with outbreaks of dangerous new infectious diseases until 2020 was that isolating symptomatic patients, and tracing their recent contacts, is the best approach.[6] This is because quarantining the healthy has health and other disbenefits which far outweigh any benefits from possible reduced transmission. Infectious disease control before 2020 was therefore focused on those who show *symptoms* of illness, and on isolating and protecting those with severe health conditions which may put them at particular risk in a pandemic. But in March 2020 the Chinese government mounted a successful media campaign, using staged videos from the city of Wuhan in which actors were shown spontaneously collapsing and shaking in the streets, to spread fear globally about the emergence of an allegedly extremely dangerous new chimeric coronavirus in December 2019. The social media campaigns were conducted with the aid of tens of thousands of 'bots' controlled from computers of Chinese government employees, and these campaigns especially targeted Italy, Germany, the United Kingdom, and the United States.[7] The Chinese also claimed that they knew of cases of asymptomatic transmission in Wuhan which, though unprecedented and coming from a communist government known for the use of propaganda, then formed the basis for an unprecedented global campaign of mass medical testing of well people although a scientific study of the Wuhan population found that no asymptomatic transmission occurred.[8]

FIGURE 5.1 Impromptu 'lockdown' bamboo barrier constructed by rural villagers in Bantul, Yogyakarta following media coverage of the Chinese coronavirus outbreak in March 2020. Author photo.

During this remarkable propaganda campaign, the WHO in a report in February 2020 changed its stance on quarantining the healthy and asymptomatic. Tedros Ghebeyesus is the first Director-General of the WHO who is not a medical doctor. He is a former Health Minister of the Ethiopian government which he joined after many years as a leader of the Tigray People's Liberation Front which was listed as a terrorist organisation by the US government.[9] Tedros was appointed as Director-General of the WHO as a result of fierce lobbying by China's President Xi.[10] It is then unsurprising that Tedros took at face value China's claims about its novel social control and mass medical testing measures, and cases of asymptomatic transmission, and that similar measures would be essential for controlling the spread of the virus in other countries.

There is now strong scientific evidence of extensive harms generated by lockdowns and other socially repressive 'public health' measures from 2020 to 2022. Health data show that large-scale suppression through 'lockdowns' of the agency of billions of humans to grow food; to gather bodily for worship, educational, hospitality, business and cultural activities; to walk outside under the sun on beaches, in parks, and forests; and to meet in public and private places with family, friends, and colleagues had multiple deleterious impacts on human health, mental and physical, while making no impact on the transmission of respiratory viruses.[11] Comparisons between nations that adopted these novel policies reveal no differences in viral transmission between countries that had 'stay at home' policies and those which did not.[12] There is evidence that countries such as Belgium and Peru that had stricter lockdowns had *higher* mortality than countries with weaker measures, and those such as Sweden and Belarus which had none. Unemployment, hunger, and extreme poverty are known causes of increased mortality and there was an inexplicable failure to investigate and model the harmful consequences of lockdowns and other measures before they were imposed.[13] It is estimated that between 400 and 600 million people were pushed into extreme poverty, hunger and unemployment as a result of pandemic social controls.[14] Social isolation is also a known killer among both the elderly and the young, and there were extensive media reports of huge increases in suicides among the young during the lockdowns

Despite the known extensive and harmful consequences of Non Pharmaceutical Interventions (NPIs) deployed for COVID-19, the 'societal shifts' they caused are said to be useful precedents for modelling and revising human impacts on the Earth's climate.[15] This claim is often linked with the claim that COVID-19 was a manifestation of the ecological crisis. Jane Goodall argued that the pandemic was caused by human environmental disturbance, and especially 'the hunting, killing, eating and trafficking of animals':

> Our disrespect for wild animals and our disrespect for farmed animals has created this situation where disease can spill over to infect human beings. We have come to a turning point in our relationship with the natural world.[16]

Goodall's argument rests upon claims made in the media, and by many scientists, that the novel coronavirus SARS-CoV-2 spread to humans from bats via zoonosis. Although scientists published *models* of a potential non-laboratory transmission route, in which the Malayan pangolin often unfairly featured, pangolins do not live in the vicinity of bats in China and though illegally trafficked for their scales they are not sold live or even dead in the Wuhan wet market which media reports identified as a possible source of zoonotic transmission. [17]

The Laboratory Origins of SARS-CoV-2

The balance of evidence concerning the allegedly 'novel' coronavirus SARS-CoV-2 from a triangle of empirically verifiable data sources indicates a laboratory origin.[18] This includes public records of research grants and patents; peer-reviewed scholarly publications; geographical evidence. I also include brief reference to a fourth perspective, which is the testimony of scientists who have analysed the genome of SARS-CoV-2:

1. **Patents:** Professor Ralph Baric of the University of North Carolina and Anthony Fauci, Director of the lead US scientific funding agency – the National

Institute of Health – for over 40 years are long-time collaborators in genetic engineering of viruses which they commenced in 1999 at Baric's Chapel Hill virology laboratory. The genetic engineering of viruses – or 'gain-of-function' research – involves creating novel *humanly adapted* pathogens by splicing together recombinant DNA from wild and human genetic material to *create* dangerous pathogens which could not have occurred in the wild. The first US-funded grant project enabled Baric to clone a novel recombinant SARS coronavirus from DNA which he spliced together with the same methods used in cloning Dolly the Sheep.[19] Baric's collaborators obtained a patent for this new method for genetically modifying SARS coronaviruses in 2002.[20] A few months before the patent was granted, the first humans were infected with a novel humanly adapted strain of SARS in Hong Kong, where Malik Peiris who collaborates with Baric, also filed for a SARS patent.[21] As Baric and his collaborators indicate, the goal of this hazardous genetic engineering research is to develop potential vaccines and therapeutics: 'Availability of a SARS-CoV full-length cDNA provides a template for manipulation of the viral genome, allowing for the rapid and rational development and testing of candidate vaccines and therapeutics against this important human pathogen.'[22] There were over 5000 patents for SARS-Coronavirus and related therapeutics and tests filed on line by pharmaceutical laboratories, virologists, and public agencies, including the US CDC and the United Kingdom's bioweapons laboratory at Porton Down, before the outbreak in 2019: the *majority* of SARS-related patent applications were for the development of potential vaccines.[23] The first such project was a DNA gene therapeutic project led and funded by Fauci in 2005.[24] Though the public were unaware that SARS-COV-2 was made in a laboratory and was the subject of intense commercial activity prior to 2020, public patent records reveal that SARS-CoV-2 was not a 'novel coronavirus' in 2020 but known to science, and to medical authorities and pharmaceutical companies, including Moderna and Pfizer.

2. **Scientific publications:** the accidental emergence of gain-of-function engineered pathogens from laboratories has precedent: in documented laboratory leaks of genetically engineered viruses and of bioweapons, from Soviet labs of a humanly adapted genetically modified version of H1N1 Swine Flu in 1977; and of a bioweaponised version of Anthrax in 1979.[25] In addition to recombinant DNA splicing, virologists use a 'sequential passage' technique when they pass viruses through many successive animal hosts. This also results in the development of gain-of-function viruses, and it gives the resultant viruses an 'artificial appearance,' as was the case with the H1N1 Swine Flu Soviet leak, and with SARS-CoV-2.[26] Virologists have used both gene splicing and sequential passage techniques extensively on bat coronaviruses since the 1990s in the United States, China, Europe, and Russia. Researchers at the National Institute of Allergies and Infective Diseases (NIAID) in Maryland, directed by Fauci, published a candidate vaccine which they trialled on mice infected with gain-of-function versions of SARS which expressed the toxic coronavirus spike protein that binds to the mammalian Ace II receptor in 2004, 18 years before the 2019 outbreaks

of SARS-CoV-2 which expresses the same toxic spike protein.[27] Fauci and Baric work with many Chinese scientists and when gain-of-function research was stopped in labs on the US mainland in 2014 they continued to work with and fund scientists engaged in this research at the Wuhan Virology Institute and the Chinese Academy of Sciences. The researchers isolated new coronavirus candidates from bats in Northern China and worked on enhancing their adaptability to mammals in the Wuhan laboratory in grant projects funded by NIAID, United States Agency for International Development (USAID), and the US Department of Defense through grants to an NGO called Ecohealth Alliance, and the Chinese Academy of Sciences. Three scientific papers arising from this research include a 2013 paper describing the isolation of the bat viral material and its gain-of-function enhancement.[28] A second paper on a different cluster of coronaviruses which were engineered for mammalian infectivity at Wuhan was published in *Nature: Medicine* in 2015 and this paper names SARS-CoV-2 as one of the novel engineered coronaviruses. The lead researchers on the project, though they put their names last as co-authors, were Baric and Zheng-Li Shi who had worked with Baric at UNC and from 2013 continued their collaborative research at the Wuhan Virology Institute after the United States declared a moratorium on US-government funding of such research.[29] A critique of the inherent dangers of this research was published in *Nature* in 2015 in which it was presciently argued that the risk of pandemics being caused by lab leaks of highly infectious and pathogenic gain-of-function viruses outweighed any putative public benefits of such research. The author, Declan Butler, cites virologist Simon Wain-Hobson at the Pasteur Institute in Paris who 'points out that the researchers have created a novel virus that "grows remarkably well" in human cells. "If the virus escape, nobody could predict the trajectory."' Butler also cites a biodefence expert at Rutgers University who said that '(T)he only impact of this work is the creation in a lab, of a new non-natural risk.'[30] Undeterred, virologists continued to work on isolating potential humanly adapted versions of SARS coronaviruses at the Wuhan Virology Institute and a second paper was published in 2017 which revealed funding by USAID and the Chinese Academy of Sciences.[31] The editor of the paper is Christian Drosten, a German scientist and key figure in the COVID-19 pandemic as publisher of a controversial and rushed paper on a new PCR test for SARS-COV-2 in January 2020 which was used globally for over two years and because of which large numbers of people were declared infected though not ill as the test was very prone to false positive results.[32] The 2017 paper from the Wuhan Virology Institute includes a five-year survey of isolated bat viral material from a cave in Yunnan. The team of scientists led by Zheng-Li, the Director of the WVI, who became known as the 'bat woman,' engaged in genetic engineering and gain-of-function of any bat viral material which showed potential for infectivity in other mammals and describe their work as follows: 'Recombinant viruses with the S gene of the novel bat SARS-CoVs and the backbone of the infectious clone of SARS-CoV WIV1 were constructed using the reverse genetic system described previously.'[33]

Given the legacy of novel laboratory-engineered SARS-Co-Vs, and the lack of evidence of a 'wild' transmission route by which the bat virus became a chimera capable of infecting humans, the balance of evidence favours a laboratory origin for SARS-CoV-2.[34]

3. **Geography of SARS-CoV-2 outbreaks and victimology:** given the lead role of the NIAID laboratory in Maryland, and of Zheng-Li and the WVI, in isolating, culturing, and genetically engineering bat viral material which shows potential for mammalian infectivity and human emergence, and especially the spike protein, it is not surprising that the reports of a highly infective and transmissible virus arose in these precise locations in 2019. The first likely victims of SARS-CoV-2, though autopsies were not conducted, were residents of an assisted-living facility a few miles from the NIAID laboratory in Maryland, which was a second laboratory, along with his UNC lab, where Baric worked on genetically engineering SARS viruses.[35] The laboratory leak most likely occurred through wastewater infecting local water supplies. The disease outbreak affected staff and residents and led to several deaths, and two weeks later the bioweapons grade NIAID laboratory, located on the US Army base of Fort Derrick, was closed.[36] The second outbreak began in the vicinity of the Wuhan Virology Institute in November 2019. The first victims were laboratory workers who were hospitalised with breathing difficulties. The laboratory was being funded by both USAID and NiH and a report had gone from the US Consul in Wuhan in 2018 to the US State Department warning that on an inspection of the laboratory they saw evidence that it was not biosecure and a leak of dangerous pathogens was feasible.[37] Two months before news of victims of a novel coronavirus began to go viral on Chinese social media, Zheng-Li took the database offline in which she had recorded the full details of her USAID, NiH, and Chinese government–funded research.[38]
4. **Investigation of the SARS-CoV-2 genome:** the likely laboratory emergence of SARS-CoV-2 was widely discussed in and beyond China and to counter it Fauci, together with Peter Daszak, who is the British Director of Ecohealth Alliance which was the principal conduit for Fauci's and the Pentagon's research funds to the WIV, coordinated a letter to *The Lancet* in which they categorically insisted that the virus had jumped from bats to mammals and thence to humans in the wild and denying the lab leak theory, although Daszak did not declare an interest in the article.[39] The first attempt by scientists with no prior involvement in SARS-CoV-2 genetic engineering, to isolate and describe the genome of SARS-CoV-2 indicated that the virus was uniquely humanly adapted and it was very unlikely the adaptation could have occurred in the wild.[40] But the pre-publication study never made it into print as Fauci used his considerable influence as the lead funder of US and Chinese medical research to suppress it. Subsequent studies by other academics not associated with Baric, Daszak, and Fauci, including by Nobel prize–winning virologist Ralph Montagnier, found that there were genetic sequences added to the bat genome of the virus which could only have got there through genetic 'manipulation':

> There has been a manipulation of the virus: at least part of it, not all of it. There is one model, which is the classic virus, which comes mainly from bats, but to which HIV sequences have been added .... It's the work of professionals, of molecular biologists. Very meticulous work. For what purpose? I don't know. One hypothesis is that they wanted to create an AIDS vaccine.[41]

The conclusion that the virus was genetically engineered, given the absence of any empirical data on a non-laboratory evolution, and the extensive trail of evidence linked to the laboratories of Baric, Fauci, Daszak, and Zheng-Li, is irresistible. And the former head of the US CDC, Robert Redfield, is on record as saying exactly that.[42]

The Pandemic Control Regime

In the socially constructed global public health regime that commenced in March 2020 to address this invisible and miniscule engineered enemy of humanity, a regime which was allegedly designed to control SARS-COV-2 transmission, the ideology of scientism, long predicted by philosophers such as Bertrand Russell as having totalitarian potential, finally realised as it became the master narrative of most of the human species.[43] From March 2020 until March 2022 a novel biopolitical regime of social control or 'biosecurity' was imposed on much of the world's population to allegedly halt the spread of a pathogen which was said to be uniquely life-threatening. Governments openly acknowledge they used fear propaganda, and mind-control techniques, to persuade people to stay indoors, not to go to work, meet friends or relatives, or even take exercise in parks or wild places, while subjecting themselves to an unprecedented regimen of mass medical testing, and travel and other restrictions which two years on had still not disappeared in much of the world.

The number of people said to have died from or with COVID-19 symptoms at time of writing is now 16 million. While not on the scale of the Spanish flu pandemic of 1918, this number of deaths over two years clearly qualifies as a pandemic. But serious questions remain about the COVID-19 pandemic including: (1) whether COVID-19 was so dangerous as to warrant health-harming lockdowns, the en masse closure of businesses, parks, places of worship, schools, and workplaces worldwide, the bankruptcies, mental health epidemic, increases in global hunger, poverty and suicides which these closures, not the virus, caused; (2) whether those counted as 'COVID deaths' died of COVID-19 or of other conditions; (3) whether the mysterious refusal of early treatment with safe generic drugs such as ivermectin and hydroxychloroquine – which were widely used to successfully treat COVID-19 infections in countries such as India and Turkey – contributed to COVID-19 deaths in the countries where these generic drugs were banned; (4) whether hospitalisation, mechanical ventilation, and injected sedatives of those alleged to have COVID-19 because of PCR tests, but who were asymptomatic, were the real cause of many deaths; (5) why novel gene-edited drugs were recommended by the WHO, the US CDC, and many other public agencies despite the very poor clinical record of such gene therapies prior to COVID-19, and the failure of previous long-term trials to establish their effectiveness as prophylactics against coronaviruses.

It would take a whole book to fully investigate these questions but on points 1 to 5:

1. The first contained study of an elderly population exposed to SARS-CoV-2 was of the passengers and crew of the *Diamond Princess* cruise ship. Despite the high average age of the passengers and that they were kept in confined spaces which tests showed were infected with the virus, numbers of deaths on the ship and in the months afterwards were not consistent with the virus being exceptionally dangerous. When the Diamond Princess case was extrapolated across the age range of the United States the Infection Fatality Rate was equivalent to a strong seasonal influenza and this is just 25 per cent of the mortality rate claimed by Fergusson and the WHO.[44] An early community study from data in Santa Clara County California in April 2020 confirmed this.[45] A 66-region survey published in January 2021 found that the IFR of SARS-CoV-2 was lower than normal flu for all age groups below 60 and equivalent to flu for those above.[46] But public health authorities ignored real-world data, preferring instead the inaccurate models of possible mortality of Neil Fergusson, who has a serial history of vastly exaggerating potential deaths from alleged pandemic diseases.
2. Numbers of alleged deaths from COVID were exaggerated in many domains by changes in procedures for recording deaths. In the United Kingdom the average age of those who were labelled as having died from COVID-19 was 82. The United Kingdom changed long-established procedures so that just one doctor, who neither knew nor had visited a deceased person, could declare cause of death based on communications with non-medical staff present with the deceased.[47] There were also no autopsies conducted into causes of death in most domains, including the United Kingdom, because the WHO claimed the disease was too dangerous. Given that the average age of death in the UK is 82, it is a reasonable supposition that most 'covid' deaths, at least in 2020, were deaths from old age relabelled as 'COVID-19.'
3. Country-by-country data on mortality indicate that countries which adopted generic drugs to treat COVID-19 symptoms in the early stages of illness had much lower mortality than those which did not. India went through a three-month period from January to March 2021 in which it stopped using Ivermectin while it rolled out mRNA vaccines. The mortality rate went up, doctors and members of the public protested, and Ivermectin was reintroduced.

4. There are reliable reports from many countries, and growing scientific data, that a leading cause of death during the COVID-19 pandemic was hospital treatment protocols commended by the WHO, the US CDC, the UK NHS, and other health bodies. These protocols subjected people who were said to be COVID-19 positive on the basis of a PCR test known to throw up large numbers of false positives, but with no difficulties breathing, to the mortality-inducing procedure of mechanical ventilation combined with sedatives and toxic drugs such as Remdesivir and Midazolam. An early analysis of use of mechanical ventilators in a hospital in Seattle found that only 4 out of a group of 20 patients escaped the hospital alive.[48] In the United Kingdom, the government stockpiled 27 months normal supply of a respiratory suppressant and hallucinogenic branded as Midazolam. Midazolam was used so extensively in elderly care homes from March 2020 that there was a significant spike in mortality in UK care homes from April to May 2020. Exceptionally, non-medical staff were licensed by the government to administer Midazolam, along with morphine, to residents without prior medical authority or the permission of relatives even though the combination is usually life-ending.[49] The UK Health Minister, Matt Hancock, gave recorded testimony on the exceptional measures leading to the widespread injection of Midazolam in care homes by non-medical care-home staff to a committee hearing in the House of Commons on April 17, 2020.[50] Midazolam is used in capital punishment in the United States, as well as in the United Kingdom's controversial euthanasia program known as the 'Liverpool Care Pathway', in which patients judged near death were intravenously injected with Midazolam and morphine, and food and water denied them.[51] The Liverpool Care Pathway was banned in 2014 as illicit euthanasia. But it was repackaged as a pandemic measure in March 2020 for community treatment of COVID-19 in care homes and hospices by the Hull and East Riding Prescribing Committee, and adopted nationwide by the UK health regulatory body the National Institute for Health and Care Excellence (NICE).[52] In the United States, Fauci led the Federal government's response to COVID-19 though he had funded the development of the creation of the pathogen and of potential vaccines in North Carolina and Wuhan. Despite the fact that Remdesivir is a highly toxic and often life-ending drug Fauci insisted it was the only one drug that was licensed for use in late-stage hospital treatment of COVID-19 patients, and that in early stages of illness no treatment should be given. The most common side effect of Remdesivir, a very expensive drug made by Gilead which has close ties to the US CDC, is *sepsis* leading to total organ failure. Robert F. Kennedy Jr. indicates that Remdesivir was commended as 'treatment' for COVID-19 because of regulatory capture of the US health care and drug licensing system.[53] Many frontline doctors ignored the official advice and used long-established safe drugs hydroxychloroquine, azithromycin and/or ivermectin, together with vitamins C and D, and zinc, in successful early

> treatment of COVID-19 symptoms.[54] And yet the US health authorities, like health authorities in many other domains, attempted to ban the use of these safe drugs in COVID-19 patients and to withdraw them from sale.[55]

The data, as well as anecdotal evidence, now show incontrovertibly that novel gene-editing therapeutics in the form of mRNA and mDNA vaccines which were rushed into emergency use in 2021 on the false basis that no other effective therapies were available, and were a financial bonanza of trillions of dollars for drug companies, have an exceptional record of injury, including death, compared to all previous vaccination programs. The rollout of the 'vaccination' program with these drugs was accompanied by significant increases in COVID-19 infections and deaths in 2021 as compared to infection and mortality rates in 2020 in most domains. In the US, which recorded the highest per person mortality from COVID-19 globally, the vaccine injury recording system – VAERS – is reliant on input of data from private doctors who are not paid for time spent recording injuries on the 28-page form required. A two-year investigation of VAERS by Harvard University scientists from 2007–9 found that VAERS recorded less than 1 in 100 of vaccine injuries the researchers recorded through community surveys in the City of Boston.[56] The scientists offered to assist the FDA in streamlining the system to make it more reliable, but the FDA refused. For the COVID-19 vaccine, the VAERS system records at time of writing 26,059 deaths, 143,554 hospitalisations, 13,360 heart attacks, 37,133 cases of heart disease, 48,342 cases resulting in permanent disability. Even if we only assume under-reporting at half the rate in the Harvard study, these outcomes would need to be multiplied 50 times. Readers may say, well if that many deaths and serious injuries ensued from the vaccines, there would be other data that shows this. And there is. US life insurers reported an unprecedented 40 per cent increase in life insurance claims for death in the second half of 2021.[57] There were similar reports from German health and life insurers, and Professor Harold Matthews conducted a study showing a huge underestimate in official statistics of vaccine damages and called for a nationwide investigation of the more than half a million serious medical incidents his research revealed to have been caused by the novel therapeutics including especially circulatory and coronary disorders, and related deaths.[58] Unexplained increases in mortality – both COVID-19 and non-COVID – are now evidenced in official statistics in many domains and show a striking correlation with the percentages of their populations who participated in the vaccination programmes.

Biopolitics and the Biosecurity State

Michel Foucault coined the term 'biopolitics' for the new regime of control over lands, soils, species, and persons which industrial societies initiated in the eighteenth century.[59] He argued that the institutions created by this novel modern State regime – clinics, psychiatric hospitals, panoptic prisons, maternity hospitals, crematoria, agricultural regulatory agencies – are unique as compared to previous polities, by the extent to which these institutions extend the claim of sovereign power of the State from political subjects to the biological conditions of life itself, including human health, mental states, mortality, natality, and sexuality. In the twentieth century, biopolitical control was extended to the atmosphere, bacteria, cells, genes, soils, and water in the more-than-human world. The regime of biopolitics is made possible by the combination of the nascent powers of new nation States with the new science of biology, including increasingly powerful techniques for studying life at the micro-scale, combined with new industrial technologies of fossil-fuelled manufacture, mechanical power, and mobility which increased speed of communication, and control of human movements en masse. Biopolitics names the new regime inaugurated first in England and Scotland, and in the West European lowlands, which was the Industrial Revolution. In the nineteenth century these new powers were mobilised by European royal families in the first 'total war' of human history, the Napoleonic Wars, in which large-scale mechanical killing, distanced war from the air via new explosives capabilities, and scorched-earth destruction of crops and villages as a technique to control terrain, were first utilised at scale.

In the twenty-first century, biopolitics became successively more invasive and controlling of human life as many nations announced 'states of emergency' to deal with alleged terror threats and pandemics such as COVID-19. In these successive states of emergency biopolitics reached a new zenith of control and suppression of human movements, meetings, and monetary exchanges, and all persons were treated as potential terrorists or biohazards and normal legal norms and rights, such as to bodily autonomy, were suspended.[60] The COVID-19 emergency regime involved the widespread use of fear propaganda, and penalties for citizens who resisted invasive government controls on their movements outside their homes. The principal and enduring economic, psychological, and social harms from the COVID crisis did not arise from the disease itself but from the constant dissemination of fear propaganda; the large-scale social isolation of individuals and households; the shuttering of huge numbers of small businesses and workplaces over an extended period; and the global rollout of states of emergency and suspensions of normal human freedoms and behaviours including family and religious gatherings, cultural, and educational activities, travel, and so on.

The social response to the emergence of SARS-CoV-2 represents a largely top-down Statist project coordinated by governments, by the World Health Organisation, Centres for Disease Control in North America, Europe, and Asia, and by central banks which created unprecedented quantities of digital money to support large corporations and stock markets, and to a lesser extent aid the billions of individuals and entrepreneurs who lost income – in many cases to the point of hunger, penury, and even suicide – during the unprecedented economic shutdown. The SARS-CoV-2 regime did not, however, shutter the activities of large corporations and indeed represented a huge economic bonanza for giant technology corporations such as Amazon, Alphabet, Apple, Facebook, and so on, while it also tilted the economic balance towards large-scale and online retailers – and particularly supermarkets – which were permitted to stay open while traditional markets, delicatessens, bakers, farmers' markets and so on were all forcibly closed for many months and in many cases never reopened.

This significant shift of economic activity from the cash economy, informal actors, and small and medium-sized enterprises (SMEs) to giant corporations is reminiscent of the global climate regime enacted by the UN Framework Convention on Climate Change which has for over 20 years focused efforts to address climate change on top-down solutions, and in particular national carbon accounts and 'carbon trading,' as favoured by nation States, investors, and energy corporations.[61] These 'solutions' have notably failed to stop States and corporations from extracting and marketing fossil fuels, or from destroying the capacity of ecosystems to absorb the resultant greenhouse gas emissions: destruction of significant carbon sinks continues, including tropical forests on land, and kelp and mangrove forests, and other critical features of the huge 'carbon pump' of the marine environment. The top-down corporately shaped climate regime is analogous to the response to the coronavirus because it suggests that only government and intergovernmental agencies acting in concert with large technologically sophisticated global corporations are capable of securing the welfare of people and planet in a global emergency such as the novel coronavirus was also said to represent.

Analogous to the State- and corporate-dominance of the climate regime, the World Health Organisation, the US CDC, the European Commission, the Rockefeller and Gates Foundations, and the World Economic Forum used the COVID-19 pandemic to advance a top-down biosecurity agenda for global health and global governance, also associated with the 'One Health' project to combine medical and veterinary science.[62] Part of the globalist technocratic vision of health that was promoted by the viral emergency was a global vaccination database accessible to governments and other agencies at borders and other sensitive locations, on the model of a system which Microsoft, in partnership with an Indian tech company, has already established to capture the identity, financial, health, and residency records of 1.2 billion Indian citizens. The system, known as AADHAR, was set up with remarkable speed under the Unique Identification Authority of India project and is also being used for voting and for tracking peoples' movements between States: it is a particular threat to minorities in States such as Kashmir who may now, at the touch of a button, cease to exist as formal citizens.[63]

In the course of the two-year-plus COVID-19 states of emergencies that governments imposed under the WHO mandate, governments and international agencies developed digital surveillance and identity mechanisms for linking the control of human movements and activities as alleged public health measures with a broader society-wide project to create new databases of citizens' digital identity, an idea that was mooted by among others the Gates Foundation, and the European Commission many years in advance of the COVID-19 pandemic. Among the principal digital devices for facilitating this project were QR codes and scanners, and so called 'green passes' or 'vaccine passports' together with 'mandates' that prevented citizens accessing public or private facilities without the requisite QR code or equivalent paperwork. Nations such as the Netherlands, Canada, Israel, Indonesia, France, Austria, and Germany attempted to set up electronic surveillance systems to scan and trace individuals' alleged 'immunity' status, and this status was judged on the basis not of their actual health and immune strength but on their willingness to undergo constant medical tests, together with their willingness to participate in the huge experimental drug trial that was initiated soon after the pandemic outbreak in Wuhan in 2020, and which led to emergency-use authorisation of novel gene-editing pharmaceutical therapeutics in the form of mRNA and mDNA 'vaccines' produced principally by American, European, and Chinese drug companies.

Gene Therapies and mRNA and DNA Vaccines

At first sight, the efforts of health authorities to suppress known and effective anti-viral and safe drugs such as ivermectin and hydroxychloroquine in the treatment of COVID-19 appear irrational, given the responsibility of doctors under the Hippocratic Oath to put the interests of the patient first. If patients report early COVID-19 symptoms and there is a drug protocol that resolves symptoms without the need for hospitalisation – as is evidently the case with COVID-19 – why would Fauci and other health 'leaders' attempt to suppress it? The answer, or at least part of the answer, is regulatory capture of these health authorities by the pharmaceutical industry. Only if there were no effective drugs to treat COVID-19 would it be possible to obtain emergency-use authorisation for experimental mRNA and mDNA therapeutics which have not been subjected to the usual five-year safety trials for new drugs and vaccines.

Fauci himself had already used this strategy in the HIV-AIDS pandemic. He promoted a very expensive and extremely toxic chemotherapy for those whom PCR tests indicated were infected with HIV in which he was personally invested. A majority of those who tested positive for HIV and took the extremely expensive AZT chemotherapy drug went on to develop full-blown immune-system disorders and eventually died from multiple organ failure. As those at risk of HIV and AIDS began to realise that their friends were being killed not cured by AZT, campaigns began for alternative drugs to be considered by health authorities. But Fauci and others strenuously resisted this approach and Fauci used his position as the lead funder of medical research in the world to prevent scientific investigation of alternative therapies for HIV. As AIDS expert and virologist Frank Ruscetti commented, 'We could have saved millions of lives with repurposed and therapeutic drugs. But there's no profit in it. It's all got to be about newly patented anti-virals and their mischievous vaccines.'[64] Celia Farber argues that methodical rules that once governed scientific investigation, and which required that researchers put findings based on empirical data above received opinion, or 'models', were set aside in medical science from the 1960s. The result was the politicisation of science, and its capture by drug companies and regulators linked to those companies, in the AIDS crisis and subsequently.[65]

The 'vaccines' for COVID-19 are not conventional vaccines since, other than the Chinese Sinovac, they do not incorporate dead elements of the virus to persuade the body to generate anti-bodies. Instead they use the same gene-editing technology that was used to genetically engineer SARS and SARS-CoV-2 as human infective pathogens. The COVID-19 therapeutics represent a genetically coded set of instructions which – in Moderna's own language – is like a software 'operating system' designed to program the human body: the instructions persuade the body first to manufacture the 'spike protein' – the principal toxic agent of SARS-CoV-2 – and then to generate an immune response to it.[66]

Ever since the expensive and international Human Genome Project to map the genes of all humans was launched in the 1990s, medical scientists and pharmaceutical companies have poured billions of dollars into developing and bringing to market gene therapies for human diseases which they believe will offer considerable, and profitable, advances over existing pharmaceuticals. But the early results of gene therapy trials on laboratory animals and on humans in clinical trials have not been notably successful. Far from it. Gene therapy trials on laboratory animals reveal a striking tendency for scientists' attempts to intervene therapeutically by inserting or shutting down genes in mammalian genomes to provoke unanticipated side effects. The most concerning of these was immune-system failure leading to spontaneous eruptions of cancers, and nervous and other disorders in laboratory animals. Many gene therapy trials, including for example a 2012 trial of mRNA coronavirus vaccines on ferrets, resulted in so many deaths of laboratory animals that the research had to be discontinued.[67]

Given the long-standing inability of scientists to vaccinate against coronaviruses, of which there are more than 60 in regular circulation, it was surprising that in 2020 most governments and the World Health Organisation insisted that the only way to end a coronavirus pandemic was via novel gene-therapy–based coronavirus vaccines which had never worked before and performed so poorly in laboratory trials that no drug company had brought one to market. Given that the 2012 SARS coronavirus vaccine trial wrecked the immune systems of other primates, experts such as Mike Yeadon, the former Research Director of Pfizer, argued on social media that this would occur in humans, but such concerns were dismissed in the media and by politicians as 'conspiracy theories,' and Yeadon was banned from Twitter and YouTube.

The most striking data on the mRNA vaccines is from the small island of Gibraltar, and the larger island of Australia. Both islands recorded rises in COVID-19 cases, and mortality, over 100 times in 2021 compared to 2020 during the vaccine rollout. There are analogous patterns of contiguity between vaccine rollout and spikes in COVID-19 infections and hospitalisations in many countries. A letter to *The Lancet* on the Pfizer trial results may help to explain this. The authors point out that Pfizer could only claim efficacy for its vaccine in preventing COVID-19 by a measure of what it called 'relative risk' which was a composite statistic applying its dubious trial results – dubious because trial participants were unblinded and many symptomatic participants were excluded from the published results. If a more appropriate measure of 'absolute risk' is taken – that is, the real-world risk of infection and illness following vaccination compared to the unvaccinated – the Pfizer results show a difference of only 0.27 per cent in likelihood of infection between vaccine recipients and those who did not receive it.[68] The Pfizer trial was also compromised by falsified data, including fictional trial subjects in Argentina; the unblinding of trial participants; and neglect in recording and following up adverse health events in the trial, as reported in the *British Medical Journal*.[69]

The bigger problem than their ineffectiveness in reducing COVID-19 infections with the novel therapeutics used on a mass trial basis from 2021 is the scale of injury and death associated with them. One study from Israel evidences significant increases in heart disease events during vaccine rollout in Israel.[70] Pfizer's own statistics on the rollout of the drugs indicate that there were 1,403 cases of serious cardiovascular events including heart attacks in the first 90 days of vaccine administration.[71] The US Vaccine Adverse Events Reporting System is cumbersome and results in huge under-reporting of vaccine side effects, as discussed above. In relation to the COVID-19 vaccines, as already discussed, VAERS reports indicate that to date the vaccines have caused 27,968 deaths, 155,633 hospitalisations, 14,326 heart attacks, 51,966 permanent disabilities, 4,615 miscarriages.[72] If this data is only 1 per cent of the actual injury rate then this would indicate a vast scale of injury and death from the COVID vaccines. The evidence for this is in actuarial reports from life insurance companies who must pay out to beneficiaries on the deaths of the insured. In the United States, life insurers report an unprecedented 20 per cent increase in death claims in 2021, and that this increase in the death rate in the year of mass vaccination was twice the 10 per cent increased mortality rate attributed to COVID-19 in 2020. And unlike COVID-19 in 2020, the increased mortality in 2021 was among working age people.[73] Scotland is the most heavily vaccinated country in the United Kingdom and Europe and its Public Health data in 2021 revealed an unexplained increase in adult all-cause mortality, most of it attributable to heart attacks and strokes.[74] Scotland also reported two major unexplained spikes in deaths of new-born babies.[75] Analogously, analysis of Our World In Data statistics reveals that in every country where vaccination rates were high, there was a subsequent increase in all-cause mortality.

There is not space here to review the evidence in detail, but the literature is now so extensive it is hard to deny something strange is occurring when governments and health authorities in many domains require citizens to participate in a hazardous experimental drug trial with a proven record of mass injury and death vastly in excess of any previous mass vaccination as condition for normal engagement in human life. As already discussed, regulatory capture is part of the explanation as pharmaceutical companies reaped unprecedented profits globally from the COVID-19 vaccine program. But there are significant indications that the COVID-19 response reaches far beyond merely the profits of drug companies, and represents a post-democratic, and even post-human, project to treat all humans going forward as potential biohazards, and to digitally track and trace all human movements, meetings, and interactions henceforth. The means for linking access to normal life and the vaccine record of individual citizens is the adoption in many domains, and at airports and ports, of QR-code scanners linked to vaccine and digital ID databases that many governments set up in concert, across Europe, North America, Australia, New Zealand, and much of Asia and Latin America.

The Digital Panopticon

That this new authoritarian and technocratic biosecurity panopticon was not merely a response to COVID-19 but a larger project long in the planning is evident in an extensive array of reports and research projects by private wealth foundations, the US and Chinese CDCs, North American and European universities, and the European Commission. One of the earliest is a report of the Rockefeller Foundation in 2010 in which a global pandemic is indicated as an opportunity to organise a new form of global government in which all national governments act in 'lockstep' and set up new forms of post-democratic technological control of citizens as means to achieving a new globally coordinated 'world society.'[76] Another advocate of such a global technocracy is Klaus Schwab whose private World Economic Forum was a training ground for many of the 'young global leaders' who were involved in coordinating the international technocratic lockstep COVID-19 response including Angela Merkel, Bill Gates, Emmanuel Macron, Matt Hancock, and Vladimir Putin.

In *COVID-19: The Great Reset* Schwab and his co-author Thierry Malaret propose a post-human merger between human and digital technologies by which humans will in future be machine-trackable 24/7. They argue that the pandemic indicates that normal bodily interactions between humans are no longer safe and that much human bodily interaction will need to be replaced by artificial intelligence, drones, and robots which will provide human services without the need for contacts between strangers or those not living in the same household.[77] The European Commission seems to have anticipated Schwab's dystopic proposal and had a project on a Europe-wide 'vaccine passport' linked to a new EU digital ID in 2015. The European Commission also led the way in proposals to link digital ID with health records and financial records. The end in view is that all monetary exchanges will be cashless and linked to a blockchain digital currency which will give to the central bankers who issue it totalitarian control on the activities and behaviours of citizens. It will be possible with this new blockchain digital financial system to restrict the activities of citizens deemed to have an incomplete vaccine record, by preventing citizens spending money beyond a predetermined distance from their home residence, or by limiting their access to all services beyond those deemed 'essential' for their survival such as food purchases; it would also be possible to limit activities of each citizen according to a predetermined 'ration' of 'carbon emissions' associated with their activities, and Mastercard announced a partnership with a carbon-tracking initiative along these lines during the COVID-19 pandemic.

Proponents of the new digital panopticon, and its post-human and post-national agendas, argue that the global challenges now facing humanity include novel infectious diseases, unprecedented environmental risks, and risks from novel technologies such as biological and nuclear warfare, and that, therefore, it is inevitable that once-independent democratic nation States will need to develop a new global biosecurity and technocratic order in which not only humans, but animals, plants, rivers, and oceans will all be surveilled and recorded in real time in computer data bases using the 'internet of things,' and the earth-surrounding low-orbital satellites put up during COVID-19, and combined with biosensors in people and all other objects. On the ground this panopticon will be facilitated by new 5G levels of electronic radiation enabling Wi-Fi to penetrate buildings and remote areas of the planet and to offer to global databases, and those who run them, total transparent visibility to every living thing on the planet. And this is despite numerous scientific studies indicating that levels of radiation at 5G frequencies and strength are hazardous to *all* life.[78] The documented harms of this much more powerful communication frequency led to a global call by physicians and scientists to stop the rollout of the technology, a call which thus far has only been heeded where citizens have actively intervened and opposed new 5G towers and related equipment in their communities. [79]

Against the allegedly unique 'risk horizon' of late modernity, and the claim that it requires a technocratic 'new normal' to reduce risk, humans as a species in the modern era have never lived longer and more healthily than they do at present. Infectious diseases cause less illness now than at any time since the origins of the Industrial Revolution which fostered living conditions, diets, and so on in which such diseases became rife. There is also lower mortality by violence in the present millennium than at any time in the last 120 years. The risks of health harms from 'global climate change' are also exaggerated by climate apocalypticists such as Greta Thunberg who, while cruising the world in luxury yachts, trains, and planes, claims her childhood was 'stolen' by those who failed to stop the production of fossil fuels on which her wealthy lifestyle, like that of all modern Europeans, Americans, and Asians depends for base energy. The World Health Organisation estimates that potential mortality from climate change is, and will remain, a small fraction of that from household and outdoor air pollution, and from malnutrition, all of which are not treated as 'emergencies' in the way that climate change is, although they are much cheaper to fix with appropriate public interventions than the huge cost of 'decarbonising' the world's economy.[80]

Human-Earth Healing

The feature of the global technocratic order which connects closely with the argument of this book is the globalist intention to capture and control the *nonhuman* environment on the same blockchain-driven and biosensor-enabled digital surveillance model as the QR-coded societies rolled out for COVID-19.[81] The merger of globalist technocratic top-down health and environmental agendas in response to COVID-19 is predicated on a well-established line of anti-human environmentalism which can be traced to the misanthropy of such diverse figures as John Muir and Edward Abbey. It also segues significantly into the modern scientistic ideology of eugenics according to which it is the business of medical science to 'control' and 'clean up' the genetic health of the human population by suppressing disabilities and undesirable traits, euthanising those judged genetically deficient, and controlling and reducing the size of the human population.[82]

At root, COVID-19 revealed the darkly atheistic and anti-human trend of modern scientism and its attendant ideologies and projects while underwriting the long-held conservationist belief that the problem with the human relation to nature is spontaneous human behaviour. It was remarkable to see police and even drones attempting to keep people out of national parks, mountains, forests, and beaches as an alleged 'public health' measure during COVID-19. And while people were being locked in their houses, and prevented from accessing nature, environmentalists were celebrating a temporary fall in human carbon emissions in 2020 and that wild animals were starting to reappear in daylight hours in the absence of humans.

I wrote at greater length in my last book, *Place, Ecology and the Sacred*, of the roots of the modern environmental crisis in the industrial alienation between humans and nature and especially the growing disconnect between food consumption and food growing. For all human history until the last 300 years, almost everyone was involved in some way in growing and procuring food. In this activity, humans and nature were not necessarily at odds. Biodiversity at vastly higher levels than that which obtains in modern industrial farms was achieved by hunter gatherers in forest gardens, and by peasant farmers using long-established crop rotation and interplanting techniques, and the symbiotic domestication of a large range of plant and animal species in complex, labour-intensive but often aesthetically beautiful, peasant agricultural systems of the kind that flourished across the Americas, Africa, and Asia before the colonial era. Now venture capitalists in the United States and other domains are investing in a future of genetic interventions in human health, and in human food the majority of which will be grown indoors in 'vertical farms' without sunlight or soil and, in the main, tended by robots.[83] New genetic engineering techniques will permit 'meat' substitutes to be grown in vats or made from factory-farmed insects, while fruit and vegetables will be grown in hydroponic systems in skyscrapers. Most foods in these novel growing environments will be genetically modified to deliver a new industrially controlled diet linked to increasingly surveyed and gene-edited post-human bodies.

Against this dystopic vision of the future of food, farming, and human health, there has for the last 30 years been growing attention by gastroenterologists, nutritionists, and complementary therapists to the human biome which primarily resides in the digestive tract as the core of an individual's health and immune system. 'Biome' therapy favours a diet that is high in unprocessed plant diversity, and low in industrially favoured staples including refined carbohydrates, refined fats, sugar, meat, and dairy products. Such a diet supports the health and functioning of 2 to 3 trillion bacteria and viruses within a person's biome and that constitute and sustain a healthy body and strong immune system in its relations with other bodies and the environment, and in its ability to resist and expel pathogens and toxins including the SARS-CoV-2 virus.[84] A healthy biome with a good diversity of healthy bacteria also improves mental states as well as supporting optimal functioning of internal organs and the body as a whole and reducing mortality from cancers, heart disease, and strokes.[85] But such a diet is inconsistent with the top-down corporately controlled monocultural production of meats, fats, sugars, and cereals, and their conversion into highly processed foods suited to the long-distance centralised supply chains used by global food corporations and supermarkets. It is also inconsistent with a top-down public health agenda which views human immunity not as a function of an individual's diet, of their internal microbial ecosystem, and of the diversity and resilience of local food cultures, soils, and species, but as a function of products and interventions by pharmaceutical companies and global health agencies in vaccination programs

FIGURE 5.2 Rice harvest, Penestanan, Ubud, Bali, May 2020. Author photo.

and other global disease control strategies including electronic databases monitoring population health records and movements.

Relationships between people and their habitats of the kind which sustain the biodiversity and resilience of ecosystems, and of healthy soils, seas, and waterways, while also sustaining the health and immune strength of persons, are 'polycentric.' Such relationships are not characterised by the two dominant institutions in modern agriculture and food provision, and in public health, which is to say, large economic corporations and governmental or intergovernmental agencies. Instead, they are characterised by a range of taboos, traditions, social agreements, rules, institutions, and customs which evolved over time in diverse ways as suited the diverse ecological conditions of different human ecological habitats, and which originated and in many cases are still sustained by religiously underwritten communitarianism. The United Nations' climate regime, and the WHO Public Health regime, are analogous globalist projects directed by international organisations whose leaders conceive of planetary and human health as top-down projects best served by powerful agencies who colonise and suppress the agency of individuals in sustaining their own health, and of communities in sustaining healthy habitats and local food cultures.

The COVID-19 'emergency' conferred enormous agency on an invisible virus which is so small that it easily passed through or around the fetishised masks many people adopted, and which many governments mandated, supposedly to ward it off despite scientific evidence from randomised control trials (not models) that masks do not reduce viral infections.[86] While the COVID-19 crisis was not a natural disaster, given the origins of the virus in unwise human practices in mixing genes and species that in the wild are not found together, it nonetheless has revealed the enduring potential of the indeterminate agency of the nonhuman still to influence, for good or ill, science-informed and technologically empowered humans.

On the top-down response to the crisis, and the broader political and economic tendencies of surveillance and control that it may portend, the Gaian and theological vision of earth, and earthly beings as reciprocally agential from the ocean's depths to the heights of heaven, that I unfold below represents both the deep past and an alternative future in which the co-creation of the health and habit of people and of nonhuman beings are co-curated polycentrically and not subjected to monopolistic control by a small number of human actors, for many of whom profit remains the aim, and not health, either human or planetary. It is to this alternative and more hopeful vision of human–earth healing which I now turn in the next three chapters.

Notes

1 Bruno Latour, 'The Pandemic Is a Warning: We Must Take Care of the Earth, Our Only Home.' *The Guardian,* December 24, 2021; see also Bruno Latour, *After Lockdown: A Metamorphosis.* Trans. Julie Rose, Cambridge, Polity Press, 2022, and for a critique see Joshua Clover, 'The Rise and Fall of Biopolitics: A Response to Bruno Latour.' *Critical Inquiry* 47 (2001) S28–32.

2 Marie Gottschalk, *Caught: The Prison State and the Lockdown of American Politics.* Princeton, NJ, Princeton University Press, 2015.
3 Jacob Kang-Brown, Chase Montagnet, and Jasmine Heiss, *People in Jail and Prison in Spring 2021.* New York, NY, Vera Institute of Justice, 2021.
4 Loic Wacquant, *Deadly Symbiosis: Race and the Rise of Neoliberal Penality.* Cambridge, Polity Press, 2006.
5 Matthew Lowen and Caroline Isaacs, *Lifetime Lockdown: How Isolation Conditions Impact Prisoner Reentry.* Tucson, AZ, American Friends Service Committee, 2014.
6 Christophe Fraser, Steven Riley, Roy M. Anderson, and Neil M. Ferguson, 'Factors That Make an Infectious Disease Outbreak Controllable.' *PNAS* 101 (2004) 6146–51.
7 Michael P. Senger, 'China's Global Lockdown Propaganda Campaign: Inside the CCP's Use of Social Media Bots and Other Disinformation Tactics to Promote Its Own Response to the Coronavirus Pandemic and Attack Its Critics.' *The Tablet,* September 16, 2020.
8 Shiyi Cao, Chao Wang, Max Bachmann, et al., 'Post-Lockdown SARS-CoV-2 Nucleic Acid Screening in Nearly Ten Million Residents of Wuhan, China.' *Nature Communications* 11 (2020) 1–7.
9 Chris Pleasance, 'How the Man Running World Health Organisation Trashed by Trump as China-Centric Is a Career Politician Who Worked for a Communist Junta and became WHO's First NON-Doctor Director-General Following Intense Lobbying from Beijing.' *Mail Online,* April 8, 2020, https://www.dailymail.co.uk/news/article-8199719/Dr-Tedros-Ghebreyesus-career-politician-running-China-centric-WHO.html accessed March 23, 2022.
10 David Horowitz, *The Enemy Within: How a Totalitarian Movement Is Destroying America.* New York, NY, Simon & Schuster, 2021, 38.
11 See for example Eleonora Iob, Andrew Septoe, and Daisy Fancourt, 'Abuse, Self-Harm and Suicidal Ideation in the UK during the COVID-19 Pandemic.' *British Journal of Pysychiatry* 217 (2020) 543–6.
12 R.F. Savaris G. Pumi, J. Dalzochio and R. Kuns, 'Stay-at-Home Policy Is a Case of Exception Fallacy: An Internet-Based Ecological Study.' *Scientific Reports* 11 (2021) 5313.
13 Itai Bavli, Brent Sutton, and Sandro Galea, 'Harms of Public Health Interventions against COVID-19 Must not Be Ignored.' *British Journal of Medicine* 371 (2020) m4074; C. Nord, I. Warnke, E. Seifritz, et al., 'Modelling Suicide and Unemployment: A Longitudinal Analysis Covering 63 Countries 2000–2011.' *Lancet Psychiatry* 2 (2015) 239–45.
14 Andy Sumner, Chris Hoy, and Eduardo Ortiz-Juarez, 'Estimates of the Impact of COVID-19 on Global Poverty.' United Nations University, UNI-Wider Working Paper 2020/43, April 2020 at https://www.wider.unu.edu/sites/default/files/Publications/Working-paper/PDF/wp2020-43.pdf accessed March 23, 2022.
15 Joshua L. Laughner, Jessica L. Neu, David Schimel, et al., 'Societal Shifts due to COVID-19 Reveal Large-Scale Complexities and Feedbacks between Atmospheric Chemistry and Climate Change.' *PNAS* 118 (2021) e2109481118.
16 Jane Goodall, Oral presentation to a webinar sponsored by Compassion in World Farming. June 1, 2020, cited Philip Lymbery, June 2, 2020, https://www.weforum.org/agenda/2020/06/jane-goodall-coronavirus-humanity-natural-animals-COVID-finished accessed July 1, 2021.
17 Beatriz do Vale, Ana Patrícia Lopes, Maria da Conceição Fontes, et al., 'Bats, Pangolins, Minks and Other Animals – Villains or Victims of SARS-CoV-2?' *Veterinary Research Communications* 45 (2021) 1–19.
18 Triangulation is a principle, derived from navigational instruments, in empirical research which indicates that data from three different perspectives, or of three different kinds, offer a stronger evidentiary base for a claim than data from one perspective: given public controversy over the origins of SARS-CoV-2, and that the claim it originated in a laboratory is a 'conspiracy theory,' triangulation is a valuable technique: V.A. Thurmond, 'The Point of Triangulation.' *Journal of Nursing Scholarship* 33 (2001) 253–8.

19 Boyd Yount, Kristopher M. Curtis, and Ralph S. Baric, 'Reverse Genetics with a Full-Length Infectious cDNA of Severe Acute Respiratory Syndrome Coronavirus.' *PNAS: Microbiology* 100 (2003) 12995–3000: the project was funded by NIAID Grants AI 23946 and GM63228.

20 US Patent 7,279,327 'Methods for Producing Recombinant Coronavirus.' For a list of over 100 patents for genetically engineered viruses, recombinant DNA techniques, and related therapeutics and vaccines owned by Anthony Fauci and agencies he controls or funds see the compilation in David Martin, *The Fauci/COVID19 Dossier* (2021) published online davidmartin.world/wp-content/uploads/2021/01/The_Fauci_COVID-19_Dossier.pdf accessed September 4, 2022.

21 Paul Elias, 'Race to Patent SARS Virus Renews Debate.' *Associated Press News,* May 6, 2003, at https://apnews.com/article/145b4e8d156cddc93e996ae52dc24ec0 accessed March 24, 2022; Renee W.Y. Chan, Ralph S. Baric, J.S. Malik Peiris, et al., 'Tropism of and Innate Immune Responses to the Novel Human Betacoronavirus Lineage C Virus in Human Ex Vivo Respiratory Organ Cultures.' *Journal of Virology* 87 (2013) 6606–14.

22 Yount, Curtis, and Baric, 'Reverse Genetics,' 12995.

23 I.B. Nikitina and E.A. Smirnova, 'SARS-CoV-2 and Patent Activity.' *Molecular Genetics, Microbiology and Virology* 36 (2021) 10–4.

24 Anthony S. Fauci, Nancy A. Touchette, and Gregory K. Folkers, 'Emerging Infectious Diseases: A 10-Year Perspective from the National Institute of Allergy and Infectious Diseases.' *Emerging Infectious Diseases* 11 (2005) 519–25.

25 Karl Sirotkin and Dan Sirotkin, 'Might SARS-CoV-2 Have Arisen via Serial Passage through an Animal Host or Cell Culture? A Potential Explanation for Much of the Novel Coronavirus' Distinctive Genome.' *Bioessays: News and Reviews in Molecular, Cellular and Developmental Biology* 42 (2020) e2000091.

26 Sirotkin and Sirotkin, 'Might SARS-CoV-2 Have Arisen via Serial Passage?'

27 Himani Bisht, Anjeanette Roberts, and Leatrice Vogel, 'Severe Acute Respiratory Syndrome Coronavirus Spike Protein Expressed by Attenuated Vaccinia Virus Protectively Immunizes Mice.' *Proceedings of the National Academy of Sciences of the United States of America* 101 (2004) 6641–6.

28 X.Y. Ge, J.L. Li, X.L. Yang, et al., 'Isolation and Characterization of a Bat SARS-Like Coronavirus that Uses the ACE2 Receptor.' *Nature* 503 (2013) 535–8.

29 Vineet D. Menachery, Boyd L. Yount Jr., and Kari Debbink, 'A SARS-Like Cluster of Circulating Bat Coronaviruses Shows Potential for Human Emergence.' *Nature: Medicine* 21 (2015) 1508–13.

30 Declan Butler, 'Engineered Bat Virus Stirs Debate over Risky Research: Lab-Made Coronavirus Related to SARS Can Infect Human Cells.' *Nature* (2015). https://doi.org/10.1038/nature.2015.18787.

31 Ben Hu, Lei-Ping Zeng, Xing-Lou Yang, et al., 'Discovery of a Rich Gene Pool of Bat SARS- Related Coronaviruses Provides New Insights into the Origin of SARS Coronavirus.' *PLOS: Pathogens* 13 (2017) e1006698.

32 On the Drosten PCR controversy, and the major scientific flaws in Corman and Drosten's PCR paper see Pieter Borger, Bobby Rajesh Malhotra, Michael Yaedon, et al., 'External Peer Review of the RTPCR Test to Detect SARS-CoV-2 Reveals 10 Major Scientific Flaws at the Molecular and Methodological Level: Consequences for False Positive Results.' Submitted to *Eurosurveillance* in September 2020, and published online at https://cormandrostenreview.com/report/ accessed March 25, 2022.

33 Hu, Zeng, Yang, et al., 'Discovery of a Rich Gene Pool of Bat SARS.'

34 Nicolas Wade, 'New Routes to Making COVID-19 in the Lab.' *Medium*, September 24, 2021, at https://nicholaswade.medium.com/new-routes-to-making-COVID-19-in-the-lab-f94318cad4de accessed February 15, 2022.

35 Fredrick Kunkle, 'Third Person Has Died after Respiratory Illness Outbreak at Greenspring Village, Fairfax Officials Say.' *The Washington Post*, July 17, 2019.

36 Tim Wyatt, 'Research into Deadly Viruses and Biological Weapons at US Army Lab Shut Down over Fears They Could Escape.' *The Independent*, August 6, 2019.

37 Marlene Lenthang, 'State Department warned in 2018.' Dailmail.com April 15 2020 https://www.dailymail.co.uk/news/article-8218817/State-Department-warned-2018-Wuhan-lab-cause-new-SARS-like-pandemic.html April 15 2020
38 Billy Bostickson and Gilles Demaneuf, 'An Investigation into the WIV Databases That Were Taken Offline.' Prepint at ResearchGate, February 2021, https://www.researchgate.net/publication/349073738_An_investigation_into_the_WIV_databases_that_were_taken_offline accessed March 25, 2022.
39 Rita R. Colwell, Bart L. Haagmans, Peter Daszak, et al., 'Statement in Support of the Scientists, Public Health Professionals, and Medical Professionals of China Combatting COVID-19.' *The Lancet* 395 (2020). https://doi.org/10.1016/S0140-6736(20)30418-9.
40 Sakshi Piplani, Puneet Kumar Singh, David A. Winkler, et al., 'In Silico Comparison of SARS-CoV-2 Spike Protein-ACE2 Binding Affinities across Species and Implications for Viral Origin.'
https://arxiv.org/pdf/2005.06199.pdf accessed March 25, 2022.
41 Nobel Professeur Luc Montagnier, 'COVID-Cine Experiment Gone Bad.' Full Interview, French TV channel CNews April 17, 2020, https://youtu.be/-a7nld57f2M accessed May 11, 2022.
42 Yaron Steinbuch, 'Ex-CDC Chief Robert Redfield Explains Belief COVID Came from China Lab.' *New York Post*, June 15, 2021, https://nypost.com/2021/06/15/ex-cdc-chief-explains-belief-COVID-19-came-from-china-lab/ accessed March 25, 2022.
43 Bertrand Russell, *The Impact of Science on Society*. New York, NY, AMS Press, 1953.
44 Smriti Mallapaty, 'What the cruise-ship outbreaks reveal about COVID-19.' *Nature* 580 (2020) 18-19;
45 Eran Bendavid, Bianca Mulaney, Neeraj Sood et al, 'COVID-19 antibody seroprevalence in Santa Clara County, California.' International Journal of Epidemiology, Volume 50, Issue 2 (2021) 410–419,
46 John Ioannidis, 'Reconciling estimates of global spread and infection fatality rates of COVID-19: An overview of systematic evaluations.' *European Journal of Clinical Investigation* 51 (2021) e313554.
47 'COVID-19: Death Certification and Cremation.' https://www.bma.org.uk/advice-and-support/COVID-19/adapting-to-COVID/COVID-19-death-certification-and-cremation accessed February 12, 2022.
48 Matt Strauss, 'Ventilators Aren't a Panacea for a Pandemic Like Coronavirus.' *The Spectator*, April 4, 2020, https://www.spectator.co.uk/article/Ventilators-aren-t-a-panacea-for-a-pandemic-like-coronavirus accessed February 15, 2022.
49 At time of writing the 'Midazolam Murders' as they have become known were the subject of a UK court case in Westminster: for a summary of the case for the prosecution see https://www.thebernician.net/overwhelming-evidence-of-midazolam-murders-by-government-policy/ accessed February 14, 2022.
50 Health and Social Care Committee Oral evidence: Preparations for Coronavirus, HC 36 Friday April 17, 2020, https://committees.parliament.uk/oralevidence/288/default/ accessed February 12, 2022.
51 B.C. Paterson, R. Duncan, R. Conway, et al., 'Introduction of the Liverpool Care Pathway for End of Life Care to Emergency Medicine.' *Emergency Medicine Journal* 26 (2009) 765.
52 'Clinical Guidelines for the Control of Symptoms in the COVID-19 Patient in a Community Setting.' Hull and East Riding Prescribing Committee, March 2020 and on line at https://www.hey.nhs.uk/wp/wp-content/uploads/2020/04/COVIDsymptomControlHERPC.pdf.
53 Robert F. Kennedy Jr., *The Real Anthony Fauci: Bill Gates, Big Pharma, and the Global War on Democracy and Public Health*. New York, NY, Skyhorse Publishing, 2021.
54 There are now numerous peer-reviewed studies on the effectiveness of hydroxychloroquine as a preventative treatment of COVID-19 symptoms and as a therapeutic drug in treatment of symptoms: see for example Xueting Yao, Fei Ye, Miao Zhang,

et al., 'In Vitro Antiviral Activity and Projection of Optimized Dosing Design of Hydroxychloroquine for the Treatment of Severe Acute Respiratory Syndrome Coronavirus 2 (SARS-CoV-2).' *Clinical Infectious Diseases* 71 (2020) 732–9.

55 Yao, Ye, Zhang, et al., 'In Vitro Antiviral Activity and Projection of Optimized Dosing Design of Hydroxychloroquine for the Treatment of Severe Acute Respiratory Syndrome Coronavirus 2 (SARS-CoV-2).'

56 Lazarus Ross, Michale Klompas, and Steve Bernstein, *Electronic Support for Public Health-Vaccine Adverse Events Reporting System (ESP:VAERS)*, Cambridge, MA, Harvard Pilgrim Health Care Inc., 2010 submitted to US Department of Health and Human Services and available online at https://digital.ahrq.gov/ahrq-funded-projects/electronic-support-public-health-vaccine-adverse-event-reporting-system accessed March 30, 2022.

57 Tyler Durden, 'Life Insurance CEO Says Deaths Up 40% Among Those Aged 18–64.' Published online at *Zerohedge* January 4, 2022, https://www.zerohedge.com/medical/life-insurance-ceo-says-deaths-40-among-those-aged-18-64-and-not-because-COVID accessed May 10, 2022.

58 Matthias Toying and Jana Olsen, 'Charité-Forscher fordert Ambulanzen für Impfgeschädigte.' MDR.DE Nachrihten Deutschland, May 3, 2022, published online at https://www.mdr.de/nachrichten/deutschland/panorama/corona-impfung-nebenwirkungen-impfschaeden-100.html accessed May 10, 2022.

59 Michel Foucault, *The Birth of Biopolitics: Lectures at the Collège de France, 1978–1979.* Trans. and Ed. M. Senellart, New York, NY, Palgrave Macmillan, 2008.

60 Giorgio Agamben, *Where Are We Now? The Epidemic as Politics.* Trans. Valeria Dani, London, ERIS, 2021.

61 I discuss the reasons for the failure of the UNFCCC treaty process more fully in Michael S. Northcott, *A Moral Climate: The Ethics of Global Warming.* London, Darton, Longman & Todd, 2007, and *A Political Theology of Climate Change.* Grand Rapids, MI, Wm. B. Eerdmans, 2013.

62 Bruce Kaplan and Mary Echols, ''One Health' - The Rosetta Stone for 21st Century Health and Health Providers.' *Veterinaria Italiana* 45 (2009) 377–82.

63 See further the description of AADHAR at the Digital India government webpage https://www.digitalindia.gov.in/infrastructure accessed June 23, 2020; for a critical perspective see Shyam Divan, 'The Prime Minister's Fingerprints: Aadhar and the Garrotting of Civil Liberties.' *National Law School of India Review* 26 (2014) 159–16.

64 Cited in Kennedy, *The Real Anthony Fauci.* E-edition, 299.

65 Celia Farber, *Serious Adverse Events: An Uncensored History of AIDS.* Hoboken, NJ, Melville House Publishing, 2006.

66 The mRNA platform is *described* as an 'operating system' on Moderna's website at https://www.modernatx.com/mrna-technology/mrna-platform-enabling-drug-discovery-development accessed March 25, 2022.

67 Chien-Te Tseng, Elena Sbrana, Naoko Iwata-Yoshikawa, et al., 'Immunization with SARS Coronavirus Vaccines Leads to Pulmonary Immunopathology on Challenge with the SARS Virus.' *PLoS One* 7 (2012) e35421.

68 Piero Olliaro, Els Torreele, and Michel Vaillant, 'COVID-19 Vaccine Efficacy and Effectiveness—The Elephant (not) in the Room.' *The Lancet,* April 20, 2021, Published online https://doi.org/10.1016/S2666-5247(21)00069-0 accessed February 15, 2022.

69 Paul D. Thacker, 'COVID-19: Researcher Blows Whistle on Data Integrity Issues in Pfizer's Vaccine Trial.' *British Medical Journal* 375 (2021). https://doi.org/10.1136/bmj.n2635.

70 See for example Christopher L.F. Sun, Eli Jaffe, and Retsef Levi, 'Increased Emergency Cardiovascular Events among under-40 Population in Israel during Vaccine Rollout and Third COVID-19 Wave.' *Scientific Reports* 12 (2022) 6978.

71 'Cumulative Analysis of Post-Authorization Adverse Event Reports of PF-07302048 (BNT162B2) received through 28-FEB-2021.' Published online

by Pfizer at https://phmpt.org/wp-content/uploads/2021/11/5.3.6-postmarketing-experience.pdf accessed May 22, 2022.
72 OpenVAERS: VAERS COVED Vaccine Adverse Even Reports Through May 6, 2022: https://openvaers.com/COVID-data accessed May 19, 2022.
73 Thomas J. Britt, Paul Correla, Mike Krohn, et al., *Group Life COVID-19 Mortality Survey Report*. SOA Research Institute. Published online by the Society of Actuaries, 2022 at https://www.soa.org/48ff80/globalassets/assets/files/resources/research-report/2022/group-life-COVID-19-mortality.pdf accessed May 19, 2022.
74 *Scotland: All-Cause Mortality Week 49 2021*. Data: National Records Scotland. Summary published online at https://scottishunity.org/wp-content/uploads/2021/12/MortReport_2021-49.pdf accessed May 19, 2022.
75 Helen Puttick, 'Spikes in Newborn Baby Deaths Breached Warning Limits.' *The Times* (of London), May 18, 2022, published online at https://www.thetimes.co.uk/article/spikes-in-newborn-baby-deaths-breached-warning-limits-mxjrgzg56 accessed May 19, 2022.
76 Rockefeller Foundation and Global Business Network, *Scenarios for the Future of Technology and International Development*. New York, NY, Rockefeller Foundation, 2010 and at http://www.nommeraadio.ee/meedia/pdf/RRS/Rockefeller%20Foundation.pdf accessed May 21, 2021.
77 Schwab and Malleret, *COVID-19: The Great Reset*. E-Edition, 374–90.
78 Cindy L. Russell, '5G Wireless Telecommunications Expansion: Public Health and Environmental Implication.' *Environmental Research* 165 (2018) 484095; Noa Betzelel, Paul Ben Ishai, and Yuri Feldman, 'The Human Skin as a Sub-THz Receiver – Does 5G Pose a Danger to It or Not?' *Environmental Research* 163 (2018) 208–16.
79 '5G Appeal: Scientists and Doctors Warn of Potential Serious Health Effects of 5G.' Press release September 13, 2017, and published online at http://www.5gappeal.eu/the-5g-appeal/ accessed May 10, 2022.
80 'The Global Health Observatory: Public Health and Environment.' https://www.who.int/data/gho/data/themes/public-health-and-environment accessed May 10, 2022.
81 See especially Bram Buscher, 'The Dangerous Intensifications of Surplus Alienation, or Why Platform Capitalism Challenges the (More-Than) Human.' Forthcoming in *Dialogues in Human Geography* (2022).
82 On eugenics in the twentieth century, and its intersections with Protestant Christianity, see Amy Laura Hall, *Conceiving Parenthood: American Protestantism and the Spirit of Reproduction*. Grand Rapids, MI, WmB Eerdmans, 2008.
83 Dickson Despommier, *The Vertical Farm: Feeding the World in the 21st Century*. New York, NY, Picador, 2011.
84 Mark L. Heiman and Frank L. Greenway, 'A Healthy Gastrointestinal Microbiome Is Dependent on Dietary Diversity.' *Molecular Metabolism* 5 (2016) 317–20.
85 For a science-informed overview of the human microbiome see Susan L. Prescott and Alan C. Logan, *The Secret Life of Your Microbiome: Why Nature and Biodiversity Are Essential to Health and Happiness*. Gabriola Island, New Society Publishing, 2017; on the gut microbiome and immunity see Javier Cabrera-Perez, Vladimir P. Badinovac, and Thomas S. Griffith, 'Enteric Immunity, the Gut Microbiome, and Sepsis: Rethinking the Germ Theory of Disease.' *Experimental Biology and Medicine* 242 (2017) 127–39; on stomach flora and the brain see Clair R. Martin, Vadim Osadchiy, Amir Kalani, et al., 'The Brain-Gut Microbiome Axis.' *Cellular and Molecular Gastroenterology and Hepatology* 6 (2018) 133–48.
86 The first randomized control trial of mask wearing by householders found no difference in respiratory virus transmission among mask wearers: C.R. Macintyre, D. Dwyer, H. Seale, et al., 'The First Randomized, Controlled Clinical Trial of Mask Use in Households to Prevent Respiratory Virus Transmission.' *International Journal of Infectious Diseases,* 12 Supplement 1 E238 (2008). https://doi.org/10.1016/j.ijid.2008.05.877.

6

THE EARTH AS GAIA

Abstract

James Lovelock is an English inventor, scientist, and author whose inventions played seminal roles in revealing the presence, and harmful impacts, of industrial chemicals in the air and water since the 1960s, and the related rise of the environmental movement, and environmental regulation. In mid-career Lovelock, while working for Shell and NASA, and in collaboration with micro-biologist Lynn Margulis, developed a theory of the connectedness of all matter and organisms, and their symbiotic roles in generating over billions of years a homeostatic Earth habitable by Life. The Gaia hypothesis in its original form involved the claims that the Earth is alive; that all life and all matter are connected; and that there is a convergence of matter and organisms on Earth towards the evolution of biodiversity, conscious living organisms, and a planet habitable to them, from the dawn of time. Margulis described the Gaia theory as a scientific revival of natural theology which replaces the deistic and mechanistic tendencies of nineteenth-century science with a theory of Life that fits the surprising evolution of biodiversity, and of conscious mammals, on a large lump of molten lava and congealed rock hurtling around a solar system, and a galaxy, that is otherwise empty of Life.

James Lovelock and the Origins of Environmental Regulation

James Lovelock is a British inventor and scientist whose most influential mechanical invention was an electron capture detector which is so sensitive it measures chemicals in air down to one part per trillion. With this device he studied haze in Southern England in the 1960s, which he believed was novel compared to his childhood, and discovered that it contains chlorofluorocarbons (CFCs), a class of industrial chemicals widely used as refrigerants and aerosol accelerants, but

DOI: 10.4324/9781003110750-7

FIGURE 6.1 James Lovelock in his laboratory in photo montage at Lovelock Centennial Exhibition, Exeter University, July 2019. Author photo.

which at the time were thought to be harmless. Later, he deployed his device with the British Antarctic Survey on the ice-breaker *Shackleton* and discovered that CFCs were also present in the Antarctic atmosphere.[1] In 1984 scientists with the British Antarctic Survey using Lovelock's device on a high atmosphere balloon discovered the ozone hole in the atmosphere over Antarctica. They also found excess chlorine in the atmosphere which they believed was the cause of ozone depletion.[2] The source of the chlorine was subsequently identified as CFCs[3] (Figure 6.1).

The discovery of the effects of CFCs in thinning the ozone layer in the upper atmosphere, and hence reducing atmospheric filtering of ultraviolet sunlight, which in excess is harmful to humans and other animals, ultimately led to the first successful international atmospheric pollution treaty in human history – the Montreal Protocol. The Protocol introduced a gradual ban on the production of CFCs, negotiations for which drew on Lovelock's data on the global dispersal of CFCs.[4] Concerted international action under the Protocol to cease producing the class of chemicals responsible for destroying atmospheric ozone has stabilised the ozone hole, which is no longer expanding and should eventually close.[5] The Montreal Protocol remains the only successful international environmental treaty.

Other international treaties have since been negotiated, including on climate change and biodiversity decline but they have not achieved their scientific goals, respectively to reduce greenhouse gas emissions and biodiversity losses. As I argue elsewhere, this failure is principally because the rights of corporate agents trump the rights of people and species to stable and safe habitats.[6] The global inability to make successful environmental treaties is intrinsic to the dominance of powerful international economic corporations, in connection with globalist agencies and most national governments which act principally to promote corporate wealth over the interests of citizens and of Life on Earth. The only extant exception is the Montreal Protocol, but in this case the treaty served the interest of one very powerful corporation deeply connected with the US Federal Government: the huge US chemical company Du Pont had already begun producing the class of chemicals that would be used as substitutes for CFCs. The United States strongly pushed for the Protocol's ban on CFCs as the United States was at the time the sole domain making the substitute.[7]

The excessive influence on political deliberation and economic development exercised by corporate agents, their lobbyists together with their public-private partnerships with government agencies, *and not human beings in general*, are the principal reason reversing ecological damage to the Earth, and to its nested ecosystems at micro- and macro- levels, is proving so intractable.[8] Individuals and small communities over thousands of years developed small-scale collective action networks for the governance of commons including coastal waters, river catchments, forests, lakes, high mountain pastures, and savannahs.[9] But the corporate-led history of Western colonialism as it spread across the planet from the sixteenth century gradually corroded and destroyed these collective action networks. This resulted in what the North American political scientist Garett Hardin called the 'tragedy of the commons.'[10] But as we have seen, Hardin was wrong that commons are inevitably tragic when utilised by humans. What he seminally missed was that commons tragedies happen when long-evolved commons governance arrangements are destroyed by the intervention of external powers.

Besides Lovelock's crucial role in the first successful atmospheric healing project, his electron capture device gave Rachel Carson the evidence on the ubiquity of pesticide residues, from mother's milk in Paris to mammals in the Congo, and their toxic effects on organic life, including human cancers and nervous and other disorders, for her highly influential book *Silent Spring*.[11] Publication was blocked for a year by the chemical and agrochemical industries but its eventual release, and Carson's testimony to Congress while she was suffering from breast cancer, sparked the modern environmental movement.

The book influenced the youth counterculture of the 1960s, and the early years of modern popular – and particularly folk – music.[12] It awakened the generation known as 'baby boomers' to growing threats to Mother Earth represented by the chemical and fossil fuel industries, and the urgent need for political and social action to address these.[13] Cultural and political pressures sparked by the book led to the formation by the Nixon administration of the Environmental Protection Agency in 1970, and the creation of similar agencies in Europe and the developing world. It also led to Clean Water and Clean Air legislation in the United States, and in Europe, though in the United States these laws are significantly weaker than when first enacted, as witness the mass poisoning of the population of the city of Flint, Michigan, by the city administration's decision to switch the water supply from Lake Huron to the highly polluted Flint River in 2012.[14]

Corporate Capture of Environmental Regulation

Lovelock's electron capture detector was a crucial instrument in the cultural awakening to the causative roles of agrochemicals in systematic reductions of biodiversity, and in modern epidemics of cancers, nervous disorders, and other diseases in humans from chemical residues in food and water. But in the United States, and globally, there is significant corporate capitalist pushback by chemical and other corporations against environmental regulations concerning biodiversity and health dangers from chemical pollution, and analogously against labour market and other social protection legislation. This pushback originated in the 1960s when corporations and investors observed that the rate of return on capital invested was declining. The blame was cast upon collective action by labour unions and other citizen organisations, and on government regulation of economic activities.[15] In response, industries began to fund a range of lobby organisations and 'think tanks' devoted to casting doubt on scientific findings about pollution and its dangers to human and planetary health. On a broader front, neoliberal economists such as James Buchanan and Milton Friedman constructed economic and legal arguments to the effect that government regulation of wealth creation, including of employment, and of pollution, were forms of theft from corporations and their shareholders. And they promulgated a new economic dogma in pursuance of this approach variously known as neoliberalism, 'public choice,' or 'rational choice' theory.[16]

Economist Ronald Coase used this approach to argue that the only form of environmental regulation that is not theft is when pollution is turned into a tradeable commodity by governmental construction of property markets in permits to pollute.[17] These novel markets in toxic power-station and factory emissions were first developed in the United States. They evolved via governmental and international architecture related to Biodiversity and Climate Treaties into a range of economic and fiduciary instruments and markets in which ecosystems, habitats, and species are all rendered by accounting devices as economic resources whose use generates costs and benefits which companies and government agencies may then write down in corporate and national accounts.[18] The financialisation of nature has spawned a clutch of novel tradeable assets, such as carbon and biodiversity 'credits.' But these have had no impact on the ongoing corporate wreckage of ocean, riverine, and terrestrial ecosystems which are the principal causes of 'natural' disasters such as flash floods, declining water tables, and increasing temperatures. The public–private construction of ecological financial instruments represents a wealth-creating opportunity for 'green capitalists,' derivative traders, and hedge fund managers. But it is a misdirection of creative and entrepreneurial energy from real-world ecological solutions, and a classic example of what Alfred North Whitehead called the 'fallacy of misplaced concreteness.'[19]

Despite the discovery of human and ecosystem harms from pollution, air pollution remains, as it has since the commencement of the industrial revolution, the biggest industrial-era killer, far greater in reducing average longevity than cancer, or infectious diseases including tuberculosis and COVID-19.[20] In the nineteenth century, coal smoke in cities killed millions prematurely and even in the twenty-first century, vehicular pollution and pollution from coal-fired power stations, biomass burning, waste incinerators, and chemical plants give rise to hundreds of millions of sufferers from chronic asthma.[21] Spikes in air pollution in cities and rural areas are associated with excess mortality days, though governments and vehicle manufacturers avoid liability because of the difficulty of proving in a court of law that failure to regulate air quality led to the death of a particular person on a particular date. [22]

There is a similar problem with insecticides and biodiversity decline. Just as artificial fertilisers originated in the manufacture of explosives, modern pesticides and herbicides originated in the development of chemical weapons in Germany and the United States during the First and Second World Wars.[23] After the Second World War, nerve agents originally designed to corrode the nervous systems of combatants, by turning off enzymes to disrupt hormone production and intercellular communication, were used instead as agrochemicals. They were also used by the United States as chemical weapons when they were deployed in vast quantities as aerial defoliants in the secret aerial bombing and spraying of Cambodia, Laos, and Vietnam during the Vietnam War: the most toxic of these – Agent Orange – left a legacy in these countries of cancers and of genetic damage in the children of adults poisoned by persistent toxins from chemical weapons in food and water.[24]

The most widespread nerve-corroding agrochemical in use today is glyphosate. Glyphosate disrupts hormone production in organic life, using the same chemical pathway to attack cellular and nervous communication systems as the broad-spectrum defoliant and chemical weapon Agent Orange.[25] There are established associations between the use of such chemicals by agricultural and park workers and birth deformities among farmers' children, and the occurrence of cancers such as non-Hodgkin lymphoma.[26] Glyphosate is now present in most water catchments on Earth because of its ubiquitous use by farmers, gardeners, and city authorities. The intended effect of defoliants such as glyphosate is to kill plants which are the homes of insects which are in turn eaten by birds, mammals, and reptiles, so reducing habitat for other creatures. Glyphosate operates by switching off crucial cellular pathways in plants in order to kill them. But the same pathways are also present in all other life forms, including soils, insects, and animals, and in the human microbiome.[27] Defoliant intervention into the hormonal signalling systems of life impacts the whole chain of life including insects, birds, mammals, aquatic species, and humans.[28] This is the origin of the large-scale chemical eradication of insects, which are close to the bottom of the food chain of life, and which has been a major factor in the decline of bird numbers since 1945.[29] Glyphosate is also significantly implicated in the decline in soil bacteria and minerals taken up by food crops, and in the decline of beneficial bacteria in the human gut and in the guts of farm and wild animals. Nicotinoid insecticides have magnified the problems with glyphosate further, since they act synergistically to corrode the abilities of bees which pollinate most human food crops to navigate from nests to plants and back.[30] The same processes that corrode the nervous systems of bees are implicated in the impacts of glyphosate ingestion on bacteria in the human gut as these are also crucially involved in many immune and brain functions and hence there is a statistically highly significant correlation between rises in human autoimmune and nervous disorders since the 1970s and levels of glyphosate in foods and water.[31] As Stephanie Seneff argues in a powerful review of scientific studies on glyphosate:

> Something terrible seems to be affecting every living thing on the planet—the insects, the animals, and the health of human beings, including children. Something hiding in plain sight. While we can't reduce all environmental and health problems to one insidious thing, I believe there is a common denominator. That common denominator is glyphosate.[32]

Despite Carson's prescient warning in 1962, the sound of bird song is in major decline across the developed world, and in many parts of the developing world.[33] Agrochemicals are the single biggest cause of the huge insect and bird die-off scientists are now uncovering, with one study in 'protected areas' in Germany finding a 75 per cent decline.[34] And it is not only on land but in the oceans, which cover two-thirds of Earth's surface, that nonhuman creatures are in decline. As we have seen, since 1945 sonar – another war-originated technology – has enabled industrial trawlers to pull unprecedented quantities of fish out of the oceans. At the same time, fishing lines miles long not only trap tuna but dolphins, sharks, and other sea creatures which are drowned by the lines and discarded when the lines are hauled in. As we have seen, longitudinal studies of fish numbers drawing on oral accounts and ships logs of fishers before scientific recording commenced indicate that in the last 100 years humans have emptied the oceans of 90 per cent of all living beings.[35] Since 1945, industrial and development corporations and agencies have also destroyed more than half of the remaining old-growth forests, including the ancient and species-rich rainforests of Southeast Asia and Central Africa, and the Amazon too is now being targeted ever more intensively by farmers, loggers, and miners, with fires set on huge new unlogged areas in 2019 after the election of President Bolsonaro in Brazil.[36]

The depletion of so much life from the oceans and land has had two Earth System–wide effects which may be hard to reverse, the first being to reduce the capacities of the oceans to turn industrial emissions of carbon dioxide into oxygen, which is a key mechanism in Earth's climate-regulating system, and the second being biodiversity decline, including species extinctions. In nigh on a century of efforts to conserve the environment through regulation of air, food, and water pollution, and through protections for endangered species such as National Parks, public regulators have failed to address the systemic roots of the problem in the overwhelming activities of corporate agencies of private and government-sponsored industrial capitalism. At the same time the conservation effort has failed to revive the connected agencies of the Earth and of Life in repairing the problems industrial agencies are creating. These failures are related *both* to the corporate and scientistic devotion to dominating and controlling nonhuman life in combination with the reductionist approaches of conservation scientists who treat air pollution, habitat destruction, and species extinction as separate problems, and neglect the extent to which these problems relate to the systemic character of the global industrial economy.

Reviving the Agency of Life: the Gaia Theory

Against the tendency of industrial civilisation, and of the machine cosmology which informs it, to degrade ecosystems and species diversity – including the nonhuman bacteria which are essential to human health and homeostasis – the Gaia theory of Lovelock and Lynn Margulis offers a holistic approach to ecological conservation and planetary repair in which the agency of the Earth and humans' fellow species are the crucial means to Earth healing. Taking this approach, each human activity needs to be assessed ecologically as to whether or not it contributes to reviving the agencies of Life, from micro-organisms in the soils to the numbers of insects, plants, and mammals on land and in the oceans, so as to revive the regulatory capacities of habitats and ultimately the Earth System. Such a holistic approach to the Earth, Lovelock calls 'planetary medicine,' which he advanced in a book on Gaian approaches to the ecological crisis based on the Gaia theory.[37] Gaian planetary medicine is analogous to holistic approaches to human health which, instead of looking at disease states as caused by particular external organisms, or pathologies within one organ, look at the whole body or 'terrain' of the body as the means to health and immunity, including the microbiome discussed above. Medicine that approaches the body this way has roots in medicine in classical Greece, and in particular the idea of 'balancing humours' favoured by Aristotle and Hippocrates which remained influential in a range of medical models into the early modern era.[38] The idea of homeostasis or the self-regulating balance of the body, and by extension of nature, also has philosophical roots in the Indian Vedas, and in pre-Hellenic vitalism, as I will show in the next chapter.

That the human body has intrinsic capacities to achieve a healthy equilibrium state, including a regular temperature, was first proposed in modern medicine by Claude Bernard and Walter Cannon, who named the way the body maintains this state *homeostasis*.[39] Homeostasis is a key feature of modern physiology, and for many nineteenth-century physicians its discovery indicated that the function of medicine is to revive the body's innate capacities to maintain a healthy equilibrium state.[40] This holistic approach became unfashionable in pharmaceutically based medicine from the late nineteenth century and was supplanted with reductionist approaches focused on microbiology, genes, cellular activity, chemistry, and micro-organisms such as 'germs,' or more recently viruses, and the use of synthetic drugs. The homeostatic approach nonetheless remained a minority tradition in the West, and therapies adopting this approach are often described as 'alternative' or 'complementary' medicine.

Drawing on the medical idea of homeostasis, Lovelock, whose principal scientific discoveries to this point related to the atmosphere, first described the Gaia hypothesis in a short 1972 scientific letter in which he argued that Life had exercised a strong influence over the atmospheric chemistry of the Earth from early in the evolution of Life, and that this influence is responsible for the subsequent evolution of an atmosphere suitable for Life. The idea of the Earth as an assembly of species which collectively and symbiotically regulate their own planetary environment in a relatively stable state, despite changing energy inputs from the sun over billions of years, came to Lovelock while working for NASA. While devising an experiment for detecting life on Mars, Lovelock began to ask how the Earth sustains the constant, but also chemically unstable, composition of gases, including 21 per cent of oxygen, that is precisely the right balance for the evolution of trees, grasslands, and mammals.[41] The result he came up with is that the unique Earth atmospheric composition of proportions of oxygen, methane, and nitrogen suitable for life could not be maintained in any kind of equilibrium – since these gases react with each other in the laboratory – without constant day-to-day regulation by some other force, and that force, Lovelock concluded, had to be Life. As he put it: 'life at an early stage of its evolution acquired the capacity to control the global environment to suit its needs': the 'biosphere' is therefore not just a 'catalogue of species' but 'an entity with properties greater than the sum of its parts':

> Such a large creature, even if only hypothetical, with a powerful capacity to homeostat the planetary environment needs a name: I am indebted to Mr William Golding for suggesting the use of the Greek personification of Mother Earth, 'Gaia.'[42]

Atmospheric scientists at the time of Lovelock's discovery believed that the Earth's atmosphere was 'an end-product of planetary outgassing and held that subsequent reactions by abiological processes had determined its present state.'[43] They were therefore highly resistant to the claim that the Earth as a 'large creature' was responsible for the balance of gases on Earth. And since Lovelock called his idea after a Greek goddess, many also dismissed the Gaia hypothesis as religious magic rather than a testable scientific theory. Others challenged Lovelock to substantiate his hypothesis with evidence of a mechanism by which Life assembled and sustained homeostasis.[44]

In answer to his critics, Lovelock teamed up with the American microbiologist Lynn Margulis and proposed in a follow-up paper that the myriad activities of microbiota were and remain the crucial mechanism by which Life sustains a life-friendly mix of oxygen, methane, and nitrogen gases in the Earth's atmosphere. According to Lovelock and Margulis, Life created and maintains the conditions for life – *homeostasis* – by a constant process of energetic exchange and feedback mechanisms between living creatures, oceans, rocks, and soils which together regulate the chemical composition of the atmosphere, oceans, and soils.[45]

Gaian Homeostasis versus Ultra-Darwinism

Not content with merely answering his critics, Lovelock, together with Margulis, also advanced another highly controversial view, this time in the field of evolutionary biology. Against the conventional Darwinian view that species evolved exclusively by competition in the process of natural selection, Lovelock and Margulis argued that species evolved by 'planetary symbiosis' between organic and inorganic life, and that species modify environment influences to enhance reproductive success: [46]

> Each species to a greater or lesser degree modifies its environment to optimize its reproduction rate. Gaia follows from this by being the sum total of all of these individual modifications and by the fact that all species are connected, for the production of gases, food and waste removal, however circuitously, to all others.[47]

This more biologically inflected version of the Gaia theory proposes symbiotic feedbacks between different life forms and the Earth herself, and that these feedbacks are the *reason* for the consistent evolutionary trajectory towards greater diversity of Life on Earth over the more than 500-million-year history of species evolution discussed above:

> (A) diverse chain of predators and prey is a more stable and stronger ecosystem than a single self-contained species, or a small group of very limited mix. An essential feature of these new Gaian models is the tight coupling between the organisms and their material environment. If these findings are true, it seems likely that the biosphere diversified rapidly as it evolved.[48]

This claim sent Darwinist and neo-Darwinist scientists, most notably Richard Dawkins, into paroxysms, and provoked fierce dismissals of the Gaia hypothesis as theology not science.

For Dawkins, the Gaia theory is a denial of the core Darwinian claim that life on Earth evolved by the principle of competitive self-interested behaviours. For

Dawkins, the crucial drivers of evolution are, in his now famous metaphor, 'selfish genes' whose relative reproductive success is the adaptive mechanism which produces species modifications which over time produce novel species best fitted to particular ecological niches. Dawkins agrees with Lovelock and Margulis to the extent that he believes that in the course of adaptation, genes do modify habitats to make them more comfortable to themselves: 'Genes manipulate the world and shape it to assist their replication.'[49] But for Dawkins it is impossible that genes modify habitats to benefit Life in general since habitat modifications are driven by genes out-competing each other in modifying the environment. If genes produce methane, it is not to benefit the Earth or other genes but for their own benefit exclusively.[50] Dawkins uses the same line of reasoning in relation to homeostatic adaptations which, he argues, do not occur originally at the level of the super-organism or ecosystem, but instead it is individual organisms which develop the best internal homeostasis and so out-compete other individuals.[51] There is a similar problem for Dawkins and other biologists about the claim that Life evolved Earth as a habitat favourable to life. Again, this claim cannot be true for neo-Darwinists since natural selection *only* works by competition and since scientists lack evidence of other planets in which life attempted such evolutionary adaptations and failed, the Earth cannot have evolved in the way the Gaia theory proposes.

Lovelock responded to his critics by observing that initial strong natural scientific rejections of the Gaia theory reflected modern science's pessimistic view of life on earth 'which sees nature as a primitive force to be subdued and conquered.'[52] This underlines the critique by Carolyn Merchant and others of the Baconian legacy of mechanistic science which by conceiving of Life and Earth as 'mere' matter treats them with such disrespect as to threaten the conditions for the continuance of the stable homeostasis of the relatively cool Holocene-era Earth, and the unique array of biodiverse organisms that the Earth brought forth during and after the warm and wet Miocene Era when most of the Earth had a tropical climate and was largely ice free.[53] But Lovelock conceded too much to his opponents when he failed to challenge their underlying, but non–evidence-based beliefs: first, that selfish individual organisms, not mutual, symbiotic, and synergistic group interaction, were the means by which life evolved; second, that life *must* have evolved by chance; and third, that there is no inner-directedness, no implicate teleology in the convergence of organisms and gases on planet Earth towards the homeostatic state.

It is notable that despite natural scientific claims that science is driven only by data and theory, all three arguments against Lovelock's position are a priori arguments. The *data* on evolution of species suggests that there are a range of mechanisms in play including competition and predation as well as symbiosis and mutual help: and in the nineteenth century several scientists argued that group symbiosis was as important as competition between individual cells, genes, or species in biological evolution, including Russian scientists Peter Kropotkin and Vladimir Vernadsky.[54]

Vernadsky's *Biosphere* is also an important theoretical precursor, along with James Hutton's *Theory of the Earth*, to the Gaia hypothesis, although neither Lovelock nor Margulis knew of these when they first published on Gaia. But along with the Russians, there have been many other natural scientists from the time of Darwin who disagree with the claim that Life evolved randomly, and that nothing remotely equivalent to an intelligent purpose influencing or operating through all organisms played any part in the surprising convergence of the Earth, despite or even because of a series of cataclysmic extra-terrestrial shocks, towards ever more biodiverse life forms, including ultimately conscious life – which is to say mammals, primates, and so humans. Among well-evidenced natural scientific arguments that such convergence is not random is the evolutionary theory of Henri Bergson, and his account of the role of time and memory in the creative processes of evolution, and the work of Simon Conway Morris on convergent evolution, to both of whom I return below.

Cybernetics and Gaian Terrestrial Dynamics

Instead of pointing out the lack of an evidentiary basis for the three core a priori beliefs of modern evolutionary science just discussed, Lovelock sought to answer his critics by demonstrating the existence of the mechanisms by which Life adapted Earth to be more favourable to life while downplaying his initial proposal that the Earth as a super-organism is alive. As we have seen, with Margulis Lovelock first turned to microbiology to explain the way in which Life generated an exchange mechanism between rocky and aqueous terrestrial habitats and the Earth's gaseous envelope in such a way as to produce an atmosphere more beneficent to organic life. Lovelock's second turn was towards cybernetics, which reflected his own prior expertise in computer modelling, instrument making, and atmospheric analysis.

Lovelock published a paper with Andy Watson which invoked a computer model of the Earth–Life–Sun relationship which he called Daisyworld. Daisyworld was an attempt to answer critics such as Paul Ehrlich who said that 'the idea that life evolves in a way to make the planet more hospitable for itself collapses for want of a mechanism.'[55] In the model there are two kinds of organism on the Earth's surface: some are white daisies, and some are black. And there is the varying heat output from the sun – it has declined over the 3.5 billion years of the Earth's present state – and the given distance of the Earth from the sun. When the software model of the daisies is run, the varying albedo of the two kinds of daisy – heat-absorbing and heat-reflecting – compete with each other, and interact with the variable of solar radiation such that they produce enough of each kind of daisy to generate relatively daisy-favouring, and stable, surface temperatures.[56] As Michael Ruse comments, the advantage of this reductionist model of how Life on the surface of Gaia regulates the atmosphere and the temperature to produce homeostasis favourable to life is that it eschews teleology and symbiosis and foregrounds the individually disconnected but, in sum, collectively influential agencies of organisms in the production of homeostasis. But it is only a model. It does not show how this might have happened.

After the Daisyworld paper, Lovelock still needed to discover a mechanism by which Life did what the model claimed – modify the atmosphere. The question posed by his critics was 'how does Life modify the atmosphere?' Feedback between the Earth's surface and the atmosphere from the varying albedo of different organisms in the real world is insufficient to explain this. Lovelock then made a third turn, this time towards marine biology in combination with atmospheric chemistry, and this was arguably his most significant evidentiary turn in proving the Gaia theory. The feedback mechanism Lovelock came up with, together with his collaborators, was interactions between clouds, oceanic condensation, and algae. Phytoplanktonic algae are among the most numerous of oceanic organisms. They excrete the chemical dimethyl sulfate (DMS) which, when it rises, interacts with nitrogen and forms particles which seed cloud formation. The white albedo of clouds is the equivalent of the Daisyworld albedo mechanism. The extent of algal blooms, their excretions, and the formation of clouds which influence solar-radiation terrestrial capture, and its dispersal into space, is the variable device, along with the reflective capacities of clouds themselves, by which Life modulates the atmosphere in interaction with the varying heat output from the sun.[57]

Further feedback mechanisms are also suggested in the paper, including the possibility of ocean–land feedbacks where sulphur from the oceans reaching inland promotes increased terrestrial productivity which benefits Life forms such as mammals, and also ultimately produces another feedback by which the sulphur runs off the land, with the aid of water condensed from DMS-promoted clouds, back into the oceans.[58] The DMS mechanism is not the only Life–Atmosphere feedback loop: there are many more, and some I have already discussed, including the action of sea creatures in bringing cool water from the depths of the oceans to the surface, and returning heat the other way. But the DMS mechanism could reasonably be said to have preceded other evolved mechanisms and have been the means by which ocean life began to regulate the atmosphere to the point that it developed the mix of gases, including oxygen and methane, that were critical to the evolution of oxygen-breathing mammals.

The original insight that sparked Lovelock's Gaia hypothesis was that atmospheres of living planets are different to dead planets. Earth has an atmosphere with a long-run stable composition of approximately 21 per cent oxygen, 78 per cent nitrogen, and 1 per cent composed of methane, nitrous oxide, carbon dioxide and other trace gases. Lovelock's original hypothesis was that such an atmosphere is not only ideal for, but produced by, Life since where Life is missing, such atmospheres do not exist in other galactic objects known to science at a similar distance from suns to that of Earth.[59] As Lovelock argues, '(T)he constancy of oxygen concentration suggests the presence of an active control mechanism.'[60] Drawing on research by atmospheric chemists, Lovelock argues that Life's mechanism for doing this is at least two-fold: first is carbon burial, as oxygen-producing creatures in the ocean and on land take the carbon they build, by producing oxygen via photosynthesis, down to the Earth's crust in the form of shells and skeletons in the oceans, and of buried biomass on land.[61] The second mechanism is methane. Methane or 'marsh gas' is not useful for life but is a 'waste' gas which much Life produces. Lovelock argued it is a crucial component of the way in which Life regulates oxygen to keep it at a life-promoting but also 'safe upper limit for life' of 21 per cent: if oxygen were above 25 per cent, forests would burn en masse.[62]

Another potential mechanism of Life–Earth–Atmosphere interactions concerns plate tectonics. Don Anderson proposes that life in the ocean interacts with olivine rocks at rifts in the Earth's crust in such a way as to promote movement of the Earth's tectonic plates. The same process may also be responsible for periodic major 'wanderings' of the Polar regions such as those which accompanied the Devonian and Cambrian explosions in speciation, and the tilt of the Earth on its axis in relation to the sun. If these interactions were proven, it would be a key mechanism, along with rock weathering, soil erosion, vulcanism, and the weight of the ocean on the Earth's crust in forming new layers of rock, for the System of the Earth which Hutton first named. It would also give scientific underwriting to the sense that the Earth is 'alive' which was believed in most cultures before that of modern Europe.[63]

Given the growing evidentiary base of the Gaia hypothesis, it gradually morphed in the 1990s from a hypothesis to a theory and gave rise to a now well-recognised field of 'Earth System' science. Atmospheric physicists and chemists, marine biologists, ecologists, and even some evolutionary biologists, have gradually come to accept that the Gaia theory is true, and a 2001 'Amsterdam Declaration on Global Change' summarised the theory as follows: 'The Earth System behaves as a single, self-regulating system comprised of physical, chemical, biological and human components.'[64]

In essence, the Gaia theory represents systems thinking applied to the evolution of Life and describes the way in which Life promoted, and still sustains, the present planetary state. Gaian theory is holistic and non-reductive. While Lovelock uses reductionist science to isolate and describe particular mechanisms – as in those just discussed – he does not, as most modern natural scientists tend to do, look for single-cause descriptions of evolution and global change. Lovelock, and Earth System scientists, recognise that Earth is an extraordinarily complex, multi-relational, mutually interactive collection of ecosystems and organisms and that no one mechanism – such as, say, the strategies genes pursue to succeed in reproduction – is sufficient to explain the whole. As Timothy Lenton puts it, 'to understand Gaia, you have to learn to think in terms of the nonlinear, circular logic of feedback systems.'[65]

Cybernetics and the Super-organism

The most notable feature of the Gaia theory which it shares with mainstream modern science is the eschewal of teleology: the habitable earth, and humans, are seen as outcomes of the 'self-regulating' system of Gaia, but the system is not understood as ultimately purposive. So the Gaia theory, while it resolves some of the problems with modern reductionist and anti-teleological natural science, still leaves open the biggest philosophical and scientific question of all which is: 'where does order come from?' Lovelock developed his answer to this question by means of the science of cybernetics.

Gregory Bateson, whose early field studies as an anthropologist were in Bali, was the leading pioneer of the modern science of cybernetics. For Bateson, as for Lovelock, the answer to the question of where order comes from in the world is a new variant of the cosmology of mechanism – namely cybernetics. In his first 'Essay on Cybernetics' Bateson built on the genetic insights of his father, William Bateson, and argued that genetic variation, and interactions between genotypes which produce new species, and build into complex systems such as a tropical rainforest, are the outcomes of communicative exchanges of information. In a cybernetic system 'everything that happens is still within the limits defined by the law of energy conservation.' Feedbacks and redundancy within complex multi-cellular organisms, and complex ecosystem assemblies of such organisms, are the principles, along with communicative exchanges of information, the availability of energy, and 'patterning contexts,' which explain how 'nature' and beings, through a constant and highly complex informational exchange system, make myriad choices from possible alternative outcomes which ultimately result in the capacity of unconscious systems, like a Dipterocarp forest, to manage reproductive processes so as to control seed predation.[66] Bateson argues that in the interaction between 'the individual human organism, the human society, and the larger ecosystem' consciousness should 'be considered as an important component in the *coupling* of these systems.'[67] So for Bateson, it is not consciousness but informational exchanges which are the principal drivers which explain homeostasis, and how ecosystems and human cultures generate meaningful and productive networks.

Lovelock adopted cybernetics to explain the *workings* of Gaia in creating and maintaining conditions suitable for complex life forms, and the evolution of conscious primates, in response to a huge range of attacks on the Gaia hypothesis by natural scientists, and especially to the claim that the Earth is a living super-organism. Although homeostasis is an established feature of the human body, the claim the Earth is a living super-organism which constantly regulates the planet's surface 'for the continuance of life' is strikingly at odds with the mechanistic cosmology which emerged in Europe from the sixteenth century and according to which the Earth, or 'nature,' is analogous to a machine rather than an animated and intelligent organism or network of organisms. In restating the Gaia theory as cybernetics, Lovelock therefore abandoned the claim the Earth is alive, and in so doing he made the theory more palatable to mechanistic science.

Gaia and God

The Gaia theory in its first iteration, rather than its subsequent cybernetic and mechanistic recasting, represents the scientific recasting of the idea present in many pre-modern religious traditions that Life on Earth is a manifestation of divine consciousness; and that Earth and her multitudes of atoms, energies, frequencies, gases, minerals, organic beings, rocks, and waters are the energetic and material manifestations of divine mind, or divinely created beings, powers, or spirits. But the Gaia theory differs from these religious perspectives since it represents a total immanentisation of pre-modern theisms. The Vedas, Lao Tzu, Pythagoras, Moses, Plato, Aristotle, Jesus, Paul of Tarsus, all envisaged that the Life Force, Tao, or Logos that animates or flows through living organisms, water, wind, and other natural powers emanates from, and is energised by, a higher transcendent Source or divine Power. But the Gaia theory, much like the seventeenth-century philosopher Benedict Spinoza, immanentises what traditional cosmologies represented as a transcendent divine force which manifests in the immanent world. In the Gaia theory, the creative and energising powers once attributed to the invisible God become a *purely* immanent gathering of forces or network of agencies implicated *within* the cosmos.

Modern scientists' resistance to the idea of a divine mind, mover, or source of Life beyond Earth is rooted in the scientific method Francis Bacon and René Descartes first described in philosophical form even though they themselves were still theists. For Bacon and Descartes, empirical data, material objects, organic beings, are amenable to scientific investigation and manipulation because they are material and not mental in their essence. Both assumed that mentality was a trait conferred uniquely by divinity on humans which privileged humans over all other life forms. This privileging is the origin of the modern scientific quest to uncover the laws which govern the behaviours of nonhuman material and organic beings, and with the knowledge so gained to *progress* the human condition by re-ordering the world in such a way as to increase the benefits it confers on humans.

This way of thinking emerged from the tendency in Abrahamic monotheism from the late Middle Ages to conceive of God as Wholly Other to the material and organic world, and to human consciousness, and to place God in a *chronological* rather than *energetic* relation to the ongoing life of the cosmos. The origins of this tendency were in the philosophy of the late Middle Ages known as nominalism, of which I have written more fully elsewhere.[68] The result was Deism, and the emergence of the modern Machine cosmology. Deists such as the seventeenth-century Cambridge philosopher William Paley held that God was like a watchmaker who set the world, and living beings, in motion at the beginning of time, and in effect wound up the mechanism so it functioned according to a pattern of laws, just as a watchmaker uses his craft to construct and ensure the continued movement of a watch. For Paley, the continuity of the cosmos is like a perpetual-motion device: it is set up at the beginning to go on without further intervention and thereafter the watchmaker may observe the continued working of the watch, but he does not intervene. It was this philosophy that informed the mathematics and physics of Isaac Newton, and the evolutionary biology of Charles Darwin. Darwin encountered Paleyism at Cambridge University and he reinvented it in scientific form in his account of the mechanistic workings of the law of competition, or struggle, between species as the device to explain the variety and distribution of species that humans observe from the fossil record and from living species.

The Gaia theory significantly revises Darwin, and Newton, because it demonstrates that the means by which the Earth produced Life are far more complex, and more *relational*, than either Newton or Darwin envisaged. The Gaia theory also resists the original scientistic dualism of matter and mentality and suggests instead that there is an *entanglement* between mind and matter which both Newton and Darwin denied.[69] In Lovelock's cybernetic iteration of the Gaia theory, the Earth is composed, and continuously evolved to its present habitable state, as a result of constant *informational exchanges* between all levels of being from microbiota in the deepest fissures of the Earth's crust to the gases of the stratosphere. This exchange of information is analogous to that which occurs in high-powered computers, but also to the flow of information and signals within the human brain and body, and between humans and other humans, and their environment. The entanglement of mind and matter in the Gaia theory means that even the laws which govern the Earth are caught up in, and evolved through, this entanglement.

In theological terms, the Gaia theory trounces Deism since the mentality which originated life goes all the way down, and up, and is continuously implicated in the ongoing and enduring ordering of life in the cosmos. But the Gaia theory is still an immanentist philosophy. There is no divine mind *before* Gaia, or beyond Gaia, as Plato, Lao Tzu, and Jesus all believed. So in its cybernetic iteration the Gaia theory is still a deistic Machine cosmology. And if this scientific attempt to escape the reductionism of the Machine cosmology, and the domineering mentality towards the more-than-human it fostered, is unsuccessful, this raises the question whether *any* scientistic philosophy, in which there is no transcendent divine Source, is able to resource resistance to the anthropocentric and Promethean direction which has characterised science-informed industrial humans since their adoption of the Machine cosmology. Without some kind of ontological recovery of the divine Source of Life, the Gaia theory cannot dethrone, or humble, Promethean industrial humans in their scientistic dominion of the more-than-human world. Instead it merely gives them a more systems-oriented and complex cosmology which may enable them, with the help of artificial intelligence and other cybernetic innovations, to achieve more sophisticated ways of re-ordering Gaia.

As if in confirmation of this conclusion, Lovelock in his latest book argues that the future of Life on Earth will ultimately lie in the hands of humanly created quantum computing powers, or Artificial Intelligence (AI), as this will result in the creation of self-learning machines which will far exceed human intelligence in their capacities to govern, repair, and restore ecosystems and even the Earth System. Lovelock goes even further when he suggests that the takeover of the cybernetic system of Gaia by AI will produce a new geological epoch which he dubs the Novacene.[70]

Gaian Theory as the Retrieval of Ontotheology

Despite his turn towards cybernetics and artificial intelligence, Lovelock insisted at several points in his career, and against his detractors, that 'Gaia, the system of all life and the Earth's surface environment, is alive.' For Lovelock, this is not only a scientific statement but a statement about how humans are to think of and treat of the Earth herself and in so doing begin to address the roots of the ecological crisis:

> Perhaps the single most important thing we can do to undo the harm we have done is to fix firmly in our minds the thought: *the Earth is alive.* Once such a thought becomes instinctive we would know that we cannot cut down the forests for farmland to feed ourselves without risking the destruction of our home planet. Farmland and tree plantations cannot serve as a replacement for natural forests that have evolved with their environment over millions of years and once served to keep the climate tolerable and the air good to breathe.[71]

The philosopher Mary Midgley, in whose book Lovelock makes this statement, notes that the attitude which has underwritten so much ecological destruction by industrial civilisation since the eighteenth century is that which considers the physical world mechanistic and *lifeless.* Modern science, far more than any religious tradition, was the principal origin of the machine mentality, and the domineering and controlling approaches to matter, the surface of the earth, and nonhuman creatures which followed. Before the Scientific and Industrial Revolutions, something like the Aristotelian attitude prevailed in Europe, as well as in Asia, Africa, and the Americas. For Aristotle and Hippocrates, both the human body, and the body of the Earth, were maintained in relative equilibrium by a balance of forces. The purpose of the physician, as of the farmer, the engineer, and the craftsman, was to work with these forces to sustain healthy habitats, for the body, and to design buildings, farms, and artefacts which worked with rather than against the laws of nature and so sustained the body of the Earth and the health of humans and other animals.

This ancient philosophical perspective also involved the belief that there is a mirroring that is cosmic in nature between the human body as microcosm and the Earth, the moon, the stars, and the sun as macrocosm: and that the human soul–body is therefore a unique composite whose organs and functions – physical, mental, and spiritual – correlate with the purposes and order of the exterior world as macrocosm. This mentality of a microcosm–macrocosm originated in the Indian Vedic tradition which the Greeks encountered by travelling to study in North India, and by bringing Vedic texts back to classical Greece. It was taken up into classical Greek philosophy and medicine and adopted by Neoplatonist theologians including Origen and Maximus the Confessor. To this day the belief that the human body is a microcosm of the macrocosm remains alive in the liturgy, spirituality, and theology of Eastern Orthodox Christians.

The ancient idea that the physical 'humours,' intentions, virtues, and vices of persons *affect* the habitat of the Earth in which they are situated finds new cogency in the Gaian recognition that humans, as well as other animals and plants, are intricately bound up in, and in turn influence, the Earth as a System. This way of thinking is reflected in both Western and Eastern philosophy before the modern era and is at the root of the idea of 'fit' or fittingness between right action and intention and cosmic and Earthly order and which I discuss more fully elsewhere.[72] Indian and Greek classical philosophers saw the core performative purpose of philosophy, religion, and the medical and natural sciences as to create and sustain fittingness and harmony with the original cosmic order of being: in the human as microcosm and in the universe as macrocosm.

Against the Misanthropic Misreading of Gaian Theory

The pre-modern and widely disseminated ontotheology of a divinely originated cosmic order and harmony between the human body and the body of the Earth is an important corrective to one reading of the Gaia theory, and one for which Lovelock himself is responsible. In a chapter in one of his books entitled 'The People Plague,' Lovelock says that humans 'have grown in number to the point where our presence is perceptibly disabling the planet like a disease.'[73] In a later book he proposes that global heating, melting ice, and extreme weather are 'the Revenge of Gaia' on humans, as Gaia attempts to repair temporary disequilibria modern humans have introduced into the system. Gaia's revenge may see the great majority of humans die off and the remnant that remain living on land near the Poles.[74] Lovelock claims that the Earth can only support around 500 million humans, and that one way or another Gaia will arrange things to bring back human numbers to something like that number, most likely by the year 2100.

This bleak assessment has a distinctly eugenicist and misanthropic overtone, and, in some ways, it is contradicted by Lovelock's own account of the material processes which have precipitated the environmental crisis, which are principally the tendency of modern industrial agriculture to radically reduce species diversity on land: and, I would add, industrial fishing and aquaculture in the oceans. It is true that the industrial revolution, and especially the Haber–Bosch process by which artificial fertiliser is synthesised from atmospheric nitrogen using oil, enabled an explosion in human numbers in the twentieth century. Nonetheless, it is not so much the number of people but the ways they are trained to live by industrial agriculture, fishing, forestry, manufacturing, mining, and consumerism, which are at the root of Earth System ecological disequilibria that negatively impact the Earth's in-built feedback mechanisms and so diminish species numbers, reduce ecosystem resilience, promote local climate changes and extreme weather events, and threaten larger destabilisation of the Earth System.

The misanthropic reading of the ecological crisis in the light of the Gaia theory is consistent with a long line of misanthropy going back to the beginning of the Industrial Revolution and especially the generative work of Thomas Malthus on the problem of scarcity which became a central organising principle in modern political economy. But empirically, Malthusian pessimism did not stand up. Against Malthus, and Lovelock's Malthusian musings, there is no global shortage of food that humans could not fix if they were prepared to alter modern economic arrangements in which non-farmers often lack sufficient cash income to provide for themselves. Modern famines are not caused by food shortages but by economic arrangements which determine that the landless, low-caste people, the urban poor and unemployed, have no cash to buy food.[75] There is nothing inevitable about such economic arrangements. Nor is the present-day highly centralised and increasingly toxic industrial food-supply system, and its preference for biodiversity destroying monocultures, set in stone. If humans gave up the corporate, governmental, and marketing procedures which have promoted an industrial diet laden with cheap cruelly reared and toxic meat, seed oils, sugar, Genetically Modified corn, and white starch from rice and wheat, they would significantly reduce the present-day burden of chronic human disease while at the same time reducing the needed land area for farming since such a large proportion of croplands are devoted to monocultures for animal feed and seed oils, and a growing proportion of the latter is wasted as 'biofuels' which, as earlier discussed, have a larger ecological footprint than fossil fuels. The real problem with the industrial food and energy system is that it replicates a social model of procurement based on the reductionist model of the machine cosmology which refuses that there is innate intelligence in how living systems, and indigenous peoples, organise themselves without top-down Statist and scientistic control and re-ordering. The machine cosmology ultimately refuses that there is meaningful purposive life apart from human scientistic re-ordering of Life on Earth because the machine cosmology lacks an account of *life*, or the Tao.

Science, Religion, and the Question 'What *Is* Life?'

The biggest issue, philosophically, *and* scientifically, raised by the Gaia theory, the claim that Gaia is 'alive,' is what is meant by the term Life.[76] Oddly, it is no use turning to Lovelock himself, or to other contemporary scientists, for an answer to this age-old question. The question 'What Is Life?' is not amenable to a scientific answer even though natural science allegedly has traced the course of Life from its origins in the nitrogen-laden fog of the pre-photosynthetic Earth to Life's present-day myriad complex assemblies of organisms. Microbiologists cannot explain *why* Life emerged from that original abiotic soup. Nor are they able to conjure up Life again from scratch in the laboratory when they mimic the conditions in which they allege Life symbiotically sprang into being.

Lovelock in later life reasserted his belief – as stated in his earliest Gaia papers – that the Earth is alive, and that this statement is not only a metaphor: 'I call Gaia a physiological system because it appears to have the unconscious goal of regulating the climate and the chemistry at a comfortable state for life.'[77] This reassertion of his earlier implicitly vitalistic scientific portrait of Gaia also influences Lovelock's recent views on planetary heating. As Michael Ruse argues, Lovelock in later life rejected global-warming alarmism as 'green drivel,' and insisted that Life, and humans, will adapt, and will find ways to moderate and slow planetary heating.[78] In resisting global-warming alarmism Lovelock finds unlikely allies among conservative Christians and Muslims who also refuse that the living creation, upheld and sustained by a Creator God, and not merely set in motion at a long-distant beginning of time, will spiral into a heat-state in which the habitat of humans, as the highest form of God-shaped consciousness in Earth's history, will be destabilised.

The recognition of Gaia as an apparently purposive living assembly of assemblies and mysteriously directed towards the reproduction of increasing levels of biodiversity, complexity, and, ultimately, consciousness over billions of years cannot but approach the boundary of science and religion. As Ruse puts it:

> Conceal and modify it as you may, if you are true to the Gaia project – to see the world as an organism – you are committed to something deeply end-directed, deeply teleological. Earth is in balance because that is its destiny, that is what it should be. That is a good thing. Modern science, shorn of teleology and value, simply will not allow this kind of vision within its borders.

Lovelock and Margulis 'were big people with a big vision. Whether science likes it or not, the vision lives on.'[79]

For many natural scientists, such as Toby Tyrell in a book-length review in 2013, the core problem with the Gaia hypothesis is that it is not possible to 'reanimate' the Earth as a living assembly of living organisms, rocks, and gases without invoking the teleological language, and arguments from design, that atheistic natural science has eschewed since the nineteenth century.[80] For Tyrell and other natural scientists, Earth will never be more than an assembly of automatons and hence it, too, is at most an automaton: it will always be wrong to say that the Earth is alive. The belief that 'nature' or the nonhuman lacks inherent purposiveness or teleology is core to the modern ideology of scientism but, as Rupert Sheldrake argues, this modern belief lacks any empirical foundation, it is a non-empirical axiom, an *a priori* assumption, and it is therefore not falsifiable or verifiable by the scientific method.[81] The philosopher Thomas Nagel argues that the eschewal of teleological language by most modern scientists renders the apparently purposive behaviour of biological organisms – including organisms such as the liver and the pancreas in the human body, which are essential to homeostasis but which perform their essential functions without conscious direction from the brain – hard to explain. Nagel's suggestion is that the purposive behaviour of organisms is best explained as 'system properties,' which is to say they acquire their abilities to express certain purposes from their biological contexts: this is analogous to Lovelock's cybernetic account.[82] But Sheldrake goes beyond Nagel and Lovelock in arguing that the purposiveness of individual organisms arises from within a field of resonances or vibrations which while hard to detect by conventional scientific instruments is as 'real' as the physical objects which are visible within material and organic systems. Sheldrake's account restates in modern scientific language the Aristotelian and medieval Christian view that all organisms express purposes because they are, on some level, *psuche* or soul-like.[83]

The axiomatic nature of their eschewal of teleology requires that modern natural scientists hold to the view that humans are the only rationally purposive beings and are therefore the *only* beings capable of imputing purpose into randomly evolved ecological systems, and assemblies of matter and organisms. Hence Paul Crutzen and other atmospheric chemists argue that scientists need to get into the planetary driving seat to address planetary heating and biodiversity decline with a range of technical interventions, even although it is precisely the large-scale application of reductionistic scientific techniques – such as sonar in the oceans and artificial fertilisers on land – that are destroying the once built-in resilience and climate stabilising features of highly biodiverse forests and grasslands, mangrove swamps, seas, and soils.[84]

Given the modern privileging of human consciousness, we might expect that at least natural scientists are able to explain what *consciousness* is. But when faced with this question, most natural scientists go with someone like Daniel Dennett, who asserts that consciousness is simply a more complex form of the electrical, hormonal, and nervous signalling systems that run through all biological life and that reach their zenith in the complexity of the human brain.[85] In his 'disproof' of the Gaia hypothesis, Tyrell ultimately uses consciousness, in the form of the Anthropic principle, as the means to escape the teleological problem. His argument is that since human beings are uniquely able to consciously scrutinise the workings of Gaia, they are uniquely able to impute purposiveness to the Earth because the only end state in which they could be observers of Earth is the Earth which gave rise to conscious observers.[86] In putting matters this way, Tyrell eviscerates the Gaia hypothesis of precisely the reverence for Earth, as a living organism, which is a feature of the origin of the Gaia hypothesis in space exploration, and of its cultural impact. By turning the Gaia hypothesis into yet another anthropocentric philosophy, Tyrell subverts the cultural and philosophical power of the Gaia hypothesis in 're-animating' Earth, and Life, and so deeply worthy of awe, reverence in its original ordering and not only as re-ordered by conscious humans.

Modern atheistic science and technology are on a collision course with Life. Modern scientists refuse that Life evolved purposively and many now argue that in the Anthropocene epoch it is the duty of scientists to impose purpose on Life anew and so prevent the Earth spiralling into a less biodiverse and over-heated state. But the Gaia theory suggests this collision course is a consequence of the *failure* of modern science to acknowledge that the Earth has purposes that humans, and even scientists, ought to study and learn from, and that to repair this failure farmers, fishers, and makers *ought* to develop, or recover, practices which respect those purposes. But for modern atheist scientists the Earth, and nonhuman beings, have no intrinsic moral purposes since they are no more than random assemblies of automatons, and humans, as the only valuing beings, cannot derive guidance for how they are to live from natural observation.

Bergsonianism and the Gaia Theory

If most modern evolutionary and system scientists refuse the core claim of the Gaia theory – that Earth is alive and that she and her assemblies of creatures express purposive goals which have, over many aeons, composed Earth in her current state of beauty and biodiversity – there are two scientists, Bergson, and the contemporary proponent of Bergsonian science Sheldrake, whose theories of the origins and unfolding of life on Earth are entirely consistent with, and add important dimensions to, the Gaia theory which on its own, as we have seen, has moved in directions which are ecologically and philosophically as problematic as the reductionist science proponents of the Gaia theory seek to displace.

In his philosophically and scientifically rich account of 'creative evolution,' Bergson argued that modern science, and modern intellectuals, by separating thought from experience, and form from duration, are unable to 'embrace life.' 'Evolutionary thought,' such as the theories of Wallace and Darwin, is at ease with inanimate objects and mechanisms but resists 'fathoming the depths of life.' And yet humans are not the only evolved lives. What if, Bergson asks, other forms of evolved conscious life were to merge with the human intellect? 'Would not the result be a consciousness as wide as life?'[87] Here it seems that Bergson, eight decades before Lovelock, has already anticipated but also transcended, the Gaia hypothesis. Decades before the discovery of the Gaia hypothesis, Bergson takes the idea of 'consciousness as wide as life' to its logical philosophical end, in which 'theory of knowledge and theory of life' are 'inseparable.'[88]

For Bergson, to be alive means to have conscious knowledge. But knowledge is only possible because of experiences over time. The infant only learns she is not her mother by repeated processual experiences of touch, taste, smell, and movement in which gradually consciousness dawns that the infant is an 'I' and her mother's body, her mother's mind, are not hers. Duration, time, are central to the *capacity* for experience and thought, as are prior experiences, from the smallest particles to objects in the solar system, and from which ultimately emerged the animated bodies which constitute sensing, thinking, feeling, comprehending minds. Evolution over time is the origin of novelty and divergent forms of life: 'evolution is a creation unceasingly renewed, it creates, as it goes on, not only the forms of life, but the ideas that will enable the intellect to understand it, the terms which will serve to express it.'[89] Bergson's description of the origin, and continuously unfolding novelty which Life has successively produced in Earth's long history, is what he called 'élan vital.'

Underlying, and preceding, Bergson's account of the essence and evolution of Life is his recognition, as expounded in his *Matter and Memory*, that experience, memory, and time are the *ways* in which the 'creative process of differentiation' proceeded in time and space to actualise the myriad biodiverse organisms, and conscious thoughts, that presently exist on Earth.[90] Bertrand Russell helpfully summarises Bergson's thought on this as follows: 'Intellect is the power of seeing things as separate one from another, and matter is that which is separated into distinct things. In reality there are no separate solid things, only an endless stream of becoming.'[91] Nonetheless, the intellect separates Life from matter and for Bergson humans do this by conceiving of Life as a process of continuously ascending possibilities and of Matter as that which descends beneath the process of Life's ascensions. Another and equally crucial distinction which for Bergson distinguishes Life from Matter is Time and Space. Space is 'the characteristic of matter' while Time is 'the essential characteristic of mind.'[92] And by this distinction we arrive at Bergson's highly illuminating definition of Life: 'Wherever anything lives there is, open somewhere, a register in which time is being inscribed.'[93]

Bergson's thought was fashionable in the 1920s in Europe and North America, but his writings and ideas fell out of favour. The idea of an animating *élan vital* behind all living things was too magical, too mysterious, too *traditional*, for moderns accustomed to the growing power of technique, and the myriad objects which gradually displaced living things in moderns' immediate environments as they migrated from farms and forests to city apartments and suburban houses. Machines and mechanistic conceptions of matter increasingly dominated the twentieth century and shaped the perceptions and experiences of so many humans, whether on the battlefield, in the factory, or in cars, trains, or planes en route to shopping malls, offices, or 'vacations.' What Alan Watts calls a 'ceramic' cosmology was in the ascendant. But the Gaia hypothesis, and the ecological crisis, challenge that ascendancy and revive the possibility that human thought and all living things are connected, as, among others William Blake and William Wordsworth, had argued, before Bergson, at the outset of the Industrial Revolution. But how do such connections across the duration of time work to bring forth extant and novel species? Without some explanation Bergson's claim, and his idea of *élan vital*, remain in the realm of magic and ideas, closer to spiritualism than technology.

Over a lifetime of novel empirical experiments and radical scientific ideas, Rupert Sheldrake has developed an answer to this question which engages, and provides further evidence for, the veracity of the Gaia theory. Sheldrake develops Bergson's attempts to define organisms via systems theory when he describes 'the nature of living organisms' in terms of 'concepts of information, communication, and control, including an emphasis on the importance of feedback':

> Organisms are living wholes, self-organizing processes of activity. Biological organisms are just one kind of organism. The earth is a far vaster

> organism, within which they come into being, develop, reproduce perhaps, and sooner or later die. She is indeed far more like a great Mother than a misty ball of inanimate rock.[94]

The best scientific account of precisely how the Earth births life for Sheldrake is the Gaia hypothesis which is 'a major new step in the direction of a new animism, which is why it is so controversial.'[95] It is the best account because it challenges both the cosmology of mechanism *and* contemporary humanism, which is why it is resisted both by natural scientists and most philosophers. Having carved up the world – including human experience – as objects for isolated disciplines to study and control, intellectuals are averse to seeing their separated and well-defended ivory towers toppled.

But if Gaia is animate, she must have an 'organizing principle with its own intents and purposes.' For a solution Sheldrake turns first not to memory of experiences across duration, which he, like Bergson, argues is part of the answer, but to the existence of 'morphic fields.' Examples of morphic fields include technical systems such as blue tooth devices, computers, and the internet; human families and friendships when they experience shared feelings which may occur across considerable distance in space; and biological and geological organisms when they influence each other by such forces as gravity, electricity, magnetism, hormonal and sonorous signalling systems, chemical and atomic exchanges. While it would be hard to imagine an experiment in which memory of experiences in time had influenced the development of organisms gathered in a particular ecosystem, it is possible empirically to test morphic resonance within an ecosystem by means of instruments that measure electrical, hormonal, sonorous, and other signals. Forestry is one area in which in recent years just such an approach has been developed.[96] Another is horticulture, as in the practice of interplanting of different species that grow better together because of pest control. Sheldrake's proposal is, then, more pragmatic than Bergson's and less idealistic. It offers the means to investigate the energetic workings of Gaia and not only her body.

A crucial field in which morphic resonance proves its explanatory worth is in the genesis of forms – or morphogenesis. This concerns the way in which embryos and seeds develop so as to reproduce fully grown examples of the species which they represent. How this happens is not explained by conventional biology, and the process suggests a far subtler and deeper life force is at work in all beings than that which most natural scientists are prepared to acknowledge. Sheldrake argues that genetic explanations of the development of cells into specific forms are inadequate, though much favoured, because they do not explain how two insects or two plants which share the same DNA nonetheless take two different species forms.[97] Similarly it is possible to intervene at the embryonic stage of a range of plants and animals and remove some of the embryo, and yet the remaining cells still succeed in going on to form the same plant or animal as would have been formed by the whole embryo. There is no adequate mechanical or organic explanation for this available to biological or genetic science.[98]

The solution to the puzzle, as first proposed by Bergson, is that the composites of cells which are living entities are constituted not only by reproductive mechanisms in which genetics and environment play crucial roles but by a non-material force analogous to memory which acts upon them from the field of beings that have previously existed in their shape and form. As Bergson puts it,

> All goes on as if the cell knew, of the other cells, what concerns itself; as if the animal knew, of the other animals, what it can utilize all else remaining in shade. It seems as if life, as soon as it has become bound up in a species, is cut off from the rest of its own work, save at one or two points that are of vital concern to the species just arisen. Is it not plain that life goes to work here exactly like consciousness, exactly like memory?[99]

For Bergson, each being is directed by its *predecessors* to realise its specific and given form even when its DNA or embryonic shape is in some way harmed. This sounds magical and yet, as Sheldrake points out, it is entirely consistent with quantum mechanics which reveal kinds of interactive relations between electrons and observers across time and space which are inexplicable according to Newtonian physics.[100]

But what is the nature of this force, this 'memory,' that acts from the others into the present? Marcel Vogel's influential research on naturally occurring bioluminescence offers a powerful answer. Vogel invented the liquid crystal display for IBM, and the metal coating used in computer hard drives. He also ran hundreds of experiments on the effects on plants, and on leaves detached from plants, when human thoughts are directed towards them. Vogel demonstrated that when his human subjects intuit themselves into the form of a living plant or direct a stream of consciousness to a leaf detached from a plant, that the force field so created is measurable by sensitive electro-magnetic devices.[101] Vogel's research led him to the view, as summarised by Tompkins and Bird, that 'a Life Force or Cosmic energy surrounding all living things is shareable among plants, animals and humans. Through such sharing, a person and a plant become one.'[102] In Vogel's own words: 'This oneness is what makes possible a mutual sensitivity allowing plant and man not only to intercommunicate, but to record these communications via the plant on a recording chart.'[103] Many scientists have, after Vogel, demonstrated thought–organism and thought–atom relationships.

The human brain is an incredibly powerful electro-magnetic signalling system and electro-magnetic signalling between cells is the central biological mechanism of all living organisms. Knowledge of this process is also turned *against* Life in the development of herbicides such as glyphosate which blocks the intercellular signalling system. But its inventors refuse that this same capacity might be harmful to humans who ingest glyphosate through spraying, or as residues in plants or water, despite the fact that humans too are constituted of cells which communicate with each other through the same chemical and electro-magnetic pathways: and despite the fact that multiple scientific studies now demonstrate a concerning array of ways in which glyphosate, even in micro-doses below the regulated dietary safety levels, harm many human bodily functions and organs, and especially the liver.[104]

The insights of Bergson, Sheldrake, Vogel, and others are a long way from conventional natural science and are very far from shaping science policy in relation to industrial technologies which are still built on the assumptions of Machine cosmology. Science in the main is still shaped by the presumed and deep divide between human consciousness, human corporeality, and the lives of other kind. But this divide is culturally of very recent origin and, as I go on to show below, it is not present in the world's religious traditions for most of their history. On the contrary, all religious traditions until the Protestant Reformation ascribed animus, or soul, to all living things. In classical Greece, Plato also ascribed this belief to the Earth as a whole with his conception of the 'world soul' which in medieval Europe was known as *animus mundi*.[105] And all pre-modern religious texts, including those of ancient Sumer, the Vedas, and the Bible, demonstrate an awareness of the multifarious capacity of all beings to respond and adapt to their surroundings and to other beings. The Gaia hypothesis confirms in scientific guise this ancient belief in the living body of the Earth, and the mysterious immaterial, but nonetheless long-perceived, connections between human and more-than-human life.

Gaia and Divinity

If Gaia is alive, and there is an *anima mundi*, or 'world soul,' as Plato and Plotinus believed, it is reasonable to ask whether William Golding's proposal that Lovelock call his theory the Gaia theory involves a similar claim.[106] Lovelock himself says that it does, in a number of his writings, both early and late. Early on he acknowledged his Quaker upbringing and that this had left him with an interest in transcendent experiences and transcendent explanations even though he devoted much of his life to making instruments. Later he suggested that it was time 'theologians shared with scientists their wonderful word, "ineffable"; a word that expresses the thought that God is immanent but unknowable.'[107]

But there is a very significant difference between Lovelock's Gaia theory, even when described as 'ineffable,' and the cosmogonies of Plato, of the Vedas, of Judaism, and Christianity, and it is that all of these held that *prior* to Earthly Life there was Consciousness, Divinity, Logos, or Spirit. Plato argued that consciousness *preceded* matter and organisms and is their transcendent Source. This is what the Vedas call Brahma, Jesus called 'Abba' or 'Father,' Paul Tillich called the 'Ground of Being,' and others call Divine Light, or 'the Universe.' In traditional cosmologies the divine Source of Being is transcendent to material and organic Life and continuously *imparts* to Earth *élan vital* or the Tao that is manifest in the 'self-organising' powers of species, and of the Earth herself in maintaining conditions favourable to Life. Energetic relations, or vibrations, between all living things and all matter in traditional cosmologies are reflections of the divine Source, and which are described in some religious traditions as *energeia* or divine energies.[108] In the next two chapters I explore the religious and ethical implications of the Gaian account of Life on Earth and how it interacts with traditional cosmologies.

Notes

1 'Something in the Air: James Lovelock and Atmospheric Pollution.' UK Science Museum, July 22, 2019, https://www.sciencemuseum.org.uk/objects-and-stories/chemistry/something-air-james-lovelock-and-atmospheric-pollution accessed June 2, 2020.

2 J.C. Farman, B.G. Gardiner, and J.D. Shanklin, 'Large Losses of Total Ozone in Antarctica Reveal Seasonal ClOx/NOx Interaction.' *Nature* 315 (1985) 207–10.

3 Susan Solomon, 'The Discovery of the Antarctic Ozone Hole.' *Nature* 575 (2019) 46–7.

4 For an overview of international environmental treaty-making, including the Montreal Protocol, see Scott Barrett, *Environment and Statecraft: The Strategy of Environmental Treaty-Making*. Oxford, Oxford University Press, 2005.

5 Jayanarayanan Kuttippurath and Prijitha J. Nair, 'The Signs of Antarctic Ozone Hole Recovery.' *Nature: Scientific Reports* 7 (2017) 585.

6 Michael S. Northcott, 'Artificial Persons against Nature: Environmental Governmentality, Economic Corporations, and Ecological Ethics.' *Annals of the New York Academy of Sciences* 1249 (2012) 104–17, and Michael S. Northcott, *A Political Theology of Climate Change*. London, SPCK, 2014.

7 'Daniel H. Cole, 'From Global to Polycentric Climate Governance.' *Climate Law* 2 (2011) 395–413.

8 On the power of the fossil fuel lobby see Naomi Oreskes and Eric M. Conway, *Merchants of Doubt: How a Handful of Scientists Obscured the Truth on Issue from Tobacco Smoke to Global Warming*. London, Bloomsbury Press, 2010.

9 Elinor Ostrom, *Governing the Commons: The Evolution of Institutions for Collective Action*. Cambridge, Cambridge University Press, 1990.

10 Garett Hardin, 'The Tragedy of the Commons.' *Science* 13 (1968) 1243–8.

11 James E. Lovelock, *A New Look at Life on Earth*. Oxford, Oxford University Press, 1979, xii, and Rachel Carson, *Silent Spring*. Boston, MA, Houghton Mifflin, 1962.

12 These and other environmental protest songs played a significant role in moving individuals towards taking action for the environment, as documented by Mark Pedelty, *Ecomusicology: Rock, Folk and the Environment*. Philadelphia, PA, Temple University Press, 2012.

13 See further Charles A. Reich, *The Greening of America: How the Youth Revolution Is Trying to Make America Liveable*. New York, NY, Random House, 1970, and Theodore Roszak, *The Making of a Counterculture: Reflections on a Technocratic Society and Its Youthful Opposition*. Garden City, New York, NY, Doubleday, 1969.

14 Anna Clark, *The Poisoned City: Flint's Water and the American Urban Tragedy*. New York, NY, Picador, 2018.

15 Capital's share of national income relative to labor fell from 1930 and began to rise in 1970: see Thomas Picketty, *Capital in the Twenty-First Century*. Trans. Arthur Goldhammer, Cambridge, MA, The Belknap Press of Harvard University Press, 2014, 429–30.

16 On public choice theory see James Buchanan, *The Limits of Liberty: Between Anarchy and Leviathan*. Chicago, IL, Chicago University Press, 1975; and Nancy MacLean,

Democracy in Chains: The Deep History of the Radical Right's Stealth Plan for America. New York, NY, Penguin Books, 2017.

17 Ronald H. Coase, 'The Problem of Social Cost.' *Journal of Law and Economics* 3 (1960) 1–44.

18 The fullest official account of what some call the 'new deal for nature' is in a UK Government official document: Partha Dasgupta, *The Economics of Biodiversity: The Dasgupta Review – Final Report,* London, HM Treasury, 2021; for an early critique see Clive L. Spash and Frederic Hache, 'The Dasgupta Review Deconstructed: An Expose of Biodiversity Economics.' *Globalizations* (2021). https://doi.org/10.1080/14747731.2021.1929007.

19 Alfred North Whitehead, *Science and the Modern World: The Lowie Lectures.* New York, NY, Macmillan, 1925, 52.

20 J. Lelieveld, J.S. Evans, M. Fnais, et al., in 'The Contribution of Outdoor Air Pollution Sources to Premature Mortality on a Global Scale.' *Nature* 525 (2015) 367–71; the paper attributes 3.3 million premature deaths annually to air pollution; on global ill health and mortality from water pollution see René P. Schwarzenbach, Thomas Egli, Thomas B. Hofstetter, et al., 'Global Water Pollution and Human Health.' *Annual Review of Environment and Resources* 35 (2010) 109–36.

21 Michael Guarnieri and John R. Balmes, 'Outdoor Air Pollution and Asthma.' *The Lancet* 383 (2014) 1581–92.

22 A mother in suburban London whose child died of an asthma attack sought a coroner's view on the cause of death who found there was a spike in air pollution on the day she died and put air pollution as cause of death: Bevan Shields, 'In Legal First, Coroner Finds Ella's Death as Caused by Air Pollution.' *Sydney Morning Herald,* December 17, 2020.

23 Eric R. Taylor, *Lethal Mists: An Introduction to the Natural and Military Sciences of Chemical, Biological Warfare and Terrorism.* Huntington, NY, New Science Publishing, 2001; see also Peter Sloterdijk, *Terror from the Air.* Eng. Trans. Amy Patton and Steve Cocoran, Michigan, Semiotext, 2009.

24 Amold Schecter, Le Cao Dai, Le Thi Bich Thuy, et al., 'Agent Orange and the Vietnamese: The Persistence of Elevated Dioxin Levels in Human Tissues.' *American Journal of Public Health* 85 (1985) 516–22.

25 Josef Unterweger, 'Ethics and Pesticides: The Precautionary Principle as Illustrated by Glyphosate.' In Laura Westra, Janice Gray, and Ranz-Theo Gottwald eds. *The Role of Integrity in the Governance of the Commons: Governance, Ecology, Law, Ethics.* Chambery, Springer, 2017, 215–24.

26 Robin Mesnage, Christian Moesch, and Rozenn Le, 'Glyphosate Exposure in a Farmer's Family.' *Journal of Environmental Protection* 3 (2012) 1001–3; Leah Schinasi and Maria E. Leon, 'Non-Hodgkin Lymphoma and Occupational Exposure to Agricultural Pesticide Chemical Groups and Active Ingredients: A Systematic Review and Meta-Analysis.' *International Journal of Environmental Research and Public Health* 11 (2014) 4449–527.

27 Stephanie Seneff, *Toxic Legacy: How the Weedkiller Glyphosate Is Destroying Our Health and the Environment.* E-Edition, White River Junction, VT, Chelsea Green Publishing, 2021.

28 C.R. Tyler and R.M. Goodhead, 'Impacts of Hormone Disrupting Chemicals on Wildlife.' In Norman MacLean ed. *Silent Summer: The State of Wildlife in Britain and Ireland.* Cambridge, Cambridge University Press, 2010, 125–40.

29 For a more detailed discussion of insect, bird species and small mammal declines in the United Kingdom see further Michael S. Northcott, 'Lament for a Silent Summer.' In Michael S. Northcott ed. *Place, Ecology and the Sacred: The Moral Geography of Sustainable Communities.* London, Bloomsbury, 2015, 47–64; and see case histories of species declines in MacLean ed., *Silent Summer.*

30 V. Girolami, L. Mazzon, A. Squartini, et al., 'Translocation of Neonicotinoid Insecticides from Coated Seeds to Seedling Guttation Drops: A Novel Way of Intoxication for Bees.' *Journal of Economic Entomology* 102 (2009) 1808–15.

31 Seneff, *Toxic Legacy*, 156–60.

32 Seneff, *Toxic Legacy*, 16.

33 David Goulson, *Silent Earth: Averting the Insect Apocalypse*. London, Penguin Random House, 2021.

34 Caspar A. Hallmann, Martin Sorg, Eelke Jongejans, et al., 'More than 75 Percent Decline over 27 Years in Total Flying Insect Biomass in Protected Areas.' *PLoS ONE* 12 (2017) e0185809.

35 Loren McClenachan, Francesco Ferretti, and Julia K. Baum, 'From Archives to Conservation: Why Historical Data are Needed to Set Baselines for Marine Animals and Ecosystems.' *Conservation Letters* 5 (2012) 349–59.

36 Herton Escobar, 'Amazon Fires Clearly Linked to Deforestation, Scientists Say.' *Science* 365 (2019) 853.

37 James E. Lovelock, *Gaia: The Practical Science of Planetary Medicine*. Oxford, Oxford University Press, 2000.

38 Sergio Pannazio, 'Homeostasis: A History of Biology.' *Rivista di Biologia/Biology Forum* 102 (2009) 253–71.

39 J. Scott Turner, 'Homeostasis and the Forgotten Vitalist Roots of Adaptation.' In S. Normandin and C.T. Wolfe eds. *Vitalism and the Scientific Image in Post-Enlightenment Life Science, 1800–2010*. Dordrecht, Springer, 2013, 271–91.

40 N. Wiener, 'The Concept of Homeostasis in Medicine.' *Transactions and Studies of the College of Physicians of Philadelphia* 20 (1953) 87–93.

41 Lovelock, *Gaia: A New Look at Life on Earth*, xvi.

42 James E. Lovelock, 'Gaia as Seen through the Atmosphere.' *Atmospheric Environment* 6 (1972) 579–80.

43 Lovelock, *Gaia*, 7.

44 One scientist publicly dismissed the Gaia hypothesis as 'fairy stories about a Greek Goddess' at a 1987 Berlin atmospheric science conference: James E. Lovelock, 'Reflections on Gaia.' Cited in Stephen H. Schneider, James R. Miller, Eileen Crist, et al. eds. *Scientists Debate Gaia: The Next Century*. Cambridge, MA, MIT Press, 2004, 1–6.

45 James E. Lovelock and Lynn Margulis, 'Atmospheric Homeostasis by and for the Biosphere: The Gaia Hypothesis.' *Tellus* 26 (1974) 2–10.

46 Lynn Margulis, *Symbiotic Planet: A New Look at Evolution*. New York, NY, Basic Books, 1998.

47 Lynn Margulis cited in Lovelock, *Gaia: A New Look*, 120.

48 Lovelock, *Gaia: A New Look*, 22.

49 Richard Dawkins, *The Extended Phenotype: The Long Reach of the Gene*. Oxford, Oxford University Press, 1981, 5.

50 Dawkins, *Extended Phenotype*, 200–36; for a summary of Dawkins' argument with Lovelock and Margulis, and the larger controversy around Gaia among biologists, see Michael Ruse, *The Gaia Hypothesis: Science on a Pagan Planet*. Chicago, IL, University of Chicago Press, 2013, 27–31.

51 Dawkins, *Extended Phenotype*, 236.

52 Lovelock, *Gaia: A New Look*, 11.

53 M. Steinthorsdottir, H.K. Coxall and A.M. de Boer, 'The Miocene: The Future of the Past.' *Paleooceanography and Paleoclimatology* 36 (2021) e2020PA004037.

54 Peter Kropotkin, *Mutual Aid: A Factor in Evolution*. London, William Heinemann, 1902; Vladimir I. Vernadsky, *The Biosphere*. Foreword by Lynn Margulis, Mauro Ceruti, Stejepko Golubic, et al., Trans. David B. Langmuir, New York, NY, Springer, 1998.

55 Paul R. Ehrlich, 1991, 21 cited in Ruse, *Gaia Hypothesis*, 152.

56 Andrew Watson and James E. Lovelock, 'Biological Homeostasis of the Global Environment: The Parable of Daisyworld.' *Tellus B: Chemical and Physical Meteorology* 35 (1983) 284–9.
57 R.J. Charlson, James E. Lovelock, M.O. Andreae, et al., 'Oceanic Phytoplankton, Atmospheric Sulphur, Cloud Albedo and Climate.' *Nature* 326 (1987) 655–61.
58 In writing this brief summary of the DMS feedback mechanism I am indebted to Ruse, *Gaia Hypothesis*, 158–62.
59 See the discussion of Earth's 'uniquely lucky' size, atmospheric composition, and position relative to the sun in Simon Conway Morris, *Life's Solution: Inevitable Humans in a Lonely Universe*. Cambridge, Cambridge University Press, 2003.
60 James E. Lovelock, *Gaia: A New Look at Life on Earth*. Oxford, Oxford University Press, 2000, 69.
61 Jennifer A. Logan, Michael J. Prather, Steven C. Wofsy and Michael B. McElroy, 'Tropospheric Chemistry: A Global Perspective.' *Journal of Geophysical Research* 86 (1981) 7210–54.
62 Lovelock, *Gaia: A New Look*, 68–70.
63 I was introduced to Dan Anderson's Gaian plate tectonics hypothesis by Lawrence Osborn in 'The Machine and the Mother Goddess: The Gaia Hypothesis in Contemporary Scientific and Religious Thought.' *Science and Christian Belief* 4 (1992) 27–41.
64 Cited in Timothy M. Lenton, 'Clarifying Gaia: Regulation with or without Natural Selection.' In Schneider, et al. eds. *Scientists Debate Gaia*, 15–25.
65 Lenton, 'Clarifying Gaia,' 24.
66 Gregory Bateson, 'Cybernetic Explanation.' In Gregory Bateson, *Steps to an Ecology of Mind: Collected Essays in Anthropology, Psychiatry, Evolution, and Epistemology*. Chicago, IL, University of Chicago Press, 2000, 287–94.
67 Gregory Bateson, 'Effects of Conscious Purpose on Human Adaptation.' In Bateson, *Steps to an Ecology of Mind*, 313–7.
68 Michael S. Northcott, *The Environment and Christian Ethics*. Cambridge, Cambridge University Press, 1996, 217-48.
69 On entanglement as a cosmological metaphor see Catherine Keller, *Cloud of the Impossible: Negative Theology and Planetary Entanglement*. New York, NY, Columbia University Press, 2018.
70 James E. Lovelock with Bryan Appleyard, *Novacene: The Coming Age of Hyperintelligence*. Cambridge, MA, MIT Press, 2020.
71 James E. Lovelock, 'Foreword.' In Mary Midgely ed. *Earthly Realism: The Meaning of Gaia*. Exeter, Societas Imprint Academic, 2007, 15.
72 Michael S. Northcott, 'Fittingness and Biodiversity: Biocultural Theory and Ethics as Praxis' in Michael S. Northcott and Steven C. van den Heuvel eds., *Fittingness and Environmental Ethics: Philosophical and Theological and Applied Perspectives,* in London, Routledge, forthcoming.
73 James E. Lovelock, *Gaia: The Practical Science of Planetary Medicine*. Oxford, Oxford University Press, 1991, 153.
74 Lovelock deploys the metaphor of virus in many interviews and publications including his *The Revenge of Gaia*, New York, NY, Basic Books, 2006.
75 On the real causes of imperial and post-colonial famines see Mike Davis, *Late Victorian Holocausts: El Nino Famines and the Making of the Third World*. London, Verso, 2001, and Jean Dreze and Amartya Sen, *Hunger and Public Action*. Oxford, Clarendon Press, 2002.
76 Osborn, 'The Machine and the Mother Goddess.'
77 Lovelock, *Revenge of Gaia*, 15, cited Ruse, *Pagan Planet*, 221.
78 Ruse, *Pagan Planet,* 221.
79 Ruse, *Pagan Planet*, 223–4.
80 Toby Tyrell, *On Gaia: A Critical Investigation of the Relationship between Life and Earth*. Princeton, NJ, Princeton University Press, 2013.

81 Rupert Sheldrake, *The Science Delusion: Freeing the Spirit of Enquiry*. E-Edition, London, Hodder & Stoughton, 2012, 307–26.
82 Thomas Nagel, 'Teleology Revisited.' In Nagel ed. *Teleology Revisited and Other Essays in the Philosophy and History of Science*. New York, NY, Columbia University Press, 1979, 275–315.
83 Sheldrake, *Science Delusion*, 308–9.
84 Paul Crutzen, 'The Anthropocene.' In: E. Ehlers, T Krafft, eds., *Earth System Science in the Anthropocene*. Berlin, Springer, https://doi.org/10.1007/3-540-26590-2_3.
85 Daniel C. Dennett, *Consciousness Explained*. New York, NY, Little, Brown and Co., 1991.
86 Tyrell, *On Gaia*, 183–96.
87 Henri Bergson, *Creative Evolution*. Trans. Arthur Mitchell, New York, NY, Random House, 1944, xxii–xxiii.
88 Bergson, *Creative Evolution*, xxiii.
89 Bergson, *Creative Evolution*, 103, and cited in Paola Marrati, 'Life, Concepts: The Newness of Bergson.' *MLN* 120 (2005) 1099–111.
90 Henri Bergson, *Matter and Memory*. Trans. Nancy M. Paul and W. Scott Palmer, London, George Allen and Unwin, 1911.
91 Bertrand Russell, 'The Philosophy of Bergson.' *The Monist* 22 (1912) 321–47, 325.
92 Russell, 'Philosophy of Bergson,' 326.
93 Bergson, *Creative Evolution*, 17.
94 Rupert Sheldrake, *The Rebirth of Nature: The Greening of Science and God*. Rochester, VT, Park Street Press, 537, Kindle Edition.
95 Sheldrake, *Rebirth of Nature*, 87.
96 Eduardo Kohn, *How Do Forests Think: Toward an Anthropology Beyond the Human*, Berkeley, CA, University of California Press, 2017.
97 Rupert Sheldrake, *Morphic Resonance: The Nature of Formative Causation*. Rochester, VT, Park Street Press, 2009, 158–9.
98 Sheldrake, *Morphic Resonance*, 163–4.
99 Henri Bergson, *Creative Evolution*. Eng. trans. Arthur Mitchell, New York, Henry Holt and Co, 1911, 184.
100 Sheldrake, *Morphic Resonance*, 169.
101 Marcel Vogel, 'Man-Plant Communication.' In Edgar Mitchell ed. *Man-Plant Communication*. E-Edition, New York, NY, Cosimo Books, 2016, Chapter 12, 216–53.
102 Peter Tompkins and Christopher Bird, *The Secret Life of Plants*. New York, NY, Harper & Row, 1973, 89.
103 Vogel, 'Man-Plant Communication' cited in Tompkins and Bird, *Secret Life of Plants*, 89.
104 Seneff, 'Liver Disease.' *Toxic Legacy*, Chapter 7.
105 On Platonic and medieval Christian conceptions of the anima mundi and its modern retrieval see Miklos Vassanyi, *Anima Mundi: The Rise of the World Soul Theory in Modern German Philosophy*. Heidelberg, Springer, 2011.
106 Vassanyi, *Anima Mundi*.
107 Lovelock, *Revenge*, 138.
108 I discuss the Orthodox Christian philosophical idea of *energeias* in Chapter 7.

7

GAIA AND GOD

Abstract

Science, as *scientism*, has become the lead religion of industrial societies. Its atheist cosmology of mechanism has rendered tyrannical humanity's interactions with other creatures and even with the Earth as a whole. Scientific technologies now threaten global tyranny in human societies as well as over other creatures and ecosystems. Against this threat, the Gaia theory, when seen as a scientific retrieval of the belief common to almost all cultures and to pre-modern religious traditions that the Earth is alive and that the forces which animate the Earth, other creatures, and humans emanate from a Supra-mundane divine Source, has the potential to revive respect for the freedom and autonomy of other beings as well as of humans in healing interactions with them. In Taoism and Hinduism, the ideas of the Tao and the Life Force are significant analogies for Gaia. Mainstream Western Christianity is more resistant to the autonomy and freedom of other beings, but there is a minority cosmology in Christianity which is well grounded in the New Testament, and which discerns divine forces continuously at work in the development and ongoing processes of Life and these are variously described as logos, Spirit, Wisdom, and *energeias*.

Science *as* Religion

The modern scientistic imaginary, and the cosmology of mechanism, originated in Western Europe. Their twin legacies are the dethronement of God, which produces, in place of God, humans as 'the measure of all things' and hence the absolutism of the modern State in which political structure most humans now dwell;[1] and the secularisation of human societies, the Earth, and her creatures.[2] The dethronement of God, and the marginalisation of divinity or Spirit in

DOI: 10.4324/9781003110750-8

modern Western culture, are the unintended but enduring consequences of late medieval scholastic rationalism, and the Reformation quest to root out the sacred from time and space and relocate it in the individual conscience.[3] The impact of this dethronement was to make way for the rise of modern science and technology as the dominant ethos, and set of practices, of the industrial West. These constitute the modern condition of 'practical atheism' which is to say, while most Western people still acknowledge that there is a God, for most intents and purposes this belief no longer impacts on their cultural and material practices.[4]

It is not only divinity, but *nature*, as the immanent sphere in which divinity or Spirit are encountered, which is secularised and marginalised in the West. Lynn White Jr. argued that the historic root of the ecological crisis is therefore in the Christian West. But as Seyyed Hossein Nasr argues, the evacuation of the sacred or Spirit from the everyday life and governing ethos of the West was not *intrinsic* to the Christian tradition, or its doctrine of creation. Rather it is the seminal consequence of the demise of Christian sacred cosmology represented by the early Christian and medieval view of nature as immanently infused with the divine *logos*, Spirit or Wisdom:

> Something of Christianity, of course, survived in modern man, but in most cases it was of little consequence as far as the destruction of nature was concerned. Equipped with a Faustian knowledge, secular in character, and based on power over the natural order, the new man began to create unprecedented havoc over the globe, for there was now no limit set by any spiritual laws upon his rights of dominion and no higher knowledge to set a limit upon his profane knowledge of the world.[5]

In his seminal body of work on the divine–human–nature relationship, Nasr argues that 'what is most significant is the Prometheanism that came to dominate Western civilisation to an ever greater degree,' and that this was the product of the substitution of a sacred with a scientistic view of the nonhuman world which ultimately also impacts anthropology and philosophy as well as ecology.[6]

Despite more than 50 years of Christian responses to White, and a growing scholarly and ecclesial engagement of Christianity with the ecological crisis, it is evident that the corporate, governmental, and scientific agencies of the post-Christian West remain studiedly atheistic and domineering in relation to nature, and increasingly in relation to humans also. Nasr argues that this is because most adherents of Western Christianity are ignorant of the long-enduring sacred cosmology of their tradition. The trajectory of industrial civilisation, which first arose in the Christian West, though it is now a global culture, is therefore still set on the path of resource extraction and the replacement of biodiverse and resilient self-organising ecosystems with industrial, extractive, and monocultural systems of production.

The twin legacies of practical atheism and the secularisation of nature combine in the modern post-metaphysical and materialistic vision of the redemption of the human condition as 'progress' in economic development by means of the industrial and technological conquest and control of creatures and ecosystems.

The European Enlightenment underwrote these twin legacies, most influentially in the philosophy of Immanuel Kant, who set aside revelation – in scripture *and* nature – as sources of true knowledge. The Romantics sought a different outcome from the dethronement of God, and the rise of modern science and technology. But their principal legacy was what Catherine Albanese calls modern 'nature religions' in which nature herself becomes the new sublime and transcendent source of human redemption: through aesthetics, through peak experiences in wilderness, and through new 'nature-oriented' pagan rituals.[7] But nature, or Gaia, cannot bear the momentous weight of the responsibility the Romantics imposed upon her and absent the grounding of the independent agency of 'nature' in her origination in the divine Source of all being, romanticism did not succeed in revalorising the agency of Earth sufficiently to restrain the Promethean industrial project to conquer Earth and her creatures.

The Great Reset, and the globalist top-down project of the Sustainable Development Goals and Agenda 2030, which I review above, represent the current elite-favoured outcome of the West European scientistic and atheistic legacy. Together, these projects represent a revolution in human affairs across the globe in which top-down totalitarian agencies in the form of central bankers, globalist organisations such as the United Nations, its partner organisations including the World Economic Forum and the World Health Organisation, and private wealth foundations, in concert with national governments and technological corporations, engineer a global digital panopticon as means towards the total control of the actions of all persons and ultimately of all Life down to the level of DNA and microbiota. In this project, both human development and ecological conservation are caught up in what some call the 'New World Order' and others the 'Fourth Industrial Revolution.' The project is in essence one of post-political and post-national global governance which will use artificial intelligence, biosensors and surveillance devices, and huge global databases to trace, track, and attempt to control all human–human and human–nature interactions.[8] In this gargantuan and Promethean globalist project, the principal conservation legacy of the Romantic project to secure 'nature' against scientistic and industrial re-engineering via protected areas morphs into a project to exclude *all* spontaneous human–nature interactions from *all* ecosystems and not only protected areas.

The totalitarian trajectory of post-Christian societies shaped by science and technique was forecast in the mid-twentieth century by among others Max Horkheimer and Theodor Adorno, C.S. Lewis and Bertrand Russell. For Horkheimer and Adorno, writing in the shadow of the technocratic German Third Reich in the 1930s, the Enlightenment, together with modern science, not only involved the eschewal of animism and magic but the redescription of the empirical world in terms of a purely immanent cause–effect nexus which facilitated its control by human *techne*:

> From now on matter was finally to be controlled without the illusion of immanent powers or hidden properties. For enlightenment, anything

> which does not conform to the standard of calculability and utility must be viewed with suspicion.[9]

On this account, the subjection of all empirical reality to the unitary prism of calculable reason in the scientific method enable a 'mastery of nature' in which humans achieve their likeness to God by taking over the sovereignty of a divine power who is no longer implicated in the unfolding of Being on the earth God is said to have set in motion, since that unfolding is purely a result of natural laws. Matter was finally to be controlled without the illusion of immanent powers or hidden properties.

After the Enlightenment, anything which did not conform to the standard of calculability and utility was viewed with suspicion.[10] Knowledge of the laws becomes the source of Enlightened humanity's divine likeness, but in claiming this power humans are estranged from the realm they would rule over and reorder:

> Human beings purchase the increase in their power with estrangement from that over which it is exerted. Enlightenment stands in the same relationship to things as the doctor to human beings. He knows them to the extent that he can manipulate them. The man of science knows things to the extent that he can make them. Their "in-itself" becomes "for him." In their transformation the essence of things is revealed as always the same, a substate of domination.[11]

This reading of the Enlightenment, and modern science, as the origin of the modern disenchantment of nature, and its being made available for human dominion, re-engineering, and resource harvesting, without regard for its given order, is at variance with White's later essay in the journal *Science* in which he argued that Western Christian teaching about creation and humanity's role in nature as described in the Book of Genesis 'de-animates' the nonhuman world in a way that other cultural traditions do not, and that this metaphysical de-animation of the world is the historic root of the ecological crisis.[12] But for Nasr, the trajectory of control and dominion over the Earth, and even over their own bodies and genes, which humans are on is not the result of Christian or any other kind of metaphysics, but of the *demise* of metaphysics since the Enlightenment.

As Nasr pointed out in a book-length reply to White, White's reading of the *religious* roots of the ecological crisis is mistaken historically because the Christian Middle Ages in Europe, and the same period in the Muslim world, was not the era in which large-scale re-engineering of the Earth took place, or nature was disenchanted.[13] The Middle Ages was the period when Christian cosmology in the West reached arguably its most impressive cultural flowering, as in the building of the great Gothic cathedrals, the lay-out of great cities which survive from that era almost intact such as those of Bologna, Florence, Rome, and Siena, and the development of perspective and new pigments in painting. But far from mirroring a de-animated cosmos, the architecture, paintings, sculptures, and tapestries of the Middle Ages display beliefs analogous to those of Aristotle that the Beings which humans encountered and took up with in the nonhuman world – whether water, stone, timber, plants, paint pigments, metal ores, insects, or animals – all carried within them, as Spike Bucklow explains, the living influence of the divine according to the medieval 'doctrine of signatures,' and were meant to be read by humans as living signs, or symbols, of the divine who created them.[14] The doctrine of signatures was one of the ways in which medieval philosophers incorporated Aristotle's account of the four-fold causation of all things – material, efficient, formal, and final – into what C.S. Lewis calls the medieval world model. At the core of this model was an array of symbols which helped to answer the question of how the Earth, and Life, are sustained in being long after the primeval event of Creation, the account of which Christians had adopted from the Israelite Book of Genesis.

In a series of radio talks and newspaper editorials from the 1930s to the 1950s, Lewis argued that the demise of the symbolic worldview of medieval Christianity accompanied and underwrote the use of scientific technique to control natural forces and materials. Lewis connected the totalitarian implications of modern scientific techniques with the demise of religion and in particular the loss of recognition of the divine Source of Life or, in Lewis' words, the 'Tao.' Lewis also presciently forecast that the techniques of domination that Baconian science turns upon nature will be used later by human elites to impose their will on the masses. But unlike Russell, Lewis argued that the repair of this situation will be a revised approach to natural science, as anticipated in the works of Goethe and Rudolf Steiner:

> The regenerate science which I have in mind would not do even to minerals and vegetables what modern science threatens to do to man himself. When it explained it would not explain away. When it spoke of the parts it would remember the whole. While studying the *It* it would not lose what Martin Buber calls the *Thou*-situation. The analogy between the *Tao* of Man and the instincts of an animal species would mean for it new light cast on the unknown thing, Instinct, by the known reality of conscience and not a reduction of conscience to the category of Instinct.[15]

Lewis' call for a regenerate science anticipates Nasr's call for a 'new sacred science.' Neither of these responses to scientism and the totalitarian tendencies of technique are anti-science. But both call for a new science that acknowledges the underlying interconnectedness of all things, and the divine power or Tao that originated, and still animates, all things.

Lewis was writing against the increasingly atheistic zeitgeist of the twentieth century while one of his contemporaries, Bertrand Russell, was fully with this zeitgeist in his *The Impact of Science on Society*. Science, Russell argued, is a revolutionary force whose novel techniques, in industry and war, have produced an historically unprecedented rate of change in human societies. And because of 'the new control over the environment which scientific knowledge has conferred, a new philosophy is growing up, involving a changed perception of man's place in the universe.'[16] The traditional view, as in Aristotle's theory of causation, that the cosmos is characterised by multiple causes and goals in the structures of being, and independent of human volition, is replaced with a simpler view, according to which causation is the outcome of purely physical events. This simplified view of causation as purely physical, and as lacking underlying teleological or spiritual purposes, underwrites the modern scientific and technical imposition of purely human goals over the physical world.[17]

The new powers over matter also give rise to oligarchic tendencies in human societies which are especially evident in the events that have occurred since the release of Sars-Cov-2 and the Great Reset. In the scientistic and technocratic regime that SARs-CoV-2 was used to usher in those with the resources and wealth to utilise the techniques imposed on the physical world turn these same techniques to acquiring more wealth and power for themselves and to dominating their fellow humans. This tendency is magnified by the effect of techniques such as electronic communication to concentrate power in fewer hands. In this way, technique increases the powers of central governments and large economic and technological corporations over communities and organisations at a distance from the centres of power. Within bureaucracies and industrial corporations, the same techniques enhance the potential for managerial tyranny which functions in many workplaces under the ideology and practices of 'human resources' – a euphemism for corporate serfdom – according to which employees are merely functionaries of an organisation and, like pieces of furniture, are in effect owned by the organisation and, if they step out of line, are disposable and replicable.[18] But for Russell, tyranny is not a concern. Instead, he sees the end goal of scientific and technical management of society as the generation of a eugenic 'struggle for existence' in which the 'race is improved' by the 'weakest going to the wall.'[19]

Russell presciently argued that of all the techniques that would be invoked to promote oligarchic power, propaganda and psychology were the most powerful. He comments on the use of propaganda in government schools, and in electronic and print mass media in Stalin's Russia, and observes that 'the evils that have grown up in Soviet Russia will exist, in a greater or less degree, wherever there is scientific government.'[20] Of all the techniques, 'the subject which will be of most importance politically is mass psychology' the study of which 'is immensely useful to practical men, whether they wish to become rich or to acquire the government.'[21] Here Russell anticipates the use of mind control and mass psychology techniques by Canadian, British, and other governments in the unprecedented imposition of 'lockdowns' and other social control measures during the COVID-19 pandemic.[22] Mass psychology and propaganda were used by allegedly democratic governments to promote irrational fear of a virus which was only a mortal threat to those who already had life-ending health conditions.

The ultimate purpose was not to reduce elderly mortality but to persuade the populace en masse to submit to a QR-coded social order in which every movement and transaction is tracked, including citizens' willingness to participate in a hazardous experimental gene-editing therapy which brought huge wealth opportunities for pharmaceutical corporations, while generating unprecedented levels of medical injury.

Russell's warnings are as prescient, and even more acutely so, than those of Horkheimer, Adorno, and Lewis. But Russell *celebrates* the achievements of science in displacing religion cosmologies and traditions in modern industrial societies, while Lewis was firmly of the view that only by recovering something of the symbolic and sacred worldview of Christian medievalism would moderns be able to resist the destructive and totalitarian tendencies of science and technique. For Lewis, as for Nasr, it is only through a recovery of the idea that the Earth and humans are divinely originated, and imbued by a force something like the Tao, or the divine *logos* or Spirit, that modern science-informed humans will recover a deeper respect, and awe, at the given beauty, diversity, and harmony of Gaia. *Because* it is divinely originated, and still continuously divinely energised, there are limits to humanity's re-engineering of Life. Lewis and Nasr also differ from Horkheimer, Adorno, and Russell in their argument for a revision of scientific cosmology with a new 'sacred science' as the way towards a more holistic and less domineering approach to the more-than-human world. But Nasr is perhaps more realistic than Lewis in his recognition that mainstream Christianity may already be too far gone to be able to recover the sacred cosmology of its medieval past.

Medieval Nominalism and the Roots of the Machine Cosmology

If we enquire into the roots of the present anti-natural and transhuman turn of Western scientistic culture, we come unavoidably, as Nasr argues, to the history of religion and in particular to the late Christian Middle Ages in which a new attitude to animals and the nonhuman more broadly began to emerge. Until that time, it was believed that animals had souls analogous to those of humans, and that they were capable of volitional acts – including malevolence – which at times led to animals being tried, much like humans, when an animal was considered to have been the cause of a human fatality.[23] But a significant modification to the traditional view that saw more analogies than disanalogies between humans and other animals was made by Thomas Aquinas, the most influential Catholic medieval philosopher and theologian, who argued that, while in most respects there are common bonds between humans and other animals, what sets humans apart is their *intellective* soul which for Aquinas are the only kind of souls which can exist apart from the material body.[24] In this, Aquinas was influenced by Augustine, and by the recovery in the Middle Ages of certain Aristotelian and Stoic texts in which the contrast between human intellective powers and animal irrationality was a central theme.[25]

The increasing late medieval focus on reason as the quality in humans that sets them apart from other animals gave rise to a new view of nonhuman beings in the nominalist and scientific innovations of William of Ockham and John Duns Scotus.[26] The nominalists adopted a sharp distinction between mental categories, such as the *names* of species and things, and physical 'accidents' in the material world. As Louis Dupré argues, this was the key philosophical innovation in Western Christian Europe which led to the erosion of traditional Christian cosmology according to which divine immanence in the cosmos was a continuing process – in Latin *creatio continua* – which connected and imbued all living things through the immanent divine Spirit in the Great Chain of Being.[27]

This traditional cosmology is powerfully illustrated, for example, in the metaphor of the vine in the Gospel of John, Chapter 15.[28] It is also illustrated in Christian art from the earliest centuries, such as the apse mosaic of the church of St Clement in Rome which shows different creatures joined by palm branches to the revivifying cross of Christ (Figure 7.1).

The apse represents the medieval Christian view of the 'Great Chain of Being' and the way in which the cross as the 'healing tree' was said to have set the whole creation on a path of restoration from the alienation emanating from the 'tree of good and evil' in the Garden of Eden, whose fruit the serpent was said to have led Adam and Eve to eat. The nominalist move tore apart this traditional metaphysical schema, inherited from Plato and before him from the Indian Vedas. In this schemea, the visible phenomena of the cosmos are an immanent veil of bodies and forms which participate, along with the mind-body-soul of human beholders, in the divine Source, and participate in the history of salvation which,

FIGURE 7.1 Apse mosaic, St Clement, Rome. Author photo.

for Christians, is focused on the revivifying effects of the life, death, and resurrection of Christ.

Intellectual developments in medieval scholasticism associated with the rise of nominalism accompanied the development of new technologies of control over natural processes – and especially the deep plough and mechanical time pieces – which were crucial tools in the development of more instrumental and control attitudes towards the nonhuman and which paved the way for the Renaissance in which man became the 'measure of all things.' The significant change in attitudes towards nonhuman creatures which followed is recorded in fascinating detail in Keith Thomas' survey of attitudes to nature in England from the late Middle Ages to the early modern era which reveals a notable increase in cruel and instrumental uses of animals and nature more broadly.[29] This gradual cultural shift was magnified in Renaissance humanism, and at the Reformation, and European political theology is subsequently characterised by a heightened emphasis on human power and control. This is evident in the political theories of Niccolo Machiavelli, John Locke, and Thomas Hobbes, and in the scientific method as described by Francis Bacon and René Descartes. These intellectual shifts underwrote the materialist and mechanistic turn in scientific cosmology as in the astronomical theories of Galileo and Kepler, the physics of Isaac Newton, and, much later, the biology of Charles Darwin.

Medieval Western Christian theology, in the form of nominalism, therefore played a seminal role in the birth of modern technologies, and in the origins of the scientific and industrial revolutions which are responsible for the large-scale transformation of human–nature relations which characterise Western, and now global modernity. It is for this reason that I devote a significant part of this chapter to a reconsideration of Christian cosmology and ontotheology. If the root of the crisis is there, then I argue that its resolution also needs to occur there, even if, because of secularisation, Christianity is much less influential now in Western culture than it was in the Middle Ages. But before turning to this needed reconsideration of Christian ontotheology, I consider the prior cosmology of Eastern religious philosophy, which as Alan Watts and Thomas Merton first argued in the 1960s, shows greater potential in its intrinsic cosmology towards repairing the modern breach with nature than Western Christian cosmology, at least in its post-medieval and subsequent iterations. Eastern religious thought also shows great potential in respect of a metaphysical view of the cosmos which is consistent with the less mechanistic iterations of Lovelock and Margulis' Gaia theory.

Dolphins, Dams, and Taoism

The Gaia theory reviewed in the last chapter is evidence that against the mechanistic and reductionist trend of modern Western cosmology since the late Middle Ages, a significant transformation is now underway in scientific cosmology which is as momentous as the rationalist and nominalist turn of the late Middle Ages. This trend is evident from the early twentieth century in Einstein's theory of relativity, Werner Heisenberg's uncertainty principle, and quantum mechanics, as already discussed. It is evident in the findings of Nicolas Tesla concerning energy, electricity, and the electrical frequencies which run through all living things: many of Tesla's controversial findings were suppressed in his lifetime but are becoming better known.[30] It is evident in Henri Bergson's philosophy of memory and consciousness, and in the related morphic resonance theory of Rupert Sheldrake.

What some call the post-modern turn in contemporary science is also evident in ethological studies of other animals, in which field the best-known pioneers were Jane Goodall and Diane Fossey, whose studies of, and companionship with, chimpanzees and gorillas, while initially criticised by zoologists, are now complemented by a large array of scientific studies showing the extent of communicative, intellective, and sentimental capacities displayed by other animals.[31] In *Dependent Rational Animals*, the philosopher Alasdair MacIntyre argues that developments in the scientific study of dolphins, dogs, and primates such as chimpanzees and gorillas reveal that in intellect there is much more in common between humans and other higher mammals than the medieval scholastic and most modern scientists believed.[32] Dolphins are among the closest of all animals to humans in their capacities for communication, group moderation and planning, nurturing capabilities, and the range of sentiments they have been observed to display. The remarkable displays of collective planning which dolphin groups have been observed to manifest when hunting shoals of fish, and their complex patterns of nurture of the young, together with the capacities of dolphins to communicate and form relationships with other primates, and especially humans, all testify to the fact that many aspects of what Aquinas, with Aristotle, called the intellective soul are not confined to humans.[33] But despite their proximity to dolphins intellectually and in other key respects, modern industrial humans show scant regard for these wondrous creatures. Huge numbers are killed every year by the three-mile-long lines of death strung out by fishers of tuna, and in the vast trawl nets of modern industrial trawlers.

The Chinese river dolphin is a unique species of freshwater dolphin whose last remaining habitat, the Yangtze River, was blocked by the gigantic Yangtze Dam in China. The Yangtze Dam is so vast that as the reservoir filled with such a weight of water it also compressed the earth's crust to the point of causing seismic disturbances. Earthquakes, and floods, which accompanied the construction of the dam led some Chinese to argue that the dam represented the refusal of the Chinese Communist Party (CCP) to honour the more-than-human world, and that by its construction the CCP had forfeited the divinely conferred 'mandate of heaven' to govern China.[34] Shortly after the completion of the gigantic Yangtze Dam scientists also discovered that the Chinese river dolphin had gone extinct since the dam cut it off from much of its former habitat.[35]

Water and water flow hold a central place in the Chinese Taoist tradition, the nature imaginary of which is closely related to nature as flow or process. Blockages of water, and of qi, are understood to cause illness in humans, and malfunctions in nature. In the *Tao Te Ching*, traditionally attributed to Lao Tzu, the Tao is an order of Being which is fluid but in its transcendent form unnameable:

> Tao can be talked about, but not the Eternal Tao.
> Names can be named, but not the Eternal Name.
> As the origin of heaven-and-earth, it is nameless.
> As 'the Mother' of all things, it is nameable.[36]

Lao Tzu envisaged embodied human life, and the bodies of other beings, as physical manifestation of an original unitary nature which is manifest in a flow of energy or qi which in Taoism means 'vital force.' As in the Gaia theory, Taoists hold that the human body is a 'self-regulating cybernetic system' which 'also functioned as a template for understanding other parallel systems.'[37] The body is understood to maintain itself in health through the circulation of qi, principally through breath and blood circulation, and through regular physical movement. Taoist related religious practices such as breath control and meditation, qi gong, and other kinds of sustained physical movement, including meditative walking in forests and other natural settings, are designed to support and increase the flow of qi. In Taoism illnesses are said to be the result of blockages of qi which can be unblocked through such practices, and through medical practices such as acupuncture.

Taoist traditions lay particular emphasis on mountains where water gathers to form rivers and flood plains below, and on the flow of water: mountains and water courses – both natural and humanly made – are central features of Taoist influenced landscape paintings, Taoist religious practices, and Taoist influenced agriculture. Taoist masters sought out mountainous places, and caves, as sites of revelation and religious practices.[38] The central place of mountains in Taoist cosmography influenced attitudes towards the care and preservation of mountainous ecosystems such as forests and springs. Even more central to Taoist cosmology are the flow of rivers, and water. As James Miller explains, the flow of waters in China from west to east traditionally represented a physical analogy for the Way of the Tao through all being which is conceived as a 'fluid vector by which the processes of life are never static but always in motion.'[39] The belief in the centrality of water to the Tao underlay the early development in China of a 'hydraulic state' in which the control of water, the prevention of floods, and the irrigation of farmland were central. Taoist beliefs influenced the development of an irrigation system in the third century BCE in Sichuan province. The weir that was constructed in the Min River was connected to water courses that to this day provide water to over 50 cities and irrigate 670,000 hectares of farmland. Unlike the Yangtze dam, the Min weir did not stop the flow of water or cause floods upstream. Rather it regulated the flow, so guiding the natural qi of water into multiple human and ecological uses that remain productive to this day.[40]

Taoists believe that all Beings are part of an interconnected whole even when the separate parts – and especially humans – imagine themselves as having identities which are materially and intellectually separate from the whole. In Taoism, humans are uniquely mindful beings, but their exceptional mindfulness is the source of tendencies to develop desires, and to dwell on Earth, in ways which neglect the given order of Being, and block or disrupt the flow of Tao, both in the human body and in the more-than-human world. The damming of the Yangtze River is a powerful example of this tendency. Another would be the spread of deserts in China north of Beijing as a result of the destruction of forests and their replacement with industrial monocultures.

The lack of fit between modern humans and the Tao, including in China, is a consequence of China's adoption of the same industrial practices which have gradually wrecked Western, and Western colonial, environments since the industrial revolution in Britain and Europe in the eighteenth century. The extinction of the Chinese river dolphin by the Yangtze dam is just one example of the ecological disasters across China which accompany its rapid industrialisation.[41] Many are now asking if China can recover its ancient wisdom, and especially the wisdom of the Tao, in revising its development model and restoring the integrity and resilience of its seriously degraded ecosystems.[42]

Decades before Chinese intellectuals began to re-examine China's indigenous traditions in order to repair its contemporary ecological destructiveness, C.S. Lewis invoked the idea of the 'Tao' from Chinese Taoism as a figure for the traditional ways of life and thought that modern technologies are displacing in the name of 'Reason.' There is a seminal divide, he argued in three talks on modern education and technology, between those teachers of children who stand 'within or without the Tao.'

> For those within, the task is to train in the pupil those responses which are in themselves appropriate, whether anyone is making them or not, and in making which the very nature of man consists. Those without, if they are logical, must regard all sentiments as equally non-rational, as mere mists between us and real objects.[43]

Once science-informed moderns step outside the Tao, and regard it as a 'mere subjective product' it becomes 'a mere abstract universal' and it is a short step towards 'the world of post-humanity' which 'nearly all men in all nations are at present labouring to produce.'[44]

Lewis invokes the Tao as the Chinese equivalent of the Western idea of 'natural law' which is a Christian cosmological conception of biological and moral universals said to be set into the structure of the cosmos and of life, and towards which humans and other animals are best fitted, and hence experience flourishing, when they train their desires and practices to honour them.[45] Lewis is prescient both in his realisation that the industrial and scientific civilisation is the first truly global form which subverts all other cultures, both East and West, while at the same time subverting both the goals and purposes of other creatures, and, ultimately, of humans themselves as their freedom and dignity is gradually sacrificed to the post-human technological society he already envisaged as being in the making in the 1930s.

In the 1960s, when the ecological crisis came to the fore in Western cultural consciousness, the Trappist monk, mystical writer, and peace campaigner Thomas Merton also turned towards the wisdom of Eastern traditions as a way of revising the orientation towards nature of his own tradition of Roman Catholicism. For Merton, the essence of Taoist thought is *wu-wei*, 'the non-doing, or non-action, which is not intent upon results' but is focused instead on being 'in harmony with Tao – the cosmic Tao':

> The true character of *wu wei* is not mere inactivity but perfect action – because it is act without activity. In other words, it is action not carried out independently of heaven and earth and in conflict with the dynamism of the whole, but in perfect harmony with the whole.[46]

Taoist cosmology recognises that 'Life is a continual development. All beings are in a state of flux.'[47] But this flux is not unstable, the development not directionless since from the least to the greatest, there is the Tao:

> Tao is Great in all things.
> Complete in all, Universal in All,
> Whole in all. These three aspects
> Are distinct, but the Reality is One.[48]

Taoism is a teleological and metaphysical tradition which to Western minds sounds like pantheism. But this is because of the tendency of Western logics to create oppositions in language, and thence in thought, which are not present in Chinese thought, in part because Chinese language is pictorial and symbolic and not merely alphabetical.

Alan Watts argues that the symbolic character of the Chinese language gives a different mindset to Chinese traditions as compared to Western culture. And of the two dominant Chinese traditions – Confucianism and Taoism – Watts argues that Taoism concerns itself more with 'the understanding of life directly' rather than with abstract thought. Taoist thought is therefore 'naturalist' in a quite different way to scientific naturalism.[49] The Taoist conception of being is of an unconscious process, similar to homeostasis by which the temperatures of the human body, and of the Earth, are regulated. But at the same time there is intelligence in the Tao, a universal intelligence which was there before the origin of all things in the 'void' and which, though 'blurred and indistinct,' is understood as a 'mental power,' the Chinese word for which is *ching*. Unlike the Western conception of mind, *ching* 'combines the ideas of essential, subtle, psychic or spiritual, and skilful':

> The important difference between the Tao and the usual idea of God is that whereas God produces the world by making (*wei*) the Tao produces it by "not-making" (*wu-wei*) – which is approximately what we mean by "growing". For things made are separate parts put together, like machines, or fashioned from without inwards, like sculptures. Whereas things grown divide themselves into parts, from within outwards.[50]

For Watts, another distinctive feature of Taoist cosmology, as compared to Western, is the understanding of the spontaneity of the processes which constitute Life on Earth: the spontaneity of the Tao means that 'it loves and nourishes all things, but does not lord it over them.'[51] By contrast, Western cosmology, and especially the view of the cosmos as being a machine of parts made by a divine maker, is more shaped by the metaphor of a divine controller. In this cosmology, humans are the divine representatives on Earth and the human function in essence is one of control rather than of going with the flow of the Tao. This leads to an intent to control and suppress the spontaneity of nature which is carried over from Western theology into Western technology as a 'mistrust of nature' and a consequent desire for control over nature.[52] Against this mistrust, Taoism teaches that the human tendency to active intervention is a function of overactive minds, and that against this, Taoism teaches that those who learn to regulate the body and unify mental attention will find that 'the harmony of heaven' comes upon them.

As in the Gaia theory, Taoism indicates that there is an unconscious tendency or flow of all life towards stable states that, while it is analogous to intelligence, is not a function of intentional thought or reason but of an underlying integrated and spontaneous power. Hence, Taoists invoke the unceasing and apparently uncoordinated groupings together and interactions of 'birds and beasts,' 'trees and shrubs,' which 'grow together without exception,' as examples for humans to follow: 'Be like these: follow the Tao, and you will be perfect.' Allowing the mind and body to function in an analogously integrated and spontaneous way is the aim of Taoist religious practices in which 'virtue' or 'power' is not so much in 'moral rectitude' as it is in effectiveness, by analogy with the healing properties of a plant.[53]

But while there are significant parallels between the Gaia theory and Taoism, the principal difference is the eschewal of teleology among advocates of Gaia theory which reflects the modern scientific preference for a metaphysics of atheism and chance over the metaphysics of divinity and deep purposiveness. Another crucial difference is the combination of humanism and naturalism of Taoism. Taoist sages such as Lao Tzu taught that when human beings align themselves with the Tao and train their desires and practices towards the unconscious but unitive flow of the Tao in all things, they contribute to sustaining the flow and ultimate goodness of the Tao in all things. By contrast, Gaia theory, for example, in Lovelock's metaphor of humans as a planetary as a virus, tends in an anti-human direction.

Lewis observes that there is a paradox in Taoist teaching for it suggests that when humans acknowledge the Tao by honouring nature and living in ways that do not overpower nature with technique, they also find a stronger and more secure foundation for honouring their own humanity, and hence resisting what Lewis foresaw as the tyrannical turn of technological control over nature into control over humanity. Hence, for Lewis the recognition of the metaphysical depth of the Tao in all things indicates, as discussed above, the need for a 'regenerate' science which resists atheism, instrumentalism, and post-humanism.[54] This call for a regenerate science finds significant echoes in the work of Nasr, and of Roy Rappaport as discussed in Chapter 8.

Religion and World Maintenance

The Taoist belief that personal growth in virtue, and especially in fitting the human mind–body towards the divinely given and sustained order of things, is at the heart of what it means to be human, and the right relation of humans towards nature, is not unique to Taoist thought but it also found in the older tradition of the Indian Vedas. The earliest written textual evidence for this belief is the Vedic classic *Bhagavad Gita*, emanating from the Himalayan region of North India three centuries before the *I Ching*. In the *Gita*, religion is in essence good acts, intentions, and practices, performed in the spirit of detachment and not for personal gain, honour, or moral prestige, which 'maintain the world' human and more-than-human.

> You should always perform with detachment the tasks that have to be done, for by performing actions with detachment a man attains the highest good; it was through action that Janaka and others achieved perfection. You should act, too, with the very maintenance of the world in view. What the good man does, other folk do also: he sets the standard, and the world follows it.[55]

In the *Gita*, as in Taoism, moral and spiritual training and related ritual actions are understood to have *theurgic* functions towards all beings: right actions involve honouring, mediating with, and hence sustaining, the networks of beings, invisible and visible, in which human life is situated and from which humans draw substance. And this has implications for the habitat of humanity, including the composition and design of human dwelling places and cultural artefacts, and the procurement of food, water, and materials for craft and making.

In Hindu perspective, humans uniquely combine elements of divine, animal, and earthly natures. This three-fold conception of human nature and the cycles of life, death, and eternity is reflected in the post-Upanishad divine Trinity of

Brahma, Vishnu, and Siva. Siva is that aspect of life that includes both creativity and destruction, death, and rebirth, and is highly venerated on Bali where this book was largely composed.[56] Brahma is the source of all life, or creator, in Hindu theology, and a rather distant figure in most Hindu cultures. Vishnu is the divine helper of human beings, and sometimes also described as the life force. Vishnu is said to become Incarnate in the figures of Krishna and Rama who are widely venerated among Hindus.

Balinese Hindus have evolved a three-fold philosophy along these lines which is called *Tri Hita Karana*, and is in effect a fusion of Balinese and European cultural ideas and informs ecological projects in Bali which attempt to repair the damaging influence of modern industrial fertilisers, plastics, and large-scale tourism.[57] The Balinese organise their human communities around this three-fold character as each Balinese village or 'banjar' is governed by three committees concerning the temple and ritual; human social life and family matters; and the environment including water and land maintenance. This Trinitarian conception informs a contemporary project in the town of Ubud which addresses Bali's serious waste problem by promoting household, business, and community-based separation of organic and inorganic waste, and creating a compost facility for organic waste which can be used instead of fertiliser on rice paddy, while the end result of the waste project is also to significantly reduce toxic waste going to landfill.[58]

Raymond Panikkar in his Edinburgh Gifford Lectures of 1989 argues that the Vedic Trinity is the predecessor of the divine Trinity in Christian theology; that this three-fold conception of modes of Being is manifested in the metaphysics of all life, human and nonhuman; and that it is especially focused in human beings. The Trinitarian pattern of Being underwrites the 'real cosmos' which is 'neither purely spiritual nor merely material.'[59] The human is a unique being in the cosmos because humans uniquely combine material, mental/soul, and spiritual dimensions and are therefore, as the Vedic texts also propose, a unique microcosm of the macrocosm. For humans, then, 'participation in the dynamism of the cosmos is a sharing in the divine dimension' and hence, 'from ancient times the sage was considered a powerful force in the cosmos.'[60] For Panikkar, therefore, the human has an intrinsic vocation to transform the world. While 'no longer inhabiting an earthly paradise,' humans retain, nonetheless, a 'heavenly spark whose task is to be the *trikarma-krt*, the doer of the triple work' as described in the Katha Upanishad.[61]

The belief that humans combine in a unique fusion the different dimensions of being that are present in the cosmos – from spiritual beings, such as angels, to the lowliest creatures – is an ontotheological presupposition present in world religious traditions until the origin of modern scientific atheism. In traditional ontotheology the prior order of the world which humans encounter on earth is a divinely originated and sustained order which, by their three-fold nature, humans are uniquely able to recognise, but also to reshape, including in ways that either degrade the harmony of the cosmos, or enhance it. Perhaps the most culturally ubiquitous form in which this ontotheological perspective is experienced, even in modern secular cultures, is that of music since musical instruments create a range of frequencies, and interactions between these frequencies are of the essence of all human musical forms. In the European Baroque era this realm of vibration or frequency was sometimes called the 'music of the spheres.' This realm is represented culturally in many sacred musical traditions, such as the gamelan orchestras of Bali, as well as the European Baroque tradition as, for example, in the music of J.S. Bach.[62] Ethnomusicological studies, as well as studies of the physiology of sound, have established that in many cultures musical harmonies, rhythms, and timbre are not only human creations but cultural reflections of cosmic harmonies and vibrations between creatures and the art of crafting musical instruments from metal, skin, trees, bone, and other natural materials is therefore also an art which engages the Earth and human culture in a unique fusion which is accessible to anyone who has strummed a wooden acoustic guitar, beaten a rhythm on a skin drum, or pressed the ivories of a grand piano made from timber and metal.[63]

In the ontotheological perspective which I am arguing is shared between Taoism, Vedic Hinduism, and indigenous cultures, there is a sense that when humans take up with the Earth by creating cultural objects such as musical instruments, food, clothing, dwellings, and temples they are required to do so in ways that engage positively with the prior animating energies of the world, so enhancing the divinely imbued cosmic harmony, and repairing any disruption that their actions may temporarily occasion. As discussed above, this recognition that humanity's unique calling is to enhance this harmony finds expression in the religious and philosophical meanings of the Sanskrit word *rta* – which means cosmic order – and equivalent words in European languages including *logos*, *pneuma*, order, and wisdom.

In this traditional religious perspective, the Earth, and all creatures, alongside humans, have exercised agency over considerable periods of time in such a way as to generate flows of energy and numbers of species which constitute the habitat, and the *surplus* production of energy and materials which humans use in curating their distinctively complex cultures and habitats. In these traditional religious cosmologies, fitting human life to the given cosmic order is not only about fitting *in* but adding something distinctively human that produces greater harmony, beauty, and fertility. This is often conceived, including in Jewish and Christian as well as Hindu traditions, as the priestly role of humans who in offering sacrifices to divinity – either in physical form or in the form of ethical or spiritual practices such as breathing techniques, fasting, and meditation – are said to complete and enhance the beauty and harmony of cosmic order, while at the same time receiving from the earth that which makes human life possible including water, food, fibre, minerals, and building, craft, and manufacturing materials. This role of sacrifice is powerfully evident in Balinese traditions which, as the case study of water temples discussed in the following chapter demonstrates, proved more successful than modern scientific agronomy in achieving a beautiful and harmonious, *and* a more efficient rice-growing culture. In Bali, as in many other enduring cultural traditions, religions' ancient and pivotal role in theurgically bringing human-intelligent agency and the intelligent agency of the Gaia, and hence the Great Chain of Being of which humans are a part, into harmony is still powerfully evident.

Christian Ontotheology

The ultimate metaphysical question raised by the Chinese philosophy of the Tao, the divine Trinity in the Vedas, and the interreligious practices of theurgic sacrifice which fit humans to cosmic order, and are said to enhance it, concerns the *agency* of divinity: how divinity is manifest on Earth and how humans may sensitively cooperate with it. As Walter Desmond observes, in modernity it is considered a scientific and philosophic error to ascribe active powers to the divine when simpler, more materialistic cause–effect explanations can explain a particular set of events or processes.[64] Modern scholars, unlike the sages of Xian or the Himalayas, are therefore hesitant to ascribe agency in the visible world to a transcendent divine Source or Being who is so beyond human apprehension as to be ultimately unknowable. The God of Moses is said to have said when asked for the divine name: 'I am that I am.' In modern English we might translate this as 'being itself' or Tao. And this answer is reminiscent of Lao Tzu's classic statement, 'the Way that can be spoken of is not the Way.' If this is the case, then it might be said that the modern view is correct that the less said about God the better so far as cause and effect, and human uses of the nonhuman, are concerned. But as Desmond argues, it is the *absence* of God from modern science and philosophy in the last 200 years which has rendered up first 'nature,' and now humans, to the technocratic urge to dominate and control life down to the minutest subatomic particles, as in in the Large Hadron Collider at Cern, and in attempts to engineer human as well as nonhuman organisms at the genetic level as in the Human Genome Project, and in bioweapons laboratories.

If there is no God, there is no given order that requires ultimate honour and respect by humans in the way in which they take up with the world. As Desmond puts it, Kant's demolition of metaphysics in philosophy produces

> an alternation in our inhabitation of the ethos of being, in which there is a changed relation to otherness, be it the other-being of nature, or of God; a massive objectification of nature, coupled with a huge subjectification of human existence; a stripping off of other-being of the qualitative signs of givenness that communicated of transcendence as other to pre-modern man. *After* it, we find a certain intensification of an innerness that, on the one hand, seems to betoken a truer wat to the ultimate and the divine but that, on the other hand, too often simply led to will to power, to the tyrannical soul.[65]

As Lewis also argues, the modern scientific and technological quest for control leads to tyranny because of the loss of the sense of the real independent agency of the Tao, and of divinity mirrored in the Tao, which Christians call God, or, in Paul Tillich's less metaphysically encumbered phrase, the 'ground of being.'[66] Consequently, even although the Gaia theory indicates that in very real senses Gaia is alive and the networks of Being which constitute her have agency in providing a habitat that enabled humans to evolve, if Gaia is ultimately no more than a complex cybernetic machine, and not divine in nature or origin, her aliveness is not so wondrous as to prevent even Lovelock himself from suggesting that 'cyborgs,' artificially intelligent machines, will be able to do better than humans, and even Gaia, in arranging her and humanity's futures![67]

The contemporary ecological crisis, and the current array of human crises, including the discernible trajectory towards post-humanism, in the midst of which this book was composed, are not so much the product of a *de-animated* cosmos as they are the product of a cosmos in which there is no God, or at least no God who makes a difference to the ongoing processes of the unfolding of material laws and organic beings and mental faculties on earth. So, against the claim that Christianity's de-animation of the earth is the root of the modern ecological crisis, the deeper root of the crises towards which the domination of being has brought the earth and 'us' are modern science, together with the Enlightenment which, having put aside a transcendent and sustaining divine Source, turned the 'man of science' into a god whose *raison d'être* becomes the conquest and control of nature.

That the 'man of science,' rather than the myth of Adam, is the origin of the pervasive and ultimately tyrannical modern urge for control – of other creatures and now of humans themselves – is most clearly evidenced by the history of environmental degradation. The modification of ecosystems at scale, to the point of causing local and regional climate changes, began during and not before the industrial era with the use of industrial-scale land clearance and habitat modification. But nonetheless, I do accept that my own religious tradition – Christianity – has a weaker conception of the agency of the Earth and of creatures in generating and sustaining life than that offered by the Chinese conception of Tao or the Vedic Trinity. This is principally because Christian theologians lay far greater emphasis than Asian traditions on the original acts of the divine *making* of the universe and Earth, and far less emphasis on the role of divinity in the *continuance* of creation. This emphasis on First Cause thinking is a critical root of the subsequent *de-animation* of the chain of beings which occurred in the Western European Middle Ages and was closely related to the gradual erosion of belief in the ongoing agency of creative powers resident in Gaia, and which Asian traditions, and Eastern Orthodox theologians, never stopped ascribing to divinity. The *metaphysical* problem, as Catherine Keller argues, is the tendency of Christian theologians in the Western traditions of Roman Catholicism and Protestantism to situate almost all generative agency in the formation of life as we experience it in the original acts of divine making.[68]

Keller's answer to this problem is reminiscent of the pantheism of Giordano Bruno and Baruch Spinoza since she quotes Alfred North Whitehead's statement with approval that: 'It is as true to say that God creates the world, as that the world creates God.'[69] Bruno was the first Western theologian to attempt a metaphysical embrace of the 'new science' of the post-Renaissance era in which the deep and generative powers of material and organic being were increasingly recognised. That his solution was to identify God and the cosmos almost completely led to him being condemned, and shockingly executed, as a heretic by the Roman Catholic church: if confirmation were needed that the wrong metaphysics leads to tyranny, that terrible ecclesiastical execution at the dawn of the modern age provides it.[70]

While I reject pantheism, I agree with Keller that the metaphysical break between Asian religious conceptions of the agency of nature, or the Tao, and Western tendencies to eschew this independent agency are connected with the domination of First Cause thinking in Western theology. As Friedrich Schleiermacher was the first to note, after the Enlightenment, the problematic dominance of First Cause thinking in Western Christianity is manifest in this tradition's radical metaphysical break between the original divine acts of creation and the continuing creative processes which are evident in the capacities of the earth to bring forth life. As Wolfhart Pannenberg puts it, First Cause thinking in Western theology prefers a conception of nature as a 'steady state' and this conception shaped both the First Philosophy of René Descartes and the cosmology of Newtonian science until Einstein.[71]

Bruno and Spinoza both sought to repair this break by fully identifying God and the world. Schleiermacher's more nuanced answer was to argue for the idea of *creatio continua* – continuing creation – and so to collapse the distinction between *creatio ex nihilo* and what Reformed theologians call Providence. For Schleiermacher, all beings, nonhuman and human, are characterised by the deep ontotheological principle of 'absolute dependence' on divinity.[72] But this dependence is not a cause–effect process in which the original cause is *creatio ex nihilo* and subsequent dependence is merely the unfolding of the effects of the original divine creative acts in time. Instead, beings are constantly in flux across time and space in a process of evolution which is *continuously dependent* on divinity and is thereby given teleological and eschatological direction and purpose by the divine Source.

The analogy which Schleiermacher uses is religious experience. Those who have the 'feeling of absolute dependence' are those who learn spiritually to dwell in relation to God, or 'abide in Christ,' which is the language the New Testament uses. But this feeling is not a once-and-for-all experience but a *continuous* spiritual state, and by analogy the creation *and* preservation of the cosmos are also continuous and inseparable:

> If we think of the creation of the world as a single divine act and including the whole system of nature, then this conception may be a complete

> expression of the feeling of absolute dependence, so long as we do not conceive of that act as having ceased, and consequently imagine on the one side, in God, an alternation of activity and rest relatively to the world, and on the other side, in the world, an alternation between a determination of the whole through God and a determination of all single individuals through each other. In the same way if we regard Preservation as a continuous divine activity exerted on the whole course of the world, covering the first beginning no less than each subsequent state, then this is a complete expression of the self-consciousness in question, provided we do not think the origin of the world is conditioned by something else before and after that activity.[73]

Schleiermacher's *The Christian Faith* from which this statement comes is often compared to Aquinas' *Summa Theologia*,[74] and is arguably the most successful attempt by any theologian to bring together what Schleiermacher calls 'the interests of piety and science on the one hand and of morality on the other.'[75] The divine quality which for Schleiermacher, as for the New Testament and especially the Gospel of John, infuses all cosmic processes, and which is more fully revealed in the Incarnation of the *logos* in Jesus, is love. Love is the moral and spiritual quality which turns the continuous absolute dependence of being on God, which Schleiermacher describes, from a potential source of control into a source of empowering agency, and which sustains the *freedom* of beings to evolve by their own agency and in networks of agency and so reach towards cosmic harmony. It is this quality of the freedom of the creature in love which is undermined by the 'man of science' who in the post-human future announces the intention to control life even down to the level of biota, cells, and genes towards engineered technocratic purposes.

If Schleiermacher is the pioneer theologian in modernity to attempt a resolution of the metaphysical problem of the continuance of creation in Western Christianity, his repair was, for the most part, not taken up. Most in the Western Christian tradition remain firmly wedded, despite Earth System and evolutionary theories of the long and continuous processes through which life emerged, to First Cause thinking, and the deistic tendency to eschew the ongoing active agency of nature, Gaia, the *logos*, or the Tao in the very long history of life. Hence, despite Schleiermacher's important innovations, the sense that there is an intrinsic conflict between scientific accounts of being and Christian accounts endured into the present, and elementary biology lessons in schools in the West are therefore sources of the practical and philosophical atheisms which have led to the dismissal of God as divine Source and active agency of love for so many contemporary Western people.

The origin of the dominance of First Cause thinking can be traced back in Christian theology to the efforts of early Christian theologians to resist Gnosticism, and the Gnostic belief that material bodies are intrinsically imperfect and incapable of holding the divine light which is the truest nature of being in the cosmos. Against the Gnostics, the second-century theologian Irenaeus of Lyons argued that Jesus Christ as the divine 'offspring' and 'First-begotten Word' had descended to the creation in order that 'it should be contained by Him; and on the other hand, the creature should contain the Word, and ascend to Him, passing beyond the angels, and be made after the image and likeness of God.'[76] But by identifying divine implication in the cosmic *pleroma* (fullness) of being exclusively with the *logos* incarnate in Jesus, a historical figure who, while he was said to have uniquely Incarnated the divine in human form subsequently was said to have ascended to heaven to 'sit at the right hand of God,' Irenaeus set up a problem that endured in Western theology. If the fullness of being is ontologically secured by its new post-Christological relation to the divine nature via the 'First-begotten Word,' then the absence of the Word from Earth after the Ascension left the Earth ontotheologically insecure in the Western Christian schema.[77]

Some of the Eastern Fathers, most notably Basil of Caesarea, sought to put the doctrine of the Holy Spirit in place of the divine *logos* as the divine power who drew life from the primeval watery chaos at the beginning and who continues to sustain life in Being.[78] But given the early Christian tradition's identification of Jesus as the full Incarnation of the divine in human form, *and* as the primeval Creative divine power, the account of the Spirit or 'left hand of God' as the agent of *continuing* creation lacked consistent explanatory power for theologians in the West.

Eastern theologians more often resolved the God–World relational problem via the Vedic and Platonic idea already discussed of the correlation between the Microcosm of humanity and the Macrocosm of all being. Sixth-century theologian Maximus the Confessor developed this idea to its fullest when he argued that there are ontotheological correlations between immanent material and transcendent spiritual planes of reality, and that these are reflected in the diversity of all species, and the different divisions of beings from angels to insensible matter such as rocks.[79] For Maximus, the Creator so arranged relations between human persons as made in the divine image, and the diversity of matter and beings that exist in the world, that the creation as macrocosm finds its correlates in the microcosm of the human person, and each species, and each division of being – including the elements – finds correlates in human character traits, virtues, and vices, and the spiritual and physical 'humours' or types, displayed by rational souls in their inner lives and outward behaviours.[80]

Maximus' theology anticipates in some ways the humanism of the Renaissance, quantum mechanics, *and* the Gaia theory since it proposes that dividing lines between forms of being as material, intelligible, and spiritual are not hard and fast but instead in humanity these realms of being find a 'little cosmos' in which they interpenetrate one another. But Maximus retains an account of what theologians call the Fall. Created beings – both rational and non-rational – were originally created for ultimate union with God but since humanity's fall, the destiny of all creatures to return to the Source is frustrated. The metaphysical resolution of this problem Maximus finds in theological anthropology according to which the Incarnation, and especially the Transfiguration, of Jesus as the divine *logos*, renews humanity's capacity to draw all the different divisions of being back toward their destiny in God. This for Maximus is the call of the Saints whose core cosmic work is the perfection of their own natures, after the perfect human nature of Jesus Christ, so that through their spiritual life, and through the theurgic power of the liturgy and sacraments of the Church, the destiny of all created order is recapitulated.[81]

Maximus' theology anticipates the anthropic version of the Gaia theory developed by some of its critics, such as Tyrell as discussed above. But it differs from the Gaia theory markedly in that it has no account of the autonomy and freedom of created beings, and the Earth herself, apart from the Incarnate *logos* and humanity more broadly: the independent agency of the Earth in deep time *before* the Incarnation of the *logos* in generating and sustaining life finds no place in his schema. To find an account of the autonomous agency of nature in Christian theology across all time and space I turn instead, as does Carolyn Merchant, to the ninth-century Irish monk John Scotus Eriugena who developed a dialectic of two natures to resolve the metaphysical problem of the agency of nature (or what, above, I called the Tao and Gaia) by reconceiving of nature as *creating* and not only *created*.[82]

Eriugena lived for most of his life in the monastery in Laon, Northern France, and there wrote in Latin and developed a five-volume cosmological schema known as the *Periphyseon – Concerning Nature* – in which he developed a major new dialectic in Christian theology between *natura creans* – nature creating – and *natura creata* – nature created. For Augustine, nature was only ever *creans*, whereas God, and the three 'persons' of the divine Trinity, were *uncreated* and *creating*. Eriugena took Augustine's categories of created and creating, and under the influence of the Neoplatonists Plotinus and Pseudo-Dionysius, he applied them to Nature. As he put it in Book 1 of the *Periphyseon*:

> So when we hear that God makes all things we ought to understand that God is in all things, that He is the Essence of all things. For only He truly exists Himself, and He alone is everything which in the things that are is truly said to be.[83]

And again, in the Third Book:

> He is the causal Beginning of all those things, and the essential Middle which fulfils, and the End in which they are consummated and which brings to rest every motion and imposes tranquility.[84]

Eriugena's *Periphyseon* was widely read in medieval Europe as it was the most sophisticated treatment by a Christian philosopher of the status of the nonhuman world and nonhuman beings that had ever been written. Although it was condemned as a proscribed work by the Catholic Church in 1225 since it was said to identify God so completely with the nonhuman world as to be Pagan, or in contemporary parlance 'pantheist,' Eriugena's ideas were influential among key medieval figures including Aquinas and Bonaventure through whom they influenced early modern philosophers including especially Spinoza and Hegel.

Against the attempt of the Catholic church to proscribe Eriugena's ideas, there is actually overlap between Eriugena's account and that of the Apostle Paul in the New Testament concerning the relationship between the Christian God and Creation. Paul revised the Jewish *Tanakh*, in which as a former Pharisee he was very well versed, and adopted elements of Pagan thought to explain Christian belief in the God who was Incarnate in Jesus Christ to Greeks and Romans. In a Pauline speech reproduced in the Book of Acts to the Athenians at the Areopagus – a statue to the 'unknown God' – the following claims are made:

> The God who made the cosmos and everything in it, this one – being Lord of heaven and earth – does not dwell in sanctuaries made by hand; needing nothing, neither is he aided by human hands – he who gives life and breath to all things.
>
> For in him we live and move and are, as indeed some of the poets among you have said: "For we too are his offspring."[85]

In an analogous passage in the Letter to the Colossians, which is traditionally also said to be by Paul, Jesus, the 'Son' of God, is described as

> the image of the invisible God, firstborn of all creation, because in him were created all things, in the heavens and on the earth, the visible as well as the invisible (whether Thrones or Lordships or Archons or Powers); all things were created through him; and he is before all things, and all things hold together in him.[86]

In claiming that God is 'the beginning, middle and end of the created universe' Eriugena seems to be agreeing with Saint Paul. Further, Paul, unlike later theologians, also says that Jesus as the Anointed Son of God was the 'first born of creation' and hence created by God to be the one who 'gives life and breath to all things' and 'in whom all things hold together.' In so doing, Paul puts the figure whom John identifies as the *logos* in the same position relative to God as Plato's demiurge to God. And so it is unsurprising that a number of the Greek fathers, and most influentially Basil of Caesarea, explicitly use the Platonic demiurge in their accounts of creation.

If Eriugena was a 'pantheist' for restating these Pauline claims, albeit at length and in systematic treatises, it is reasonable to argue that the Apostle Paul was too. But nonetheless, passages such as the following led the church authorities to condemn Eriugena's work:

> There is a most general and common nature of all, created by the one First Principle of all; and from it, as from a very copious fountain, corporeal

> creatures, like rivulets, are channelled through hidden passages and break forth into the different forms of individual things. That force, coming forth through different seeds from the secret recesses of nature and first emerging in the seeds themselves, then mixed with the different fluids, bursts out into the individual, sensible species.[87]

In this passage, Eriugena comes closer than any of the Greeks, or previous Christian theologians, to explaining the relationship of the divine origination of all things – the 'First Principle' – to the Life force, and the continuous development of Nature in different material forms and varieties of living species since the beginning of time. In essence Eriugena's position is that the Earth is not autonomous from the Creator but intricately related to the Creator at every level of existence from the highest to the lowest.

The Quickening of Earth by the Divine Breath, Sophia or Wisdom

Although Eriugena is sometimes represented as a pantheist who was out of step with the general tenor of European Christian culture, most Christian, Jewish, and Muslim theologians in the Middle Ages believed as he did that the Earth and all living creatures are inspired by the divine breath or Spirit; moved by divine energy; pursuing goals and instincts which are guided by divinely originated Natural Law; and on the great cosmic journey of emanation and return to the One which Plotinus and the Neoplatonists, who influenced Augustine in the Roman West and Orthodox theologians such as Maximus the Confessor in Byzantium, proposed four centuries after Aristotle in a Vedic-inspired revision of Aristotle's schema.[88] There will be some Christians, however, who at this point will think that I at least have abandoned any attempt at orthodoxy in this book and have allowed Lovelock or Gaia to lead me into Paganism. For those readers, if they are still with me by this point, I think it is worth a short digression to look at further biblical evidence, in Genesis, the Book of Proverbs, and the Christian Gospels for the position Eriugena proposed.

Against the dominant Jewish and Christian reading of the Genesis creation account, a literal reading of the first account of creation in Genesis 1, while it indicates an external creator who produces Earth, and life on Earth, from an original primordial void and watery chaos, nonetheless also ascribes creative, life-bearing roles to *Earth* and the *sea*. The divine agent who sets the creation of life in motion, in Hebrew is *ruach* – variously translated wind or spirit. This divine wind or Spirit is said to have been sent by God, and is described, rather like a spaceship, as 'hovering' over the lifeless waters of a primordial earth (Genesis 1: 2). Subsequent to this quickening of the waters, the divine Word or command is given – 'and God said' – and the land is separated from the sea and sky. The same divine Word commands several succeeding events that lead to the emergence of life: 'let the Earth sprout' fertile plants and trees with seeds; let the waters 'bring forth' fishes and birds. God is said to create 'sea monsters' directly indicating perhaps an ancient sense of the closeness of whales and dolphins to humans in their mental and social capacities; and then the Earth 'brings forth' land-dwelling creatures including mammals, insects, and reptiles (Genesis 1: 7, 11–12, 20–21).[89]

This active agency of the Earth in the first biblical creation story in bringing forth life, though missed by most Jewish and Christian commentators, is closer to Vedic than Greek cosmogony. The Genesis account is clear that the primordial origin of all things is a 'God of gods' who sends forth a divine Spirit who imbues the primordial waters, and the land, with the powers to bring forth life in their turn. But the agency of the divine Spirit, and the living Earth, in bringing forth nonhuman life are not said to be responsible for humans. Instead, like the 'sea monsters,' 'man' is said to be directly made by God: 'And God created man in His image, in the image of God He created him: male and female He created them.' And God then put humans in charge as the divine representative or 'vice-regent' on earth to 'be fertile and increase, fill the earth and master it' (Genesis 1: 26). There are, however, two things to note here about who is doing the creating. The first is that the creator is plural and not singular, and the second is that the creator is male *and* female.

In an insightful essay on the divine feminine, Queen of Heaven, and Wisdom (in Greek *Sophia*), in the Old Testament, Margaret Barker argues from a range of textual and archaeological evidence, including Jeremiah, the first book of Enoch, Proverbs, and the Wisdom of Solomon, that the Israelites in Judah, and in exile in Egypt, worshipped male and female representations of God in the first Temple. Hence there were two cherubim in the Temple of Solomon, not one, and at the Exile, against Jeremiah, who blamed idolatry for their exile, the Israelites argued back that the reason the favour of God had departed from them, and the Temple in Jerusalem was destroyed, was because they had *stopped* worshipping 'with libations, incense and moulded loaves, the Queen of Heaven who protected Jerusalem and Judah.'[90]

Elsewhere in the Old Testament, the Queen of Heaven is spoken of as the feminine figure of Wisdom, and most clearly in Proverbs. As Barker notes, in Proverbs 8 Wisdom cries from the hilltops and outside the gates of the city whence she was banished by the foolish people within for reasons of power and wealth. The chapter goes on to speak of Wisdom as a feminine Being who was created before all things, and who, like Plato's *demiourgos*, was the craftsperson who was beside God in the making of heaven and earth and all that is in them.[91]

The feminine figure of Wisdom has been largely erased from scholarly commentaries on the *Tanakh* by modern Jewish and Christian exegetes. But the idea of the 'Queen of Heaven' as source of the fertility of the Earth, and hence of the capacities of the Earth as Mother of all Beings to generate Life, persisted into the late Middle Ages in Europe and acted as a cultural restraint on invasive activities such as deep mining.[92] It is arguable then that the idea of Earth as Mother, which was revived in official Vatican teaching by the Papal Encyclical *Laudato*

Si', represents a cultural and folk link between the Genesis account of a quickened Earth as Life Bearer, and Christian culture.

In a reconstruction of the history of the figure of Sophia in the Christian tradition, Michael Martin argues from a range of sources that the figure of Sophia did not disappear in early Christianity, and instead that Sophia represents a submerged metaphysics.[93] In support of this claim, the first textual recognition of Sophia in Christianity is the description of the child Jesus who is said to have been 'filled with wisdom' from an early age, and actually to *be* 'wisdom from God' (Luke 2: 40; I Cor. 1: 30). Wisdom is also referred to in the important second-century text *Apostolic Constitutions* in which God is described as the Father of the female figure of Wisdom.[94] But nonetheless, most Christian theologians from the second century preferred the Greek idea of *logos*, as identified with Christ in the Prologue of the Gospel of John, as the organising principle of life. Sophia lived on, as in First Temple Judaism, but more in popular folk imaginaries, for example, in the Medieval figure of the Fairy Queen in Britain than in official Western Christian theology.[95] But in Eastern Christendom what many consider the holiest of all Eastern Orthodox churches – Hagia Sophia in Istanbul which is now a mosque – is named after the figure of Sophia, as are many churches in Russia, Georgia, and the Ukraine.

Related to this Eastern Orthodox veneration of the divine Sophia, in the nineteenth and twentieth centuries the Russians Vladimir Solovyov and Sergei Bulgakov attempted a theological revival of sophiology, but their accounts never achieved widespread adoption. Martin's account of Sophia as a 'submerged metaphysics' is therefore accurate. The principal way in which it was submerged was through the identification of Jesus as not only the anointed one, the Messiah, or in Greek the Christ, but as a quasi-divine figure. From the New Testament on, Jesus is spoken of variously as the 'begotten,' and yet divine, Son of God who came from and returned to heaven; as divine Wisdom; as the *Logos* or divine Word who was with God at the beginning of all things; and as the divine means, within and beyond earthly history, of the renewal of all things a 'New Creation' remade, reconstituted, in a new and ultimate harmony of all Being with the divine source of all. In the Synoptic Gospels of Matthew Mark and Luke, as well as the more Greek-influenced Gospel of John, there are numerous pointers to the early Christian belief that in Jesus the divine power which had originated Life in the cosmos was uniquely present and available on Earth for a brief period, and especially as source of deliverance, healing, and renewal for humans, and that he was so recognised not only by some of the Jewish people and their teachers, but by angels, daemons, and a number of Gentiles.

There is not space to reprise all the passages which so indicate this but to those who know the Gospels, the numerous stories of healings – both physical and psychological – will be familiar. In addition, there are striking passages, such as the story of the Gerasene demoniac, in which the daemons recognise Jesus; there are many passages in which the people, the teachers in the Temple, some of the Pharisees, and even Roman centurions, legionnaires, and Pontius Pilate acknowledge the unique intellectual and spiritual powers of Jesus. For analogies with the Pauline passages about the cosmic Christ already cited, perhaps the most striking evidence in the New Testament is the short reference in Mark to the period of the temptation of Jesus in the wilderness when it is said Jesus 'was in the wilderness, tempted of Satan; and was with the wild beasts; and the angels ministered to him.' (Mark 1:13). As Richard Baukham argues, this is strongly suggestive that the animals knew him, befriended him, and ministered to him.[96] The other cosmically significant moment – the most significant of all in the Gospels – is the point at which Jesus is said to have stopped breathing during the crucifixion. At that time, the Gospels say that 'the veil of the temple was torn in two,' which symbolically indicates that the prophecy of Ezekiel had come true, and that God was now present to Life on Earth in a way that God was only said to be present, in the era of the First Temple, in the holy of holies.[97] And, as if to confirm this, the Earth is said at the death of Christ to have undergone both a brief eclipse and an earthquake. In the culture of the early Christians, no stronger symbolic references could be made than these to indicate that Jesus was not only a human being, but a divine figure, the Life principle, divine Wisdom, the divine Word or *Logos*, Incarnate in human form and that Earth *knew*.

There is also evidence that the Christians in the post-apostolic era believed that Jesus had inaugurated the promised eschatological restoration of relations of peace between humanity and other animals, as in the famous 'lion lying down with the lamb' passage of Isaiah 65. One significant text along these lines is the second-century text of Pseudo-Matthew, according to which lions and panthers accompanied the Holy Family in their flight into Egypt 'showing them the way, bowing their heads; and showing submission by wagging their tails' because 'they adored Him (Christ) with great reverence.'[98] The early theologians interpreted this and other texts as indicating that in Jesus the original peace and harmony that there was in Eden between humans and other kind would be progressively restored after the Ascension of Jesus, and the most extensive discussion of this claim in the first Christian millennium is in the writings of Isaac of Nineveh, also known as Isaac the Syrian.[99]

There are also numerous accounts of saints developing remarkable companionable relationships with wolves, birds, and other animals. The two best-known stories are those of Saint Jerome removing a thorn from a lion's paw, and of Saint Francis befriending a wolf.[100] In both cases, animals conventionally associated with predation lose their predatory characteristics, while also losing their fear of humans. Instead, they become naively familiar with humans, just as the animals who are described as coming before Adam to be named by him in the first creation story in Genesis. Both stories are also suggestive that there is an eschatological telos in Judaism and Christianity, as Walter Brueggemann argues, towards a progressive extension of the 'circle of neighborliness' to a growing range of other beings whom, in the course of Christian history, come to include other kind.[101] In similar vein, that most eschatological of modern Protestant theologians, Karl Barth, argues that the progressive realisation of Christian redemption as

FIGURE 7.2 *Saint Jerome and the Lion*, Fra Filippo Lippi and Workshop, 1455–60, National Gallery, London. Author photo.

a distinctive form of human ethical life would tend towards vegetarianism over time since the original pre-fallen creaturely ethic of Adam and Eve in Paradise is said to have been vegetarian and hence one in which there was no killing:[102] although Barth himself was a joyful meat eater!

A striking feature of many of the narratives of the saints rediscovering peace with other animals is that it is not so much human as *animal* agency which is the key to these new relationships: animals are drawn to the saints because in their humility and poverty they radiate compassion towards all creation, and creatures respond in kind by losing the fear of *Homo sapiens* said to have begun with the fall of Adam and Eve. The restoration of peace is reciprocal and involves mutual compassionate recognition. This progressive unfolding, from the life of Jesus through the lives of the Saints, of an extension of the circle of neighbourliness to include other animals, also admits of the agency of other creatures in completing the eschatological extension of the circle of compassion exemplified in Jesus' Parable of the 'Good Samaritan' to include all creatures.

It is also reasonable to make a connection between this extension of the circle of Christian compassion to include other animals and the contemporary great leap forward in primatology in which chimpanzees, gorillas, and other apes are increasingly recognised by scientists as being much closer to humans in a remarkable range of cognitive, communicative, and empathic behaviours. As discussed above, the first contemporary scientist to go into the field and begin making observations of primates based upon a lack of fear, and an expectation of empathy, was Goodall, who is a lifelong Christian. [103] In the title as well as the narrative of her autobiography, Goodall describes how her Christian faith is a mainstay of her compassionate regard for primates and other creatures, and her charismatic and extensive campaigning efforts – in the movements she has founded – to spread compassion and care for primates, and other kind, around the Earth.

The evidence many scholars have assembled in recent years of distinctive Christian attitudes to other kind does not gainsay however another strain – and it is fair to say, the majority view – in the Christian tradition and Christian history according to which animals, and creation as a whole, are created by God instrumentally to serve humanity rather than as deserving of independent moral regard as beings who are also caught up in, and sometimes agents of, the story of salvation. I reckon the root of this belief is to be found in the apostolic argument in the New Testament between the apostles Peter and Paul. Peter intended that Gentile Christians should adopt Jewish ritual laws: this would have included the circumcision of Christian males before baptism, and their adoption of Jewish dietary practices as prescribed in Leviticus including proscription of eating of certain creatures – particularly pig and reptile – and the requirement that all animals and birds that are to be eaten should be slaughtered in such a way that the blood is ritually drained and returned to the earth which reflects the belief that blood is sacred because animated by the divine Spirit and therefore must be returned to the ground because 'the Earth is the Lord's' (Psalm 24: 1). Paul resisted the idea that Gentile Christians need adopt Jewish dietary and ritual laws and this became a major contest in the churches Paul founded, including those at Corinth, Galatia, and Ephesus. The conflict came to a head in Jerusalem at what was the first Apostolic Council where Peter softened his stance and agreed an 'ecumenical accord' which indicated that Gentile Christians would be purified 'by faith' and therefore had no need of circumcision.

As for dietary laws, the judgement, which was written down in a letter that Judas and Silas were to take in person from Jerusalem to the Gentile churches, was to the effect that Gentile Christians should not be bound by ritual and dietary laws that some from Jerusalem had attempted to impose upon them but, being faithful to the name of the Lord Jesus Christ they need only 'abstain from meats offered to idols, and from blood, and from things strangled, and from fornication: from which if you keep yourselves, you shall do well.' (Acts 15: 29). Paul suggests to the Christians at Corinth that such dietary liberty may even be extended to meat offered to idols provided that it is not consumed in the meetings of the Christians for worship, and provided that their Christian neighbours are not offended by this liberty: but if this liberty does give offence, Paul suggests it is better to abstain from eating meat (1 Cor.: 7–8).[104]

Jewish dietary law rests upon what many modern liberal exegetes see as a 'primitive,' quasi-magical, belief in the sacredness of the life blood of animals. In the Tanakh, it is clear that in Paradise there was no killing; violence, including meat eating, is said to have caused the primeval Flood: and *after* the flood meat eating is permitted by an apparently reluctant Creator in a new covenant with Noah because of the weakness of men and women. But this permission is only given with a number of strict regulations and prohibitions. The most important of these is that the life blood – which is described in Torah as 'inspired by the breath of the Spirit' since the same Hebrew word *nephesh* is used of blood and spirit – is returned to the earth 'which is the Lord's.' Returning the blood to the ground, and sacrificing a portion of every animal and bird eaten – first at the Ark of the Covenant and later in Solomon's Temple – were therefore requirements to give respect to the creator spirit who breathes through all being. And when the Israelites settled in cities in the Land, butchering animals – because of the moral and polluting dangers of the act of killing – was to be conducted by only one group or tribe – the Levites – who were to live separately from the other tribes so as not to infect them with the potential evils, moral and physical, of killing and slaughter. The Jewish apostles, it is reasonable to argue, maintained much of this ritual understanding of the significance of blood and of the theological and ethical rationale for the Jewish proscription on eating blood. And therefore, it was the sole dietary requirement they imposed upon the Gentiles. But while a minority of Gentile Christians from the first century abstained from meat eating altogether because of its potentially profane associations, vegetarianism was more often associated with Gnosticism than with orthodox Christianity, and most Christians also did not follow the dietary rule of Acts 15.

The New Testament proscription on eating blood is a critical link with the sacred view of life that was sustained – however complexly – in the dietary rules of the Israelites, and subsequent Jewish culture. The loss of this complex of rules means that Christians – uniquely among those of the Abrahamic faith – do not have any formal ritual dietary laws except those governing the Eucharist. It is even possible, though the evidence chain is too long and multifarious for definitive proof, that this helps to explain the distinctively instrumental relations with other animals which developed in medieval Christianity, and ultimately into the secularised forms and practices of modern industrial capitalism, though as I argued above, these attitudes only came to the fore under the influence of scholastic rationalism and nominalism in the late Middle Ages.[105] Hence, from the late Middle Ages there are significant non-biblical sources for an instrumental view of creatures, of which Thomas Aquinas is the most influential in the Western tradition. His instrumental view is illustrated in his argument in the *Summa Theologiae* that the reason that cruelty towards other animals is morally problematic is because of the effects it has on humans and not on the creatures themselves.[106] But there is now a significant and growing effort to revise this tradition, among several Protestant theologians, some of whom are referenced above, and now in Roman Catholicism.

In his encyclical *Laudato Si'*, the Argentinian Pope Francis challenges the mainstream Roman Catholic tradition for having a degree of instrumentalism towards other creatures. After his namesake, Saint Francis of Assisi, Francis indicates that this tradition ought now to be transcended and set aside. In so doing,

Francis has arguably begun to embrace what McLaughlin and others call the minority tradition of compassion to other animals, apart from their service of and uses to humans. In line with this minority tradition, Pope Francis effectively channels Saint Francis in his 'naive' relations with other animals when he indicates in *Laudato Si'* that all creatures, and not only persons, have 'intrinsic value' and that all creatures are redirected by the Christ events toward the 'end of time' when 'the Son will deliver all things to the Father' and thus 'the risen one is mysteriously holding them to himself and directing them towards fullness as their end.'[107] In this way Francis indicates that creatures, and not only persons, are part of the eschatological transformation of all things towards their future consummation: and that this salvific status of creatures is the theological ground for Catholics to honour the divinely given worth of animals, and also by extension of plants, rivers, forests, oceans, and the atmosphere.

From Gregory Palamas to Post-modern Science

If the Western Christian tradition in the second millennium gradually lost the belief in the *logos* as the life principle within all cosmic and material relations, the case is often made that Eastern Orthodox Christianity did not follow the West in this regard. Byzantium was notably also more resistant to the scientific reductionism of Bacon and Descartes. And most Orthodox philosophers also reject the European Enlightenment view that modern science requires a new theory of knowledge, and a new ethics, in which the nonhuman is strictly demarcated from the human. The core question that Eriugena raised in the West was how creatures participate in the divine nature. If God is ultimately One, as Christians believe, and if the *logos* was in Christ who is now ascended and has returned to the Father, then what is the continuous means of the participation of creatures in divinity which animates and upholds them?

The Orthodox theologian who came up with a unique answer to this question, and who, unlike Eriugena, was not accused of pantheism, was Gregory Palamas. Palamas, more than any early modern Christian theologian, anticipated the cosmological implications of what some call 'post-modern science,' including the Gaia theory, quantum mechanics, and Bergsonianism as in the work of Rupert Sheldrake. I therefore conclude this chapter with a brief consideration of Palamas' teaching and of parallels between that teaching and post-modern science.

Palamas was a monk for most of his life after 1320 at the famous Greek monastery on Mount Athos. Palamas practised the distinctively Eastern Orthodox approach to the contemplative life which involve ascetic practices of prayer, and especially the prayer of the heart which is a form of continuous inner praying with the Jesus Prayer, in combination with what is known as *nepsis* which, as Constantinos Athanasopoulos explains, is 'a spiritual and bodily process of cleansing the passions and transforming them, so that a meaningful compunction is achieved, with endless tears and love for Christ.'[108] The ultimate fruit of the contemplative life is what the Eastern church knows as 'deification,' in which the one who prays is gradually transformed so as to become closer in being to the divine nature.

Palamas realised that deification raises an issue concerning the being of God. If humans can participate in the divine nature, this seems to suggest the divine nature is many and not one, and therefore verges on pantheism. In order to safeguard belief in the Oneness and transcendence of God, Palamas appealed to the fifth-century Cappadocian theologian Basil the Great when he said that created beings participate in the divine names, including divine goodness, light, life, and wisdom, which exist 'around God.'[109]

To explain the difference between God's being as One, and the divine influences or forces that issue forth from God into the cosmos and creaturely beings, Palamas introduced a new philosophical distinction between divine *energeias* and the divine *essence*. To explain this distinction, Palamas uses the analogy of the sun and the sun's rays: 'For one applies the word "sun" to the rays as well as to the source of the rays; yet it does not follow that there are two suns.'[110] According to the distinction between the divine essence and divine *energeias*, the participation of created beings in God, and especially the heightened participation of contemplatives who are on the way to deification, does not mean that God is not One. Rather the energies which animate all beings in the cosmos are divine in origin and in potentiality, but not *equivalent* to the transcendent divine Source of these energies, who is the One God.[111]

Palamas' valuable distinction between *energeias* and *essence*, just because it was novel, led to a considerable theological dispute which became known as the Hesychast controversy in which ultimately Palamas prevailed against those who accused him of heterodoxy. At issue was not only the distinction but the claim that Palamas was able to make with it which is that when the mystic sees the divine light, the light she sees is truly divine light. Palamas' opponents argued the experience of illumination in prayer was symbolic and not a real phenomenal perception of the divine light of God. The dispute recurred centuries later, albeit in secular language, in the rise of phenomenology in Western philosophy as Maurice Merleau-Ponty and others rejected the rationalist orthodoxy of post-Kantian philosophy by reconnecting mental perceptions with bodily sensory experiences (Figure 7.3).

Palamas was the first philosopher in either Eastern or Western Christianity to explicitly challenge the mentality–materiality distinction that became such a strong feature of Christian philosophy from Augustine onwards, and which in post-Kantian philosophy and mechanistic science became the humanist orthodoxy of modernity according to which humans are the only conscious and intelligently purposive beings. Modern humanism leaves material and organic being beyond the human available for re-engineering by humans since it is said to lack intelligent purposes of its own.

Despite the claims of his opponents, and many since who resist him to this day, Palamas' position is well rooted in the Christian ascetic tradition. From the third

FIGURE 7.3 Icon of Gregory Palamas, Mount Athos. Photo made available under Creative Commons CCO 0.1 by Monastery ofVatopaidi, Mount Athos, Greece.

century onwards there are well-attested traditions among the desert fathers, who were the originators of Christian monasticism, that bodily healing, life extension, and new harmonious relations with other creatures were fruits of the ascetic life among the desert fathers.[112] The traditions of companionship between the desert saints and wild animals earlier discussed, and of exceptional plant growth in their desert gardens, are also strong evidences of this belief which endured into the second Christian millennium among Orthodox monks. In this tradition, the mystical life is the way that leads to a restored creation, and the recovery of the material and organic relations that existed in Paradise. The endurance of this belief in the East is powerfully represented not only in Palamism and Eastern Orthodox monasticism more broadly but in the art and symbolism of Eastern Orthodox churches.[113] And it lives on today in Eastern Orthodox liturgy and theology, and in distinctive theologically shaped cultures of food, family, and other practices in Eastern Christian lands.[114]

Western theologians have in the main continued to hold that Palamas was heterodox in what they allege was a speculative innovation in theology. And they also resist Palamas in his claim that divine illumination is a genuine bodily as well as intellectual and spiritual participation in divine being and can influence and manifest in bodily phenomena such as physical healing and life extension.[115] This is indicative of a deeper cosmological division between Western and Eastern Christianity. In the West from the tenth century onwards there is a gradual resiling from the early cosmological claims of Saint Paul, and later Patristic teaching, that there is a continuous and ongoing divine animation and illumination of created beings which gives to them capacities of agency and relational powers to engage with other beings, and which evidence divine energies active within created beings. Instead, in the West, as we have seen, the divine influence in creation is said to primarily reside in the divine as First Cause having imbued certain law-like qualities into created beings which then develop and are generated in a quasi-mechanistic fashion and without continuous and divine influence. This is why Eriugena was ultimately condemned in the West. But for Palamas, divine influence, or potentiality in created beings, is a continuous process, as it is for Schleiermacher, and not primarily something which only occurred at the origin of the cosmos. Just as there would be no life on earth without the continuous and energising rays of the sun, for Palamas there would be no life in the cosmos, and hence no *sun*, without the energies which issue from the invisible God, who exists beyond what an English mystic called the 'cloud of unknowing.'

Unlike Eriugena, who never precisely distinguished between the divine *forces* which animate life and the divine *Source*, Palamas in Orthodox Europe developed a cosmological theory of the inherence of divine energies in matter and in all living things which represents a significant speculative advance on predecessor patristic and medieval theologians. Palamism represents an early modern development in Christian theology which takes a different path from Baconianism and understands the cosmos as divinely infused, but not in a static way as so much of Christian creation theology tends to do. Instead, Palamism anticipates the tradition recovered by Friedrich Schleiermacher of *creatio continua*, and aspects of the Process Theology of Whitehead. In this science-informed but still deeply metaphysical cosmology, the ongoing and evolving processes of creation are manifest in the autonomous and goal-oriented agencies of the more-than-human world which are divine in origin and yet expressed in the freedom of creatures to pursue their own goals and purposes, and in relations with others to contribute to the beauty and harmony of the larger created and cosmic order. That all creatures, and not only humans, participate as agents in the divine process of continuous creation also finds testimony in another key concept in Orthodox thought, namely that of the *logoi* which are said to animate all creatures, and in *mystagogia*, as Orthodox theologian Tamara Grdzelidze lucidly expounds in the thought of Maximus the Confessor and the Orthodox tradition more broadly.[116]

In sum, Palamism, and analogous cosmological perspectives in Eastern religious traditions, anticipate the findings of what some call post-modern science in the form of the Gaia theory, the Bergsonian concept of creative evolution, and the related theory of morphic resonance as developed by Sheldrake and quantum mechanics. These scientific approaches represent the gradual modern scientific awakening, first made explicitly by Einstein, to the fundamental role of *light* in the cosmos which Einstein discovered performed like particles in interaction with other atoms and electrons.[117] Einstein himself speculated that light at the quantum level represents a measurable field of divine influence on the constitution, development, movement, and interaction of earthly and heavenly bodies, material objects, *and* minds. But there was disagreement on how the particles that constitute physical reality did this at the quantum level between Einstein and Niels Bohr. Bohr proposed that all particles, including photons, are entangled at the quantum level and that this explains how subatomic particles are mysteriously linked even when separated by space and time. But Einstein called this 'spooky' and resisted Bohr's account by arguing that Quantum reality only exists when it is observed.[118] In 1964 physicist John Bell finally confirmed Bohr's view that at the quantum level the inexplicable or 'spooky' interactions that scientists detect and yet cannot predict – Heisenberg's Uncertainty Principle – do really exist in the quantum realm, that these interactions can be demonstrated in a repeatable experiment, and Bell's resolution of the problem became known as 'Bell's theorem'.[119]

Quantum mechanics revolutionised modern physics and it led to significant developments in atomic energy, digital technologies, and the study of what used to be called 'psychic phenomena' such as Extra Sensory Perception and Remote Viewing. That interactions between minds and particles across time and space - quantum entanglement – are real and yet still not really understood was underwritten in a further breakthrough experiment by Freedman and Clauser in 1972.[120] In the last five years Quantum entanglements are also being discovered in the behaviours and evolution of biological organisms as well as in the behaviours of photons and electrons. It has been demonstrated for example that birds navigate through chemical responses in their nervous systems to the Earth's magnetic field which are physical evidence of Quantum entanglement.[121] Quantum entanglement is evident in chemical and nervous system interactions with molecules that are manifest in the human and other animal sense of smell and which are responses to *vibrating* bonds between chemical molecules analogous to musical vibrations.[122] Quantum entanglement is also now being discerned as crucial to embryonic and morphological development such as the change of a tadpole into a frog. Quantum entanglements of these kinds underlie relationships between *all* beings across time and space, and it is likely in the future that quantum entanglement will eventually explain how single and multi-cellular organisms over 500 million years generated, and now sustain, a Gaian environment which is conducive to ever more diverse forms of life, including conscious humans. Quantum entanglement and the Gaia theory are modern scientific ways of describing interconnections between all beings across time and space that were first described thousands of years ago in religious beliefs: as the Tao, as the world soul, as the animating divine Spirit, and as *energeias*.

In the light of the Gaia theory and Quantum mechanics the current array of ecological crises may be said to arise from the closing down, misdirection, or obstruction of Quantum entanglements that, because modern humans cannot control or predict them, they disregard as mere superstition. The closing of the Western mind and of modern science since Bacon to divine cosmic *illumination* and influence is on this account the critical metaphysical root of the refusal of the autonomous, free, and purposive agencies, visible and invisible, that constitute the more-than-human world. This refusal is accompanied by the ongoing imperial and industrial assault on the complexly diverse, free, and autonomous agencies of the vast multitude of other beings with whom our indigenous ancestors shared the Earth. The resolution of the ecological crisis therefore requires a recovered ontotheological metaphysic of the divine energies and entanglements which animate and connect all beings across time and space. Such a recovery will, if it acquires cultural force, restore to other beings their seemingly autonomous, divinely inspired, and relationally entangled agencies in making Gaia habitable for them and us. In the final chapter I describe four cases of Earth–human relations which exemplify this approach.

Notes

1 Alan Watts, *The Way of Zen*. E-Edition, New York, NY, Vintage Books, 1989, 67.
2 Syyed Hossein Nasr, *Religion and the Order of Nature*. Oxford, Oxford University Press, 1996, 23.
3 On secularisation and the Reformation see Brad S. Gregory, *The Unintended Reformation*. Cambridge, MA, Harvard University Press, 2012, 121–8.
4 According to Etienne Borne '(P)ractical atheism is not the denial of the existence of God, but complete godlessness of action': Etienne Borne, *Atheism*. New York, NY, Hawthorn Books, 1961, 35.
5 Nasr, *Religion*, 179.
6 Nasr, *Religion*, 177.
7 Catherine Albanese, *Nature Religion in America: From the Algonkian Indians to the New Age*. Chicago, IL, University of Chicago Press, 1991.
8 See further Klaus Schwab, *The Fourth Industrial Revolution*. London, Penguin Books, 2017.
9 Max Horkeimer and Theodore W. Adorno, *Dialectic of the Enlightenment*. Trans. E. Jephcott, Stanford, CA, Stanford University Press, 2002, 3.
10 Horkeimer and Adorno, *Dialectic of the Enlightenment*, 3–4.
11 Horkeimer and Adorno, *Dialectic of the Enlightenment*, 6–7.
12 Lynn White Jr., 'The historical roots of our ecologic crisis.' *Science* 155, 1967, 1203-7.
13 Seyyed Hossein Nasr, *The Encounter of Man and Nature: The Spiritual Crisis of Modern Man*. London, George Allen & Unwin, 1968.
14 Spike Bucklow, *The Alchemy of Paint: Art, Science and Secrets from the Middle Ages*. London, Marion Boyars, 2009.
15 C.S. Lewis, *The Abolition of Man: Reflections on education with special reference to the teaching of English in the upper forms of schools*, Oxford, Oxford University Press, 1943, 79.
16 Bertrand Russell, *The Impact of Science on Society*. New York, NY, Simon & Schuster, 1953, 1–2.
17 See the account of Aristotelian causation, and the implications of its demise, in Bucklow, *Alchemy of Paint*. E-Edition, 124–6.
18 I first encountered the dehumanising impacts of industrial management theories and practices in my doctoral project on urban industrial mission in chemical and

steel works in North East England: it was later published as Michael S. Northcott, *The Church and Secularisation: Urban Industrial Mission in Northeast England*. Frankfurt, Peter Lang, 1990.

19 Russell, *Impact of Science*, 17.

20 Russell, *Impact of Science*, 49.

21 Russell, *Impact of Science*, 29.

22 See further Matias Desmet, *The Psychology of Totalitarianism*. Trans. Els Vanbrabant, White River Junction, VT, Chelsea Green Publ. 2022.

23 John D. Salisbury, 'Animals as Humans.' In Salisbury ed. *The Beast Within: Animals in the Middle Ages*. London, Routledge, 2010, chapter 5.

24 Jason T. Eberl, 'Pomponazzi and Aquinas on the Intellective Soul.' *The Modern Schoolmen* 83 (2005) 65–77.

25 Richard Sorabji, *Animal Minds and Human Morals: The Origins of the Western Debate*. Ithaca, NY, Cornell University Press, 1993, 64–86, 113.

26 Louis Dupre, *Passage to Modernity: An Essay in the Hermeneutics of Nature*. New Haven, CT, Yale University Press, 1993, 40–81.

27 Lovejoy, *Great Chain of Being*; on *creatio continua* see David W. Congdon, '*Creatio continua Ex Electione:* A Post-Barthian Revision of *Creatio Ex Nihilo*.' *Koinonia* XXII (2010) 33–53.

28 Jan G. Van Der Watt, *Family of the King: Dynamics of Metaphor in the Gospel According to John*. Leiden, Brill, 2000, 25–55.

29 Keith Thomas, *Man and the Natural World: Changing Attitudes in England*. London, Penguin, 1991; on the Greek origins of the Western rationalist turn against animals see Richard Sorabji, *Animal Minds and Human Morals: The Origins of the Western Debate*. Ithaca, NY, Cornell University Press, 2018.

30 Nikola Tesla, *My Inventions and Other Writings*. Mineola, NY, Dover Publications, 2016.

31 Jane Goodall, *In the Shadow of Man*. New York, NY, Houghton Mifflin, 1971; Dian Fossey, *Gorillas in the Mist*. New York, NY, Houghton Mifflin, 1983; on the philosophical implications of these developments see Frans de Waal, *Good Natured: The Origins of Right and Wrong in Humans and Other Animals*. Cambridge, MA, Harvard University Press, 1996.

32 Alasdair MacIntyre, *Dependent Rational Animals: Why Human Beings Need the Virtues*. London, Duckworth, 1999, 6–8.

33 See further Michael S. Northcott, 'Do Dolphins Carry the Cross: Biological Moral Realism and Theological Ethics.' *New Blackfriars* 84 (2003) 540–53.

34 Audrey R. Topping, 'Ecological Roulette: Damming the Yangtze.' *Foreign Affairs* 75 (1995) 132–9.

35 Samuel T. Turvey, Robert L. Pitman, Barbara L. Taylor, et al., 'First Human-Caused Extinction of a Cetacean Species?' *Biology Letters* 3 (2007) 537–40.

36 Lao Tzu, *Tao Teh Ching*. Trans. John C. Wu, New York, NY, Barnes and Noble, 2017, 3.

37 James Miller, *The Way of Highest Clarity: Nature, Vision and Revelation in Medieval China*. Magdalena, NM, Three Pines Press, 2008, 17.

38 James Miller, 'China: Landscapes, Cultures, Ecologies, Religions.' In Willis Jenkins, Mary Evelyn Tucker, and John Grim eds. *Routledge Handbook of Religion and Ecology*. London, Routledge, 2017, 184–5.

39 Miller, 'China,' 185.

40 Miller, 'China,' 185.

41 For a first hand account of environmental issues in contemporary China see R. Edward Grumbine, *Where the Dragon Meets the Angry River: Nature and Power in the People's Republic of China*. Washington DC, Island Press, 2010.

42 See further the collected essays in James Miller, Dam Smyer Yu, and Peter van der Veer, eds., *Religion and Ecological Sustainability in China*. London, Routledge, 2014.

43 Lewis, *Abolition of Man*, 21.

44 Lewis, *Abolition of Man*, 74.
45 I give an extended account of the ecological potential of the Greek and Christian tradition of natural law thinking in Northcott, *Environment and Christian Ethics* though I now think that that tradition had problems of its own, as discussed by Carolyn Merchant in her *Autonomous Nature: Problems of Prediction and Control from Ancient Times to the Scientific Revolution*. London, Routledge, 2016, 52–112.
46 Thomas Merton, with an Introduction by George Woodcock, *Thoughts on the East*. Tunbridge Wells, Burns & Oates, 1996, 22–3.
47 Merton, *Thoughts on the East*, 23.
48 Merton, *Thoughts on the East*, 32.
49 Alan Watts, *The Way of Zen*, 10–12.
50 Watts, *The Way of Zen*, 16.
51 Watts, *The Way of Zen*, 18.
52 Watts, *The Way of Zen*, 20–21.
53 Watts, *The Way of Zen*, 25–6.
54 Lewis, *Abolition of Man*, 79.
55 *Bhagavad Gita* 20, in John D. Smith, *The Māhabhārata: An Abridged Translation*. London, Penguin Books, 2009, 356.
56 Fred B. Eiseman Jr., *Bali Sejaka & Niskala: Essays on Religion, Ritual, and Art*. Singapore, Tuttle Publishing, 1990, 23.
57 Jan Hendrik Peters and Wisnu Wardana, *Tri Hata Karana: The Spirit of Bali*. Jakarta, Kepustakaan Populer Gramedia, 2013.
58 Michael S. Northcott, 'Rubbish, Recycling and Religion: Indonesia's Plastic Waste Crisis and the Case of Rumah Kompos, Ubud, Bali.' *International Journal of Interreligious and Intercultural Studies* 3 (2020) 1–19.
59 Raimon Panikkar, *The Rhythm of Being: The Unbroken Trinity: The Gifford Lectures of Edinburgh University*. Maryknoll, NY, Orbis Books, 2010, page numbers in E-edition.
60 Panikkar, *Rhythm of Being*, 1676–7.
61 Panikkar, *Rhythm of Being*, 1679.
62 On the theological and spiritual significance of the pervasiveness of music in nature and in human cultures, and as reflected in the music of J.S. Bach, see Christoph Schwoebel, 'Mutual Resonances: Remarks on the Relationship between Music and Theology.' *International Journal for the Study of the Christian Church* 20 (2020) 8–22.
63 See further my discussion of ethnomusicology in Michael S. Northcott, 'Anthropogenic Climate Change and the Truthfulness of Trees.' In Sigurd Bergmann and Dieter Gerten eds. *Religion and Dangerous Environmental Change: Transdisciplinary Perspectives on the Ethics of Climate and Sustainability*. Berlin, LIT Verlag, 2010, 104–12.
64 William Desmond, *God and the Between*. London, Routledge, 2008, 1–2.
65 William Desmond, *The Intimate Strangeness of Being: Metaphysics After Dialectic*. Washington, DC, Catholic University of America Press, 2012, 92.
66 Paul Tillich adopted the ontological metaphor 'ground of being' to describe the 'permanent creativity of God in everything that is' in Paul Tillich, *Biblical Religion and the Search for Ultimate Reality*. Chicago, IL, University of Chicago Press, 1955, 74.
67 James Lovelock with Brian Appleyard, *Novacene: The Coming Age of Hyperintelligence*. Cambridge MA, MIT Press, 2020, 29.
68 Catherine Keller, *Face of the Deep: A Theology of Becoming*. New York, NY, Routledge, 2003.
69 Keller, *Face of the Deep*, 181.
70 For a fine summary of Giordano Bruno's cosmology see Merchant, *Autonomous Nature*, 74–5.
71 Wolfhart Pannenberg, 'The Doctrine of Creation and Modern Science.' *Zygon* 23 (1988) 3–21.
72 Friedrich Schleiermacher, *The Christian Faith*. 3rd Edition, Trans. Various, London, Bloomsbury, 2016, 147–55.

73 Schleiermacher, *Christian Faith*, 147.
74 Brian A. Gerrish, 'Foreword to the Second Edition,' Schleiermacher, *Christian Faith*, xiv.
75 Schleiermacher, *Christian Faith*, 193.
76 Irenaeus of Lyons, *Against Heresies*, V. 36. In *Ante-Nicene Fathers, Vol. 9.* Trans. Alexander Roberts and William Rambaut, Ed. Alexander Roberts, James Donaldson, and A. Cleveland Coxe, Buffalo, NY, Christian Literature Publishing Co., 1885, revised and edited for New Advent by Kevin Knight, http://www.newadvent.org/fathers/0103.htm accessed July 7, 2020.
77 For an insightful discussion of this cosmological problem see Douglas Farrow, *Ascension and Ecclesia: On the Significance of the Doctrine of the Ascension for Ecclesiology and Christian Cosmology.* Edinburgh, T&T Clark, 1999, 41–3.
78 On Maximus see Andrew Louth, *Maximus the Confessor.* London, Routledge, 1996.
79 See further Torstein Theodor Tollefsen, *The Christocentric Cosmology of Maximus the Confessor.* Oxford, Oxford University Press, 96–107.
80 On the Vedic and Greek origins of the microcosm-macrocosm conception see Alex Wayman, 'The Human Body as Microcosm in India, Greek Cosmology, and Sixteenth-Century Europe.' *History of Religions* 22 (1982) 172–90.
81 I rely in this brief account on Louth, '5. Cosmic Theology.' In Louth *Maximus*, 61–74.
82 Merchant, *Autonomous Nature*, 46–7.
83 John Scotus Eriugena, *Periphyseon* (*De Divisione Naturae*), Book 1 Ch. 72, 209 cited Merchant, *Autonomous Nature*, 58, note 11.
84 Eriugena, *Periphyseon*, Book III, 621–2a cited Merchant, *Autonomous Nature*, 59, note 12.
85 Paul, 'Speech at the Areopagus' in *Acts of the Apostles* 17: 24–25 and 28 in Hart, *New Testament*, 260.
86 Colossians 1: 15–17 in Hart, *New Testament*, 398.
87 Eriugena, *Periphyseon*, Book 4, 19 cited Merchant, *Autonomous Nature*, 58.X
88 On Plotinus and the Neoplatonist predecessors, and their influence on Eriugena see Merchant, *Autonomous Nature*, 33–7, 47–8; that Plotinus was influenced by the Vedas is evident in the proximity of his thought to the Vedas, and that Vedic texts and ideas grew in influence in the Mediterranean as trade routes were strengthened in the first centuries of the Christian era; see further Daniel Regnier, 'One and the Possibility of Many in Greek and Indian Philosophy: Plotinus and Ramanuja.' *Philosophy East and West* 67, (2017) 825-840.
89 Quotations are from *The Jewish Study Bible: Tanakh Translation*. Eds. Adele Berlin and Marc Zvi Brettler, Oxford, Oxford University Press, 2004.
90 Margaret Barker, 'Wisdom: The Queen of Heaven.' *Scottish Journal of Theology* 55 (2002) 141–59.
91 Barker, 'Wisdom,' 149.
92 Carolyn Merchant, *The Death of Nature: Women. Ecology and the Scientific Revolution.* San Francisco CA, Harper & Row, 1980.
93 Michael Martin, *The Submerged Reality: Sophiology and the Turn to a Poetic Metaphysics.* Kettering, OH, Angelico Press, 2015.
94 *The Apostolic Constitutions*, 7.35.10 cited Barker, 'Wisdom,' 146.
95 Ronald Hutton, 'The Fairy Queen.' In Hutton, *Queens of the Wild: Pagan Goddesses in Christian Europe: An Investigation.* New Haven CT, Yale University Press, 2022, 75 – 109.
96 Richard Bauckham, 'Jesus and Animals II: What Did He Practice?' In Andrew Linzey and Dorothy Yamamato eds. *Animals on the Agenda.* London, SCM Press, 1998, 49–60.
97 On this reading of Ezekiel see further Michael S Northcott, *Place, Ecology and the Sacred: The Moral Geography of Sustainable Communities.* London, Bloomsbury, 2015, 18.

98 Pseudo-Matthew 18, quoted in Ryan Patrick McLaughlin, *Christian Theology and the Status of Animals: The Dominant Tradition and Its Alternatives.* New York, NY, Macmillan, 2014, 72.
99 For a fuller discussion of Isaac of Nineveh's eschatological theology of peace between humans and other animals see McLaughlin, 72–3.
100 McLaughlin, *Christian Theology and the Status of Animals,* 71.
101 Walter Brueggemann, *Reverberations of the Faith.* Louisville, KY, Westminster John Knox Press, 2002, 143: see also the thesis of my doctoral student Daniel Miller, *Animal Ethics and Theology: The Lens of the Good Samaritan.* London, Routledge, 2013.
102 Karl Barth, *Church Dogmatics III.2.* Trans. Geoffrey Bromiley, Edinburgh, T&T Clark, 1989, 457–9; see also David L. Clough, *On Animals: Volume II: Theological Ethics.* London, T&T Clark, 2019, 216–7.
103 Jane Goodall with Philip Berman, *Reason for Hope: A Spiritual Journey,* New York, NY, Warner Books, 1999; see also Jane Goodall with Thane Maynard and Gail Hudson, *Hope for Animals and Their World: How Endangered Species Are Being Rescued from the Brink.* New York, NY, Grand Central Publishing, 2009.
104 For fuller discussions of this apostolic argument and its import see Michael S. Northcott, 'Eucharistic Eating, and Why Many Early Christians Preferred Fish.' In David Grummet and Rachel Muers eds. *Eating and Believing: Interdisciplinary Perspectives on Vegetarianism and Theology.* London, T&T Clark, 2008, 232–46; and Michael S. Northcott, 'Killing for Philosophy and a Creaturely Theology of Nonviolence.' In Celia Deane Drummond and David Clough eds. *Creaturely Theology: On God, Humans and Other Animals.* London: SCM Press, 2009, 231–48.
105 Northcott, 'Eucharistic eating', 240–1.
106 Thomas Aquinas, Summa Theologia II.II, Question 159, 'Cruelty'. Online edition published at New Advent.org at https://www.newadvent.org/summa/3159.htm accessed May 31, 2022.
107 Pope Francis, *Laudato Si': On Care for Our Common Home.* Rome, Vatican Press, 2015, 100, 140.
108 Constantinos Athanasopoulos, 'Introduction.' In Constantinos Athanasopoulos, ed. *Orthodox Mysticism and Asceticism: Philosophy and Theology in St Gregory Palamas' Work,* Newcastle upon Tyne, Cambridge Scholars Publishing, 2020, xxii.
109 Norman Russell, *Gregory Palamas and the Making of Palamism in the Modern Age.* Oxford, Oxford University Press, 2019, 23.
110 Gregory Palamas, *The Triads,* III, iii, 11. In John Mehendorff ed. *Gregory Palamas: The Triads.* Trans. Nicholas Gendle, NJ, Paulist Press, 1983, 87; see also Patricia Calvario, 'The Problem of Deification.' In Constantinos Athanasopoulos ed. *Triune God: Incomprehensible but Knowable – The Philosophical and Theological Significance of St Gregory Palamas for Contemporary Philosophy and Theology.* Newcastle upon Tyne, Cambridge Scholars Publishing, 2015, 222–231.
111 Norman Russell, *Gregory Palamas and the Making of Palamism in the Modern Age.* Oxford, Oxford University Press, 2019, 9–11.
112 Bodily restoration and slowing ageing among the desert fathers was the subject of the research of my doctoral student Todd Daly, now published as Todd T. Daly, *Chasing Methuselah: Theology, the Body, and Slowing Human Aging.* Eugene, OR, Wipf and Stock Publishers, 2021: see especially 156–203.
113 See further the fine account of this tradition, including a travelogue of old Eastern Orthodox churches in Eastern Europe, as well as of its demise in the West in Rita Nakashima Brock and Rebecca Ann Parker, *Saving Paradise: How Christianity Traded Love of this World for Crucifixion and Empire.* Boston, MA, Beacon Press, 2009.

114 See further my discussion of bread-making in Greece in Northcott, *Place, Ecology and the Sacred*, 133–4.
115 See further Daly, *Chasing Methuselah*.
116 Tamara Grdzelidze, 'The Church.' In Michael S. Northcott and Peter M. Scott eds. *Systematic Theology and Climate Change*. London, Routledge, 2014, 141–57.
117 Albert Einstein, 'Concerning a Heuristic Point of View about the Creation and Transformation of Light.' Trans and ed. Anna Beck, *The Collected Papers of Albert Einstein Vol. 2. The Swiss Years: Writings 1900-1909*. Princeton NJ, Princeton University Press, 1989, 86–103.
118 See further Pierre Marage and Gregoire Walleborn, 'The Debate between Einstein and Bohr, or How to Interpret Quantum Mechanics. In Marage, P., Wallenborn, G. eds. *The Solvay Councils and the Birth of Modern Physics. Science Networks · Historical Studies, vol 22*. Basel, Birkhäuser, Basel, 1999.
119 John Stewart Bell, 'On the Einstein Podolsky Rosen paradox.' *Physics Physique Физика*. 1 (1964) 195–200.
120 Stuart J. Freedman and John F. Clauser, Experimental Test of Local Hidden-Variable Theories.' Phys. Rev. Lett. 28, 938 (1972) – Published 3 April 1972.
121 Peter J. Hore, and Henrik Mouritsen, The Quantum Nature of Bird Migration" in *Scientific American* 326 (2022) 26-31.
122 Jennifer C. Brookes, 'Quantum Effects in Biology: Golden Rule in Enzymes, Olfaction, Photosynthesis and Magnetodetection.' *Proceedings of The Royal Society A Mathematical Physical and Engineering Sciences* 473 (2017) 20160822.

8
GAIAN ETHICS

Abstract

Much contemporary focus on environmental action centres on greenhouse gas emissions from humans and other animals. But this reductionist focus fails to address the core causes and features of the ecological crisis, and especially the declining capacity of ecosystems to support diverse life forms and to respond to external and climate shocks. Industrial land use and industrial fishing practices are wrecking forests, savannah, soils, and oceans and ending long-evolved symbiotic relationships between indigenous people, small farmers, other animals, and microbiota, fungi, and phytoplankton in soils and oceans. They are also reducing the capacities of creatures on land and in the oceans to utilise the excess carbon the industrial economy is emitting into the atmosphere. A Gaian approach to planetary medicine involves reviving the creaturely agencies of the other animals, plants, and microbiota with which humans share the Earth. In this final chapter, case studies are described from Indonesia, Southern Africa, Siberia, and England which demonstrate the potential for Earth healing when humans work towards enabling other beings, and especially large mammals such as herbivores and sharks, to regenerate ecosystems by selective reintroductions of keystone species and guarding them from destruction by industrial fishers, farmers, and foresters. The potential role of religions in valorising this approach is considered in conclusion.

Introduction

Many argue that Gaia, and humans, are facing an immanent climate apocalypse. But the gravest crisis Gaia faces currently is not climate change. Gaia has gone through climate perturbations throughout her history. The most recent

DOI: 10.4324/9781003110750-9

was in the seventeenth century when an extended solar minimum led to a decline in heat energy reaching the Earth. In temperate zones the resulting collapse in harvests was accompanied by a die-off of between one and two thirds of human inhabitants between 1615 and 1680. Many died from infectious diseases as hunger reduces disease immunity. But even more critical in population declines was the exceptional number of wars that occurred in that century, from Ireland to China.[1]

In the present century, humanity is also said to be facing a global climate crisis, not from cooling but heating. But presently, as in the seventeenth century, the leading causes of human suffering, mortality, and out-migration from many regions are not hunger, or heat, but violent conflict and war. From the First World War through to the many conflicts in the Middle East, Eastern Europe, and Africa which are in the main associated with the United States' 'war on terror', tens of millions of combatants and civilians died, while many other millions took to the roads and seas in efforts to escape disease, famine, and violence that accompany and follow in the wake of war. In the US- and UK-initiated Iraq war, which began in 2003 on a false pretext, a community survey found 655,000 died violently in the subsequent three years and the death toll has continued.[2] But apart from the devastating Iraq invasion, the numbers of civilians directly killed in conflicts have declined significantly since 1945. Instead, the greatest human effects of war have been environmental. As residential areas, energy, food, and water sources are destroyed, or rendered inaccessible, by modern indiscriminate aerial weapons, and land mines, out-migration from conflict zones has become the principal enduring effect of contemporary warfare.

Since 1990, tens of millions of people have migrated from conflicts associated with the United States' post-9/11, unliterally declared, and endless, 'war on terror.' The long list of over 90 such conflicts includes Afghanistan, Iraq, Niger, Mali, Bosnia, Rwanda, Palestine, Lebanon, the Congo, Eritrea, Ethiopia, Angola, East Timor, Pakistan, Darfur, Iraq, Somalia, Sudan, Syria, Libya, and Yemen.[3] The resultant migration crisis is often presented by academics, and in the media, as a consequence of climate change. But unlike in the seventeenth century, climate change to date has produced a net increase in food production because of a longer growing season in temperate areas. Food shortages at the present time are not caused by climate change but by disruptions to food production in war zones, and by global supply-chain disruptions. The fragile and global nature of many food supply chains are a consequence of concerted actions by governments, food corporations, and intergovernmental agencies to reduce national food security, and increase dependence on food imports from the 1970s as part of the broader neoliberal globalisation project which was led by the United States and the World Trade Organisation.[4] At time of writing disruptions to the newly global nature of much food production are not a result of climate change but of the unwise actions of States in relation to the COVID-19 pandemic, and especially of those States which prevented farmers growing food, or even paid them to destroy produced food, allegedly because of transportation shortages resulting from government restrictions on human movements.

Despite the major roles of war, and of globalisation-induced food insecurity, globalist organisations and governments argue that increased human migration and food shortages are caused by climate change. Thus, a UK government report on the 'economics of climate change' quotes with approval the claim that there will be up to 50 million 'climate refugees' by 2030.[5] There are analogous claims in the media and by governments that climate change is causing food shortages, and that it will eventually bring about a reduction in human numbers. On the basis of these alarmist predictions, government actions said to be aimed at mitigating climate change, and especially raising the price of energy while limiting the production of fossil fuels, are generating an economic crisis for businesses and households.

Against climate change alarmism, James Lovelock observes, correctly, that since 2000 'global average surface and atmospheric temperature has hardly warmed at all' and this is despite a huge increase in global carbon dioxide emissions.[6] Computerised Global Change Models on which the United Nations' Intergovernmental Panel on Climate Change bases its predictions of future apocalyptic temperature increases assume a steady linear increase in temperatures but the empirical evidence does not support this. The decline in temperature increase since 2000 could be because Gaia has worked out how to make use of the additional atmospheric carbon since carbon is a fertiliser and not a toxin, and this is underwritten by satellite data since 2000 which show greatly increased greening of the Earth since 2000 despite ongoing destruction of old-growth forests.[7] It could also be other factors, such as an unexplained but major increase in upwelling of cold waters to the surface in the Pacific Ocean.[8] As Lovelock says, the climate system is extremely complex and there is no straightforward coupling between greenhouse gas emissions and the Earth's temperature.

There are highly perverse outcomes from what Björn Lomborg calls the global false alarm of climate change, and its representation as the gravest near-term threat to human wellbeing and the future of life on earth. The first set of perverse outcomes relates to misguided actions allegedly to mitigate climate change which, far from reducing human suffering, significantly increase it. This is because the main focus of government and activist climate action has been energy sources. In the United Kingdom, the price of energy has trebled since 2011 as investments in oil and gas have declined under pressure from anti–fossil fuel campaigns. At the same time government taxes on energy, allegedly to address climate change, include 'renewable energy' and 'biofuel' surcharges even though biofuels are, as we have seen, a *source* of ecosystem breakdown and local heating. Government and investor actions to raise the price of energy disproportionately harm the poor who in rich countries pay 10 per cent or more of their income in energy costs.[9] There is already far greater risk of lost life years in Northern Hemisphere countries from increased energy costs, which force poor and elderly people to avoid heating their houses, and to spend more time in bed, than there is of lost life years from climate change.[10]

In developing countries, forcing 'climate mitigation' policies which stop the availability of cheap energy will also have far more devastating consequences in lost life years than climate change since cheap energy is the principal driver of improvements in infrastructure, such as the accessibility of potable water, adequate nutrition and electricity which are closely associated with increased longevity.[11] The behaviours and developmental changes proposed as climate 'fixes' by climate change alarmists, such as using electricity to power vehicles, substituting electric heat pumps for gas boilers, burning wood instead of coal in electricity generating stations, and eating less meat, are not climate fixes. And in the case of substituting living forests for decayed forests in power generation this 'fix' considerably increases ecological harms and increases carbon dioxide emissions since burning wood produces more carbon dioxide per unit of heat than burning coal.

These alleged 'climate fixes' also divert attention from the real contemporary causes of human ill health which remain, as they have since the beginning of the industrial revolution, lack of secure access to non-toxic air, food, shelter, and water. Pre-industrial peoples – subsistence farmers and indigenous peoples – did not need to rely on global energy, water, and food corporations and government handouts to subsidise inaccessible prices charged by energy, food, and water corporations, since their farming and gathering systems derived the essentials of life directly from sun-fuelled ecosystems. This is not to say that these sun-fuelled provisioning systems always produced better outcomes in health and longevity than industrial-based global supply systems. But large-scale famines, and violent conflict, all grew significantly in the course of the global expansion of European companies and their colonial extractive system. And the forced migration of peoples from zones of European colonial corporate extraction and warfare were, and remain, the leading causes of death and disease in the colonial and postcolonial eras.

The second set of perverse outcomes from climate alarmism are ecological. While proposed lifestyle changes have little or no impact on atmospheric levels of carbon dioxide, they divert attention and resources from the deeper causes of the ecological crisis in industrial harvesting and production methods[12] The real threat to the *fertility* of Gaia, and to the in-built capacities of Gaia to maintain Earth's homeostasis, which is to say, temperatures equable for life, including humans, comes not from the air but from changes on land and in the sea. In the last 120 years industrial farmers and fishers, and the food corporations that increasingly control them, have destroyed vast myriads of life forms in the soils and the oceans. In so doing, they are systematically destroying the capacity of Gaia to respond to climate shocks as she has done in the past. Huge declines of microbiota in soils, and of algae, seagrass, fish, and phytoplankton in the oceans, are reducing the capacity of Gaia to recycle atmospheric carbon. And they are reducing the capacity of the Earth to generate and continue to support biodiverse ecosystems on which ultimately human health, as well as climate resilience, depend.

From a socio-political perspective there is a third perverse outcome of present-day and proposed climate change and 'carbon' policies, or what some call the 'climate leviathan,' and this concerns control, freedom, and human agency.[13] The majority shareholdings of the companies that control global energy and food supplies, and many water sources, are owned and controlled by just three US investment entities – Blackrock, Vanguard, and State Street which together own 70 per cent of all shares in the US Stock Market, and significant proportions of European and Asian stocks as well.[14] These three investment houses are therefore responsible for most of the carbon emissions put out by the global industrial economy. If governments and financiers really believe that carbon emissions are a grave threat to the future of humanity, and even to life on earth, the logical thing to do which would have the least harmful consequences on ordinary humans would be to address this problem through the corporate governance protocols of the owners of most of the world's companies. Instead of doing this, the climate regime of the United Nations Framework Convention on Climate Change (UNFCCC) and the Intergovernmental Panel on Climate Change (IPCC), and governments enacting it, are pushing farmers, governments, householders, and small and medium-sized enterprises to change their behaviours in myriad ways. Since the origin of the UN climate regime, or 'climate leviathan,' it has had zero impact on rising carbon emissions from the extractive land and marine practices of industrial corporations. But the principal instrument of the climate leviathan – energy taxes – are already seriously impacting human freedoms and farmer and householder self-sufficiency.

An even more invasive project to 'control' the carbon emissions of ordinary people is to link digital money – in the form of credit and debit cards and proposed Central Bank Digital Currencies – to estimates of individuals' carbon emissions. On the basis of such estimates, individuals will be given carbon scores in a system of top-down economic control analogous to social credit scores already linked to digital money, and Statist political control of citizens, in communist China.[15] While current systems linking carbon emissions to Fintech in the West are voluntary, in a cashless economy central bankers and governments will be able to use carbon footprinting to determine how small businesses and individuals spend their money.[16] This will be a climate leviathan on speed, and a technocratic dystopia since fundamental freedoms and human agency – cultural, personal, and political – depend upon individuals' capacities to support themselves. [17]

In his book *Regenesis*, George Monbiot notes the growing control since the 1970s of the global food supply by a small number of corporations, and that the farming practices they sponsor wreck soils, destroy biodiversity, and pollute rivers. But he argues that traditional peasant-based and small-holder farming, based on a mix of animal farming, cereal growing, and horticulture, need to be brought to an end, along with industrial agriculture, and that humans will increasingly eat food produced by large food corporations in the form of novel fermented proteins such as Quorn, and other artificially manufactured meat and dairy substitutes.[18] An earlier book of the same title does not provide confidence that the ecological or human outcomes from these new synthetic biological techniques will be benign.[19] And, as discussed above, the record thus far of biological engineering using gene editing techniques is negative both for ecological and human health outcomes. But Monbiot argues that to allow more space for wild animals and ecosystems to revive, animal farming must end. He also argues that most fruits and vegetables will need to be grown in artificially lit, hydroponic, and computerised systems, many of which will be located in city tower blocks.[20] That such a system offers possibilities for enhanced corporate control, including genetic engineering, of human foods does not, however, trouble Monbiot who argues that agriculture, and especially domestic animals, are wasteful forms of land use, and that the abolition of agriculture across much of the surface area of the habitable land is the only way that ecosystems and threatened species can regenerate.

Like many environmentalists, Monbiot believes there is an intrinsic conflict between humans and nature, and that only by enforced top-down governmental restraint can this conflict be reduced and resolved. But what this argument does is to tar all agriculture with the same brush as modern industrial monocultures and intensive animal rearing systems. I entirely agree with Monbiot that industrial agriculture, and fishing, are intrinsically in competition with the health of the planet. But this does not mean that all agricultural systems involving other animals are ecologically problematic. On the contrary, there is important historical and contemporary evidence that ecologically benign farming systems evolved over thousands of years which, while they impacted the numbers of wild animals, did not systematically reduce the biodiversity and resilience of ecosystems.

In previous chapters I show that archaeological and historical data evidence a long-established relationship between humans and other animals which was represented aesthetically and ritually in a range of forms. From ancient cave paintings and religious stories of friendships between humans and other animals to animistic dances in which animal identities are acted out by humans, there is a cross-cultural and interreligious sensibility for the critical role of other animals and creatures in co-creating and sustaining the habitat of humanity including food and water sources and materials for shelter, clothing, and cultural artefacts such as musical instruments. Humanity has also long hallowed mountains, caves, waterfalls and springs, lakes and coastlines; and sacred ritual spaces have often been located in proximity to these.[21] Post-Reformation Europe is arguably unique historically in the extent to which such hallowed relationships were abandoned and suppressed. Monbiot's environmentalism, like almost all modern environmentalism, is rooted in that same Protestant culture even although he himself eschews any religious affiliation.[22] I have argued throughout this book that the modern quest to control nature, and now humans, originated in that same culture. That Monbiot commends a farming system going forward in which most protein is grown in technologically controlled vats and hydroponic artificially lit systems is an example of where this culture is headed, and it is not in a direction which will heal relationships between modern humans, other animals, and the Earth.

In the remainder of this chapter, I describe four case studies in which cultural symbiosis between humans and other creatures, of the kind that is found extensively in religious traditions from the Indian Vedas to the stories of the Christian desert fathers, is front and centre in agricultural systems which are both productive of human foods and which are ecologically benign. Lovelock and Margulis' account of Gaia as a system of systems which are self-organising and yet directed over more than 500 million years towards finely balanced complexities of interrelations and diversity of life forms is a scientific theory which, as I argued above, mirrors in significant ways the creaturely symbiosis which is characteristic of most religious texts until the Enlightenment. In these texts, and in the Gaia theory, order and harmony are not imposed from outside the Earth by a constantly intervening divine engineer or maker but rather emerge from the self-organising and relational entanglements of myriad creatures. When most religious texts describe a guiding hand or *nomos*, it is more like an emergent and immanent life force, such as *chi* or *pneuma*, than a colonising and controlling Promethean power. The case studies that follow describe approaches to agriculture and ecological management in which multiple beings are enabled to express agency in such a way as to build up, rather than break down, the productivity, resilience, and species diversity of ecosystems.

Bad Karma and the Green Revolution in Bali

I wrote this book on the Indonesian island of Bali which Balinese call the 'island of god.' Bali's beauty and cultural richness attracted artists and anthropologists

FIGURE 8.1 Water temple, Gianyar, Bali. Author photo.

from Europe and North America from the 1930s. The Austro-Russian artist Walter Spes established an artistic community on the edge of the town of Ubud at Campuhan where two rivers meet, and the water temple is much venerated (Figure 8.1).

Since the 1960s, Bali began to attract larger numbers of tourists, in part as a result of the actions of the post-independence President Sukarno of Indonesia. Sukarno's mother was Balinese, he maintained a presidential palace in the hills north of Ubud, and he decided to build the first large concrete multi-storey tourist hotel, the Grand Inna Bali Beach Hotel, in the sheltered Balinese coastal village of Sanur. His aim was to turn Bali into an international tourist destination and dollar-earner for the Indonesian government. To celebrate the opening of the new hotel Sukarno planned to attend the Eka Dasa Rudra ceremony, held once in 100 of Balinese years, at the most sacred Besakih Temple on the slopes of Mount Agung in March 1963. Sukarno invited hundreds of foreign guests including diplomats and tourist business owners from the United States, Europe, and Australia to the ceremony. But the Temple priests were unhappy about Sukarno's plans, and apparently, they were not alone. Gunung Agung began to erupt in February 1963 and the eruption continued through the Eka Dasa Rudra ceremony in March to which, consequently, Sukarno did not show up. The first casualty of the eruption was the ceremonial arch at Besakih Temple which Sukarno had ordered built to honour his attendance, and his guests, at the ceremony. The Temple itself escaped with only its grass roofs burned, but the eruption went on to destroy hundreds of thousands of hectares of rice fields, hundreds of villages, and there was extensive loss of life and famine which endured into 1964.[23] But there was worse to come. The years 1965–6 saw a cascade of political turmoil and violence across Indonesia, beginning with the downfall of Sukarno's Presidency, and the extra-judicial killing of up to a million people across Bali and much of Indonesia. The coup which unseated Sukarno was fomented by the United States, and the United States and United Kingdom were also complicit in the mass murder which followed.[24] Nonetheless the eruption of Gunung Agung was the chronological event which set Sukarno's downfall in motion.

Sukarno was not the only outsider to Bali to disrespect Bali's religious culture, rituals, and ways. The Green Revolution took off in India in the 1960s and sparked a new more intensive phase in the monocultural and technologically dependent character of modern industrial agriculture. In India the introduction of fast-growing hybrid seeds, chemical fertilisers and pesticides, mechanical tillage and harvesting machines produced a significant uplift in rural incomes at least in the early years of the development.[25] But in Malaysia, James C. Scott describes how the Green Revolution led to landlessness and poverty among formerly self-sufficient rice farmers in his classic study of forms of peasant resistance *Weapons of the Weak*.[26] Vandana Shiva is strongly critical of the end result of the Green Revolution in India a few decades on: the chemically doused fast-growing seed system raised the cost of farming, devasted the natural fertilising effects of microbiota and minerals in agricultural soils, and pushed millions of former peasant farmers into debt. The new technologies are expensive, and as initial government subsidies declined, many peasant farmers, laden with unpayable debts, became urban migrants forced to work as wage slaves in factories or the houses of high caste urbanites, while many others used pesticides to kill themselves.[27]

The impacts of attempts to improve agricultural productivity from the 1950s in Indonesia were investigated by Western anthropologists, including Clifford Geertz, who describes a process he called 'agricultural involution' by which the Javanese found ways to incorporate new techniques into existing cultural practices which lessened their deleterious societal impacts.[28] Green Revolution techniques were, however, more culturally disruptive than earlier agricultural improvement techniques and they were introduced by Swiss-based agricultural engineers into Bali in the 1970s. The new farming methods radically disrupted the centuries-old temple-managed hydrological systems which linked Bali's mountain watersheds, lakes, rivers, rice terraces, coastal mangroves, and beaches via a hugely complex system of weirs, tunnels, channels, and rice terraces to Bali's religious calendar and ceremonial practices. Bali's water temples underwrote calendrical schedules for rice planting and long-established cooperative farming practices, including decisions of neighbouring farmers about when to harvest, and whether to burn or flood fields after harvest, and these decisions are crucial in the control of pests.

The engineers who introduced the Green Revolution techniques knew nothing of the evolved ways in which the religion of Bali controlled the water on Bali, sustained cooperation between farmers, and so also controlled pests. They knew that Bali had a 'rice cult' which they dismissed as superstition, and they forced the farmers by fines and regulations to abandon centuries-old traditions of seasonally related double-cropping patterns and coordinated fallow periods for pest control. The engineers also thought they knew better than the water temple priests how to improve the availability of water to rice farmers and introduced over 50 new concrete weirs. The Swiss agricultural consultants and Jakarta bureaucrats who engaged them envisaged that their Green Revolution techniques would enable the Balinese to grow three crops of rice a year with the more 'efficient' fast-growing hybrid seeds, and chemicals.[29]

Problems with the Green Revolution techniques emerged within a few years of their introduction in Bali as the farmers found the new hybrid seeds susceptible to insect plagues which no amount of the new pesticides could control. The new techniques and planting regime also led to significant water shortages as well as the loss of thousands of tons of rice to pest plagues: this led to lower all-island rice outputs than were achieved following traditional methods. To attempt to mitigate the pest plagues, the foreign agricultural engineers introduced successively newer hybrid seeds which would resist one pest, only to find another pest plague would take over within one or two plantings.[30]

Water-Temple Priests versus Development Economists

The problems with Green Revolution techniques were so bad that Balinese farmers started to campaign for a return to the traditional system of water-temple–controlled planting and harvesting. While this was dismissed as 'religious conservatism' by the agricultural and development experts pushing the Green Revolution techniques, the Bali Department of Public Works commissioned Balinese academics at Udayana University in Denpasar to investigate the problem. Their report found that the problems of pests and water shortages were 'a result of the policy of continuous uncoordinated rice planting' and recommended further investigation of 'the connections between the hierarchy of the subak (water) temples and cropping patterns.'[31]

The story of how the Green Revolution failed in Bali, and how the Balinese returned to their own religiously managed hydrological growing system is told in *Priests and Programmers* by the American anthropologist Steven Lancing. His field work in Bali in the 1980s coincided with the height of the problems caused by the Green Revolution. Based on the Udayana University report, and his field work, Lansing proposed in a report to the Asian Development Bank overseeing the 'Bali Irrigation Project,' as it was known, that the religiously governed calendars and cooperatively coordinated planting systems were more efficient than the new techniques and it would be best to return to the ten-century-old system of control by the Subak water temples:

> (T)he rituals of the water temples were not a template for an outmoded cultivation system but a system of ecological management with deep historical roots in Balinese culture. Agriculture was a social as well as a technical process, dependent on the 'hydraulic solidarity' achieved by the temple system. Continuous rice cropping threatened both the ecology of the terraces and the social infrastructure of production.[32]

But the foreign consultants ignored Lansing as well as the Balinese farmers.

The guiding force of modern economic development is neither human nor planetary wellbeing but what Roy Rappaport insightfully calls 'monetary epistemology' which is 'the dominant mode of assigning value in modern societies.'[33] Monetary epistemology puts the logic of money, in the form of short-term cost and benefit analysis, above all other considerations. It neglects costs, including ecological and social costs, that are not on the balance sheets of businesses, or governmental and intergovernmental agencies.

As described in more detail in Chapter 4, underlying monetary epistemology is a modern secular faith in technological progress, and belief in the presumed superiority, and greater wealth-generating powers, of modern industrial systems and technologies over all predecessor traditions and ways of life. Monetary epistemology is a poor substitute for more traditional cultural value systems because it is binary, it only values quantity, and it does not recognise value as *qualities*, such as those present in non-monetised human cultures and ecosystems.

That there is an unacknowledged but quasi-religious devotion to the measure of money, and the suppression of alternative value systems, in development agencies is well illustrated in a case study of Protestant development agencies in Zimbabwe. That case study also reveals that genuine religious faith and spiritual awareness on the part of development workers can transform a narrow perspective on material uplift into a more holistic approach.[34] But in Bali the Hindu religion was viewed by European development experts, and Jakartan Muslim agronomists, as irrational and anti-scientific and so for 20 years they continued to deny that it might offer a solution to the problem their modern scientistic techniques were generating.

Lansing's view of religion when embarking on his research in Bali was not so different from that of the agronomists. The standard modern anthropological perspective on religion is that it has an exclusively humancentric significance. As discussed above, in a 'scientific' perspective religion has ethical functions, and may promote social solidarity, but it does not affect nonhuman agencies such as ecosystems or spirits. But Lansing's field work in Bali led him to suspect that religion might influence rice production, and so he turned to investigating the problem of whether water temples play a role in the productivity of Bali's artificial agricultural landscape (Figure 8.2).

With a systems ecology colleague Lansing built a Lovelock 'daisy-world'–type simulation model of the Green Revolution techniques and compared them with a model using the traditional Balinese techniques.[35] The modelling revealed that traditional Balinese annual double-cropping methods using a range of traditional Balinese rice varieties and coordinated by the water temples to ensure careful sharing of water throughout the watershed, caused 'the least amount of water stress.' Water-temple–coordinated cropping patterns also prevented pest plagues which cannot be controlled at the village level as if fallow periods are not planned over large areas pests simply migrate between neighbouring fields.[36] And they concluded:

> The substitution of the 'high technology and bureaucratic' solution in the event proved counter-productive and was the major factor behind the yield

FIGURE 8.2 Rice terraces and water temple, Gianyar Bali. Author photo.

> and cropped areas declines experienced between 1982 and 1985 … The cost of the lack of appreciation of the merits of the traditional regime has been high. Project experience highlights the fact that the irrigated rice terraces of Bali form a complex artificial ecosystem which has been recognized locally over centuries.[37]

Lansing's daisy-world–type simulation translated the potential of Bali's traditional religiously governed rice-growing and water-sharing system into scientific language. And this translation played a role in finally persuading the non-Balinese managers of the Bali Irrigation Project that the Balinese could do better using their traditional methods. And so, the agronomists permitted informal control over rice planting to be returned to the water temples. Although the new hybrid seeds, and artificial fertilisers, were still widely used, attempts to force the Balinese to plant rice three times a year were abandoned, and the Balinese were allowed to return to their ten-century-old 'hydraulic solidarity' of water-temple–coordinated rice growing, and the pest plagues and water shortages declined. At the present time there are several projects underway to return to traditional seeds, and traditional methods of fertilising the soil, since the artificial fertilisers create algal blooms which negatively impact water flow, and the island's coastline, and traditional seeds are not subject to the expensive licenses of the hybrid varieties.

The complex cultural system of Bali is Hindu in origin and its humanly evolved landscape is more religiously shaped than even the Balinese, including Balinese agricultural 'experts' as well as foreign development consultants, realised. As Lansing comments, 'the idea of religion playing an important role in agriculture was inconceivable to development planners.'[38] But in a contest between development programmers, pushing modern industrial chemicals, hybrid seeds, and machines, and traditional beliefs and ceremonies which invoke and engage divine energies in the waters, seeds, and soils of Bali, the priests beat the programmers. The eventual abandonment of the attempts of bureaucrats in Switzerland and Jakarta to take over control of Bali's ten-century-old water-management system was a good outcome for Bali. But across much of the developing world, globalist sustainable development continues to displace long-evolved and indigenous cultural and production systems with one-size-fits-all industrial development practices.

The Case of African Drylands

Africa is the continent more than any other which is currently being subjected to global sustainable development projects that are undermining the capacity of people to feed themselves from their own resources and in ways that do not degrade the resilience of their habitats and ecosystems. In many countries, African government agencies work with overseas development agencies to take land off peasant farmers and convert it either into large agricultural projects or large conservation projects. In both cases, the shared assumption of government bureaucrats and foreign 'developers' is that indigenous people and their domesticated animals need to be cleared off the land in order for it to either be made more agriculturally productive, or for it to be conserved. One of the biggest conservation issues in sub-Saharan Africa concerns the use of drylands and the tendency of these lands when misused to turn to desert. The conventional scientific view of the problem is that these areas suffer from a surfeit of domestic animals belonging to peasant farmers and indigenous peoples and that the answer is for the land to be 'rested' by the farmers and their animals excluded.[39]

Allan Savory was a scientist and land manager, born in Zimbabwe, and grew up with a deep love of the African bush. In his scientific training he was taught, and practised, land management on the basis of the established belief that desertification is caused by deforestation, and overgrazing by domestic animals, and that the solution is to eradicate, or radically reduce, the numbers of people and domesticated animals on the land. But in Southern Africa, and in Texas after he took up teaching and land and ranch management consultancy there, Savory realised that desertification and soil erosion were occurring at scale on warm drylands on both the American and African continents, and that, despite the mainstream scientific view of the causes, the more empty landscapes were of humans and animals, the more likely they were to be desertifying.

Savory's field observations in Zimbabwe led him to the view that the problem is not caused by too many animals, as conventional land managers and conservation scientists assumed, but too few, and he later found the same problems in Texas:

> (V)ast areas of rangeland that had once sustained immense herds of bison and later cattle were hardly distinguishable from the most degraded rangelands on the fringes of the Sahara Sand dunes that were beginning to form in West Texas. The water table was falling too, and rivers once filled with fish had been dry for decades except for the occasional flash flood. The Texas government had spent hundreds of millions of dollars to fight insects, weeds, and brush that were blamed for ruining the land, but they continued to thrive. All the while, the people, as in Africa, were leaving the land for the big cities, where crime and poverty continued to grow.[40]

Savory realised that the problem of desertification was not too many animals but grazing patterns determined by humans rather than predators. In pre-historic times, before humans learned to use fire and to make more advanced hunting devices, billions of large hooved animals roamed much of the Earth's surface, but they did not graze in concentrated fashion, as modern farmed animals do, because they were constantly predated by wild animals such as lions and tigers. Predation encouraged the herds of bison, wildebeest, and so on, to graze in bunched patterns which led to rapid eating down of greenery to the roots at the same time as fertilisation from animal faeces. Animals do not like to feed on areas they have polluted so these huge bunched herds kept moving to new areas. In so doing, the animals would have prevented the hard soil crust from forming which characterises soil surfaces in areas with little vegetation and undisturbed by animals and causes runoff of rainfall, and which are therefore prone to flash floods. By breaking up the soil surface grazing animals that keep moving allow rain to replenish soil moisture and promote subsurface microbial activity including worms and numerous other creatures. As Savory observes, the majority of Earthly creaturely activity on land not farmed and denuded of most life by industrial humans occurs below the surface of the soil.[41] Holistic ecological management requires land managers to make decisions which allow creatures both above and below the surface to restore ecosystems and so reverse desertification, stop species loss, repair micro-climate changes and reduce natural disasters such as flash floods, dried-up rivers, soil erosion, and so on which are usually attributed to global climate change.

Building on these observations, Savory developed a radically new method of farmland management in 'brittle' areas – subtropical and semi-temperate lowland regions – which is Holistic Land Management (HLM). 'Brittle landscapes' are climatic zones north and south of the Equator characterised by long periods without rain during which land temperatures tend to escalate significantly. These regions constitute 40 per cent of the global land surface and are the locations of the modern phenomenon of spreading deserts, which climate and other natural scientists in the main attribute to the global process of 'climate change.'[42] Managing drylands is crucial to re-greening the Earth and Savory's HLM paradigm offers an answer to this which does not require excluding indigenous and peasant farmers and their animals.

HLM is based on four key insights. First, is that the Earth

> is not a machine made up of parts that could be isolated for study or management. In reality the world is composed of patterns – of matter, energy, and life – that function as *wholes* whose qualities cannot be predicted by studying any aspect in isolation.[43]

Instead HLM requires a holistic perspective, analogous to the Gaian perspective on the Earth System developed by Lovelock and Margulis, in which land-management decisions are based on an understanding of the total system in which they are being made, including the respective roles of people, animals, soil microbiota, plants, water catchment, and so on.

A classic modern agronomic example of the tendency to manage 'nature' as a machine was the decision of the Chinese Communist Party in the 1960s to eradicate Chinese sparrows, which consumed a modest fraction of 1 per cent of agricultural grain, in order to help address the Maoist 'Great Leap Forward' goal for China to be completely self-sufficient in food. Peasants duly followed Communist Party instructions so successfully that sparrow numbers were reduced to near extinction of the species. But within a year, a vast famine overtook China as crops failed due to huge outbreaks of pests, and especially locusts, which the sparrows had formerly eaten. The locusts did not just take a fraction of 1 per cent from rice and wheat fields, however: they ate everything, including even tree-leaves and vegetable plants, leading to the deaths of ten million Chinese from starvation. One small human collective decision to simplify an ecosystem as if it were a machine reducible to its parts, with the intention to increase food production, devastated China.

Savory describes a similar event in North and South Rhodesia in the 1950s. Both regions were at that time replete with large animals. Savory regularly counted 40 lions or more in a day's walk, and there were huge populations of elephant, buffalo, zebra, antelope, hippo, crocodile, and so on. People had lived in these areas for thousands of years in villages with vegetable gardens which they protected from wild animals with drums in the night. And in the daytime, they hunted the large animals. But the governors of both regions decided that they needed to turn much of the land into 'National Parks' and set about excluding humans from the parks to 'protect' the animals. As Savory comments, 'the results were quick and dramatic':

> Within a few decades miles of riverbank in both valleys were devoid of reeds, fig thickets, and most other vegetation. With nothing but the change in behaviour of one species these areas became terribly impoverished and are still deteriorating seriously as I write.[44]

The key process which led to this rapid degradation was the exclusion of humans who were the animals' chief predators. With no humans to fear anymore, animal behaviour changed; they stayed put instead of moving around, 'which in turn led to the damage to soils and vegetation.'[45]

The second key insight in Savory's HLM approach is that land habitats 'may be classified along a continuum of two principal categories from nonbrittle to brittle according to how well humidity is distributed throughout the year and how quickly dead vegetation breaks down.' Lands at either end of the continuum respond differently to land-use decisions, and especially to decisions concerning 'fallow' or 'rest' periods: resting the land damages brittle environments while the same procedure may restore land in continuously humid areas. This insight overturns thousands of years of belief that all environments respond the same way to human interventions. Among the oldest beliefs about land management, going back before Herodotus, is that when land is damaged

> the best remedy is to rest it from any form of human disturbance, including domestic stock. But despite the application of this wisdom, the croplands and rangelands of Libya and its neighbors have desertified anyway, as much of America is now doing.[46]

Savory points out that brittle lands all over the Earth once sustained vast numbers of animals, and of humans who, as in Zimbabwe, also lived among them and hunted them for food and hides. These animals' grazing habits, and digestive tracts, were critical to maintaining vegetation in drylands since the animals helped break down waste plant matter, even in dry seasons, and in combination with huge numbers of organisms in their stomachs, they turned plants into dung, so speeding up the cycle of decay which is much slower in dry than moist land areas.[47] Failure to understand this cycle led to the abandonment of whole cities and regions in dry areas such as Central Mexico, the Sahara, and the Yemen thousands of years ago. Can this insight be used to restore deserts today? Savory shows that it can, and this is indicated in his third key insight:

> In brittle environments, relatively high numbers of large, herding animals, concentrated and moving as they naturally do in the presence of pack-hunting predators, are vital to maintaining the health of the lands we thought they destroyed.[48]

This is because, as just indicated, large numbers of animals that graze in bunches, and keep moving, surface-plough the soil, trampling and breaking up dead vegetation, so creating ground cover for the soil, and for new seeds, while eating growing vegetation and defecating, so creating ideal conditions for plants, insects, and soil microorganisms to reproduce and regenerate in regions with both long dry hot seasons and seasonal heavy rains:

> This relationship between pack hunters, their herding prey, and the soils and plants they trampled and grazed developed over millions of years, long before humans themselves became pack hunters with the aid of fire and spear. There was no other influence that could realistically have both created the necessary soil disturbance to provide a good seed bed for new plants and protected bare soil by trampling old plant material.[49]

The fourth and final key insight of HLM that follows from this is that 'timing is everything.' Conventional land management wisdom has ignored *time*. But in all environments, 'overgrazing and damage from trampling bear little relationship to the number of animals, but rather to the amount of time plants and soils are exposed to animals.'

Savory first came to this realisation observing the differences between land which was empty of almost all large animals, land with large herds of wild animals in the presence of predators, and herds of domestic animals managed by humans. Empty land was the driest and most lacking in vegetation; naturally grazed land was the healthiest. Savory realised that 'the disturbance created by the hooves of *herding* natural wild animal populations was vital to the health of the land and that humankind had lost this benefit when we domesticated cattle, horses, sheep, and goats and protected them from predators.'[50] But the benefit was only conferred when predators kept the animals in concentrated grazing herds which never stayed in one place for more than a day or so. This was how the vast herds of wildebeest, elephant, buffalo, zebra, and other hooved animals, which once 'vastly outnumbered the cattle and sheep herds that came later' had for millennia 'enjoyed an environment more abundant than anything' that modern farmers in Africa, and in similar drylands in the Americas, Asia and Austronesia, have known.[51] In Southern Africa, areas with large numbers of animals, and natural springs, were named by the early Dutch settlers after the animals that once lived there, and sustained the springs, such as 'Sprinbokfontein' and 'Buffelsfontein.' But as the animals were hunted to near extinction, and replaced with farms, the springs – in Dutch, *fontein* – also disappeared.[52]

Having put together the four insights of HLM, Savory set to applying them among the farmers and ranchers he advised as a land consultant in the Southwest United States, and in sub-Saharan Africa. After early experiments, which failed, Savory realised how critical timing is to HLM; and that records of the time animals spend in each area, with the aid of fences, were crucial to balancing animal trampling, and vegetation regeneration. If domestic stock animals were herded too long in one area, it did not recover, and the animals would also often suffer from malnutrition as they were forced to eat plants that were not part of their preferred diets. Keeping the herds moving, in a pattern analogous to that created by predators of herds of wild hoofed animals, was the key and this could be achieved with careful written records of spatial management and the timings of animal movements. Savory's consultancy subsequently achieved considerable success in managing and restoring brittle landscapes, while introducing more animals into these areas than convention allows.

HLM provides a crucial and urgently needed solution to the global problem of spreading deserts and declining water tables which afflict drier lands on every continent and island, and which climate scientists and land ecologists have until Savory's pioneering work blamed on global climate change, and the related monocausal account of such change as being caused by industrial and agricultural greenhouse gas emissions, including from the digestive tracts of animals. Far from *reducing* the numbers of domestic animals on the planet, as Monbiot and many climate scientists propose, HLM indicates that *increases* in the numbers of hooved animals is the means to reversing desertification over large areas of the planet. As Saylor puts it:

> The discovery of the brittleness scale and the vital role of herding animals and their predators in maintaining the health of more brittle environments showed why these environments were prone to desertification. This is significant because desertification has ultimately destroyed more civilizations than war and currently, according to the United Nations, adversely affects over 900 million people in over a hundred countries.[53]

I have described Savory's HLM paradigm at length because it offers not only a solution to many of the local ecological changes that are presently, and wrongly, attributed to global climate change, but because his paradigm shows the potential ecologically transformative power of the ontological and ontotheological turns discussed above. Ancient humans lived in the midst of huge numbers of other animals, and their cave paintings, such as those in Sulawesi and Southern France, show that they did not only see the animals as fellow creatures but as keystone Beings whose co-agency with heavenly, human, and subterranean beings was crucial in sustaining the mutually dependent chain of being in which the ancients believed that they were situated. Analogously, in more recent times, when Christian, Hindu, and Muslim saints sought the mystical way, they found that it led not only to their own spiritual renewal but to the renewal of their relations with other creatures and the land.

Gaian Planetary Medicine

Savory's HLM is analogous to Lovelock's Gaian approach to 'planetary medicine' in his monograph of that title in which he argues that 'established science is ill-suited to cope with the problems of global change' because it looks only at individual symptoms of the ecological crisis rather than the physiology of the body of Gaia as a whole system, and 'its record of handing environmental problems is far from good.' In particular, he notes the distortions that arise from misguided scientific funding, which has focused most climate science on upper-atmospheric measurements and models, but failed to look at the climate impacts of land-based changes such as deforestation.[54] Contemporary science 'lacks wholeness of vision' and most practitioners are in 'isolated disciplines'; hence they have an entirely inadequate view of the Earth as a functioning system, and most are unable even to give an account of what Life is![55] Lovelock argues that looked at as a functioning organism, the great systems that create homeostasis – that regulate the temperature of the Earth by analogy with the human body – are still insufficiently understood and studied. And he argues that the roles of soil physiology, of heat absorption of the oceans (especially in the tropics), of methane generation from rock weathering, and of soil respiration, are hugely influential but little-understood processes in the regulation of atmospheric chemistry.

Another critical but little understood process is 'lowly bacteria,' quadrillions of which are in three groups of organisms: photosynthesisers including plants and plankton; animals, and especially their stomachs; and fermenters, and especially fungi, which mainly live on forest and ocean floors: 'These three bacterial groups form the most powerful part of Gaia's body, and have sustained her for some 3.8 billion years.'[56] Lovelock compares Gaia's procedures for temperature regulation, which in relation to the alleged near-term threat of 'climate change' is the most critical function of the interaction of all these life systems, by analogy with the human body. There is no single process by which the human body regulates its temperature at 38 degrees centigrade but instead all the organs of the body work together to achieve this temperature. Similarly, there is no Gaian thermostat. Myriad life forms work together to achieve temperatures which work for them, and which, from the tropics to the Poles, and from the deep ocean to surface waters, together produce global average temperatures that are the most benign to most Life forms. This is the result of 'geophysiology' in which 'the organisms and their environment are so tightly coupled that they constitute a single system, Gaia.'[57]

Lovelock's account of Gaian physiology indicates that the mistakes agronomists, biologists, climate scientists, fishery, forest and land managers are all prone to make is to discount the roles of organisms in ecosystem productivity, and in habitat and temperature regulation. Instead, they measure the success of their management, and conservation, methods through the reductionist lens of mechanistic science and quantitative, or monetary, epistemology: they focus on how much of a particular species they can harvest from, or conserve in, a managed ecosystem so as to maximise species quantity versus the monetary costs of inputs. It would be unreasonable to expect all farmers, foresters, and fishers to consider global inputs and outputs while going about their production activities. And yet there is good historical evidence that for thousands of years, until the introduction of government and corporate industrial agronomy, subsistence farmers acquired sufficient knowledge from their forebears of how to tend the land, and their animals, as to be able to farm it in ways that did not degrade it for their heirs. But technocrats now propose that the only way to prevent the ecological devastation industrial agronomy caused as it replaced indigenous farming practices is *more* central planning, more top-down control, more industrialised food production.

Globalists at the UN's Food and Agriculture Organisation (FAO) and the World Economic Forum propose that to fix climate change – which they assume without sufficient evidence is principally caused by greenhouse gas emissions from humans and their animals – most farmers and their animals should be removed from the land and artificial fertilisers banned.[58] Those farmers that remain are supposed to be digitally tracked with biosensors and other devices and so rendered visible, countable, and regulatable in the public-private partnerships which will dominate food and farming going forward under the Biodiversity and Climate programmes, and Agenda 30, of the United Nations.

The first nation to adopt FAO and WEF proposals to 'control' farmers and reduce their alleged impacts on the climate was Sri Lanka. In 2021 the President of Sri Lanka, following advice from the WEF, banned the import of artificial fertiliser, and by July 2022, Sri Lanka was facing drastic food shortages, and the government was overthrown.[59] In the same month, farmers in the Netherlands, who are the second largest global exporters of food, began large-scale protests on the highways and in the cities of the Netherlands at attempts to shut down their operations by the Dutch government under the aegis of 'climate targets.'[60] Similar plans are being rolled out in other European countries. The government of Ireland has announced that over a million farmed animals will need to be culled for Ireland to meet its climate targets by 2030.[61]

Against this top-down planned 'Climate Leviathan,' Savory offers an approach to reviving the world's deserts, savannah, and forests through a Gaian whole-systems approach applied to the physiology of each managed habitat in which management is more like a partnership between the manager and the other constituent beings of the habitat, including the humans *and* the other living creatures within it. Instead of environmental exclusion, as in the 'Protected Areas' model, this Gaian approach to restoring Earth habitats does not assume an original alienation between humans, other animals, and the health of the land, rivers, and oceans. Rather, as in Bergsonian cosmology, it is built on the perception of the web of consciousness which runs all the way down from humans to the invisible biota that inhabit the deepest volcanic ocean trenches and the near-frozen waters of Antarctica.

Savory has found a way to work with and revive the biggest anaerobic systems on land, which is to say soils and animal guts, which in their pre-industrial state harbour more microbial life forms than the densest tropical forests. As James Hutton first worked out, two centuries before Lovelock, soil deposition from the slopes of living and extinguished volcanoes into rivers and ultimately the ocean floor is the ancient and evolved means by which the 'System of the Earth,' or Gaia, regulates atmospheric carbon and keeps it at a level suitable for oxygen-breathing animals, including humans and other primates. The drawing down of carbon into soils and its eventual burial in sediments on the ocean floor is critical to the Earth's climate. And so, industrial farming, forestry, and fishing, and urban-industrial settlement design, which neglect the roles of soil and maritime microbes in Gaia's achievement of homeostasis, are vastly more important drivers of the ecological crisis, and of weather extremes, than are carbon emissions of post-consumer activities such as driving cars, heating homes, and eating meat.

The misdirection of climate policy by national governments, and global organisations such as the UN's FAO, the IPCC, and WEF, towards attempts to manage or 'nudge' farming practices and citizen consumption activities as means to 'control' planetary temperatures is, in this light, a classic example of what Whitehead called 'misplaced concreteness.' There is no planetary thermostat and there is no closely coupled connection between household and citizen consumption activities and land and ocean temperatures. Getting people to imagine that by driving an electric car powered by batteries of lithium and nickel from mines that are contributing to desertification in Latin America, Africa, Australia, and China, and built with metals smelted with coal and plastics made in oil refineries, they are limiting the spread of deserts on land and in the oceans is a powerful instance of anti-scientific propaganda by industrial and consumer corporations and their partner governmental and non-governmental agencies. Despite the 'net-zero' badges displayed on many electric cars, in reality the huge mineral and energy inputs in their construction, together with their heavier weight, means that pollution from electric vehicles is equivalent to that of modern low-emission gasoline-powered vehicles.[62]

In contrast to misguided globalist attempts to end animal farming, and to control post-consumer activities, to stop the spread of deserts and related climate changes, Savory's Holistic Land Management Model and Lovelock's Gaian approach to planetary physiology both attest to the coupled relationships between the numbers and varieties of organisms on land and in the oceans and the capacity of the Earth System or Gaia to regulate local, and thence global, temperatures to conditions favourable to all life including humans.

Reviving Ecosystems by Empowering Other Creatures

An intriguing case of symbiosis between humans and other creatures is the Knepp Estate belonging to Sir Charles Burrell in Sussex, England. On a 2000-acre farm 20 miles south of Gatwick Airport, Burrell decided to abandon industrial farming on his ancestral estate of Knepp in 2002. He removed the fences over much of the farm, sold his farm machinery, and introduced ancient breeds of animals onto the land, and for the most part left them to their own devices. The introduced species included Tamworth pigs, British longhorn cattle, and red deer, all of which were chosen as animals closer to their wild ancestors – such as aurochs – than to domesticated animals. In the 15 years that the animals established themselves on the land, they returned to living in a relatively 'wild' state, and adopted patterns of foraging, interaction, and reproduction which have created significant niches for the abundance of species that are now found on the farm. Without industrial tillage the soils have become much richer, and species such as hawthorn and sedge have re-established themselves, in turn providing protected cover for hardwoods such as English oak to re-establish naturally. The result after just 15 years is that many bird and small mammal species are now present on the farm that are rarely found elsewhere on British farmland. They are attracted by an abundance of insects, and a variety of land types including watering holes, as well as thorn bush and shrub, that are usually levelled out and ploughed up on conventional farms (Figure 8.3)[63]

The biodiversity that now exists at Knepp exceeds in its different kinds of present and visiting species *any other area* of mainland Britain, including scientifically

FIGURE 8.3 The Knepp Estate, Sussex. Author photo.

managed 'protected' conservation areas such as Sites of Special Scientific Interest. The project at Knepp was partly inspired by a Dutch rewilding project at the Oostvaardersplassen to the east of Amsterdam, that was designed by the Dutch ecologist Frans Vera, and has had analogous ecological results.[64] Rewilding is often represented as a project to exclude humans and domesticated animals from nature. The popular climate activist Greta Thunberg argues in a promotional film for the American non-profit Mercy for Animals that to stop climate change and even prevent future pandemics humans need to stop keeping and eating farm animals, and their milk and eggs, and that veganism, and even 3-D printed synthetic meat, are 'climate solutions' analogous to driving electric cars.[65] But as the founder of Survival International, Stephen Corry, points out, this argument has been used by environmental NGOs such as the World Wildlife Fund and the International Union for the Conservation of Nature, to evict millions of indigenous peoples and their animals from the lands their ancestors inhabited for tens of thousands of years.[66] As Savory's work shows, such evictions are usually followed by the ecological degradation of the lands indigenous people, and their animals, had successfully managed as highly biodiverse and complex ecosystems over tens of thousands of years.

The Knepp Estate demonstrates that rewilding can be undertaken in ways that do not buy into the narrative that humans and domesticated animals are best kept off the land if it is to be ecologically 'restored.' It also shows the powers domesticated mammals are able to exercise when farmers and land managers give them agency in ecosystem restoration and allow them to re-establish patterns of behaviour their ancestor species would once have exercised in the presence of animal predators.

A second case study that is analogous to Savory's HLM, and consistent with Lovelock's account of planetary medicine, is a marine area in West Papua, Indonesia, known as Pulau Misool. Indonesia is the world's top region for sharks, and it is therefore the prime destination for Chinese shark-fishing fleets who wastefully catch sharks for their fins and throw the rest of the animal back into the sea. The term for shark fin in Chinese – *yu chi* – indicates that the fish soup is high in the *chi* or 'life force' which, in Taoist tradition, energises all living things including mammals, birds, fish, plants, and flowing rivers. *Chi* is said to flow through the human body and to be focused at certain nodal points – or chakras – in the body, such as the forehead, the throat, and the abdomen. Increasing the 'flow' of *chi* is seen as therapeutic to humans, and Chinese health practitioners often identify 'blocked *chi*' as a cause of illness that can be unblocked through acupuncture, massage, herbal medicine, heat, and other interventions. It is not hard to see why the flow of *chi* would be thought to be especially enhanced by eating the fin of the shark. The shark is the ocean's fastest creature: sharks are very agile, highly intelligent, and powerful hunters. Chinese traditionally believe that eating certain parts from such creatures confers the same qualities on those who consume them. It is a similar spiritual logic which drives the continued hunting of rhinoceros for their horns, and lions and tigers for their bones and meat.

Chinese trade in wild animals is the biggest threat to biodiversity in Southeast Asia and it is of concern to the government of Indonesia, in part because tourism is rapidly replacing natural resource extraction as the more valuable industry across the archipelago. Significant numbers of high-paying tourists are attracted to dive sites and islands where there is abundant marine life, including sharks. Consequently, the Indonesian government, in concert with Provincial governors, have declared a number of sites off-limits to Chinese shark fishers, and designated them Marine Protected Areas. At Pulau Misool, a shark-fishing camp was evicted in 2007 and less than a decade later, scientific surveys reveal that species of all kinds have revived in huge abundance in the lagoon which is now a shark nursery filled with thousands of sharks and smaller fish, and the coral reef also has the prospect of revival.[67]

The revival of creaturely life in the Pulau Misool region is a function of two processes. First, industrial extraction practices were ended. Second, as large mammals at the top of the ocean food chain moved back into the area, so the other species that were once abundant there returned and revived in numbers also. The abundance of fish and the revival of corals in the area is back to pre-industrial levels and those who snorkel and dive in the area testify on social media to a sense that they are travelling back in time to an era before the industrial-era ecological crisis began to degrade marine environments.

As with the Knepp Estate, Pulau Misool demonstrates the capacity of large mammals, when left to their own devices, to bring back ecosystems and species numbers from a seriously degraded state to pre-industrial levels. Both case studies contradict the claim that 'it is too late' to reverse the damage industrial fishers, farmers, and foresters are doing to ecosystems. They also demonstrate that conservation scientists are not the best agents of such restoration. Instead, ecosystems can be brought back from seemingly terminal degradation when other animals are given back the agency they once exercised in co-managing such ecosystems by excluding *industrial*, but not indigenous, humans.

A third case study that powerfully illustrates the potential roles of other animals in reviving ecosystems and reversing ecological damage, is from Siberia. Permafrost is melting in Siberia across a vast area and global climate change is blamed for this. Permafrost is also a significant store of carbon and if it melts it will produce a pulse of up to 1,600 gigatonnes of CO2, which is equivalent to the existing quantity of carbon dioxide in the atmosphere.[68] Sergei Zimov, together with his son Nikita Zimov, discovered that the demise of mammoths, and contemporary reductions in the numbers of other large herbivores, are the origin of the accelerating permafrost melt in Siberia since herbivores preserve the permafrost in winter by trampling the snow on the surface in obtaining food, and by pulling up tree saplings.[69] Herbivore actions in breaking up surface snow and melted soils bring cooler air from the frozen ground to the surface, so cooling the surface; and by foraging tree saplings and preventing tree growth they revive mosses which are the evolved cover for permafrost and which cool the land in winter because mosses, when snow covered, reflect more sunlight than trees. When significant numbers of large mammals all engage in this foraging behaviour their presence keeps the permafrost from melting, and in the 36,000-acre area the Zimovs established as the 'Pleistocene Park' their reintroduction has already demonstrated that they can stop permafrost melting.[70] Thus far, the Zimovs have purchased and shipped in over 1,000 large herbivores from the subarctic, including bison, muskoxen, moose, yaks, horses, and elk which they purchased and shipped in from Denmark to Alaska. The animals have spread beyond the original park area and the land is transforming back from forest and melting soils to permafrost tundra. A second park has been opened near Moscow to try to awaken more Russians, and the Russian government, to the potential of this approach in preventing a cataclysmic meltdown of Russia's vast permafrost.

Scientists who have studied the Pleistocene Park project confirm that increasing the density of large mammals on the Russian steppe is a means to prevent permafrost melt and to keep the land from thawing.[71] The Zimovs would like to reintroduce woolly mammoths to Siberia if scientists in the future are able to revive this extinct species by means of genetic engineering as they believe this largest of all subarctic herbivores is the keystone species which will spread their project beyond its current area throughout the frozen Russian Steppe.[72]

The Siberian case illustrates the role of indigenous creatures in the ecological restoration approach I am commending here. It also suggests that modern conservation projects miss the potential role of *indigenous people*, as well as other indigenous animals, in reviving highly biodiverse ecosystems modelled on past eras. One such example in which indigenous humans play a central role is the Penan Peace Park in the Malaysian State of Sarawak, Borneo as discussed in Chapter 2. In Sarawak, the Penans were historically, and remain, the most successful conservers of the unique biodiversity of Borneo's rainforests, 90 per cent of which are now degraded by industrial activities sponsored by the governments of Indonesia and Malaysia, in partnership with logging, paper and palm oil companies, and globalist agencies such as the World Bank which funded a human transmigration programme that was hugely damaging to Indonesian Kalimantan. Established informally as means to resist the Sarawak government's destruction of Borneo, the Penan Peace Park covers an area of land in the Upper Baram District of Sarawak and has evolved two forms of income generation for Penan communities while preserving the intact rainforest. The first of these is the sustainable harvesting of non-timber forest products including valuable fruits such as durian, plants with medicinal uses, and sustainable blow-pipe hunting of wild animals for meat and skins. The second use is community-based tourism in which tourists are brought up-river in Penan-owned boats; stay in what are now settled Penan villages; and are guided in visits to the forest, waterfalls, and other natural features, and in Penan cultural activities. The resultant flows of cash income to the Penan have enabled them both to conserve the remnants of tropical rainforest from industrial loggers and palm oil companies and to obtain livelihoods which enable more Penan children to stay in Penan villages rather than migrate to Malaysian cities for work.[73]

The Penan Peace Park case indicates that indigenous people are not only capable of setting up protected areas in which animals and plants are less at risk of extinction while deriving an income from these areas, but that they may be more successful at doing this than government and non-governmental development agencies. This success is evidence for the argument made at several points in this book that contemporary models of 'sustainable development,' driven forward by globalist agencies, and the scientific concept of 'protected areas' as places from which indigenous people and domestic animals are excluded, are both wrong. They are wrong in the eugenicist assumption that humans in general are the source of the ecological crisis, rather than industrial humans. And they are wrong in the allegedly needed remedy, which is increasing levels of technocratic control, facilitated by industrial technologies such as biosensors, computer networks, drones, and artificial intelligence, of human activities down to the micro-level of their diets, their food growing and animal husbandry practices, their means of travel and of monetary exchange.

Religion and Ecology: Towards a New Synthesis

From the Vedas to the Books of Gilgamesh and Genesis, many religious texts witness to the fact that humans have the capacity, unlike all other species, to 'go wrong,' and to dwell in and use Earth's habitats in ways which cause ecological breakdown. As Hans Jonas argues, this is the outstanding feature of the human condition that has come to the fore as a result of the greatly increased material powers modern science and industrial technologies have conferred upon humans.[74] And this observation leads to the question of how such technologically empowered but increasingly post-religious modern humans may be able to recover the kinds of respect for other beings, and for particular 'natural' places and habitats, that are such core features of the cosmologies, ontological conceptions, and sacred practices, of indigenous peoples.

In his magnum opus *Ritual and Religion in the Making of Humanity*, Rappaport argues that long before the evolution of techniques and technologies, it was languages that enabled humans to adapt to, and dwell and hunt in, all the habitats the Earth offered to them and so successfully to predate other species. Languages gave humans powers of collective action and planning far in excess of other species, though scientists are still learning about the extent of such powers in fellow mammals such as dolphins. But languages also conferred on humans powers of individuation and rationalisation which significantly reduced the effectiveness of group- or species-think in restraining behaviours which have such undesirable collective outcomes as habitat destruction and pollution. Religion, Rappaport argues, was the evolved response to this species trait, providing humans with the means to create community customs, moral codes, myths, and rituals which restrain individualism and set apart certain beings and places and procedures as 'sacred,' while rendering problematic behaviours and practices as taboo.[75]

Over much of the modern Earth, as Charles Taylor charts in *A Secular Age*, the sacred cosmologies and practices which protected religiously evolved humans from destructive behaviours and extreme individualism have frayed to the extent of collapse as effective sources of ethical and social restraint.[76] As Taylor earlier argued in his *Sources of the Self*, the rise of the modern conception of the self as an autonomous rational agent was a central feature of this fraying.[77] But Jonas, and his teacher Martin Heidegger, are surely right that equally influential are the cultural influences of mechanistic science and technology which foster the idea that the Earth and even other creatures are insensible machines, and that humans are the only rational and conscious beings in a cosmos otherwise empty of meaningfully conscious beings – whether gods, other animals, or Gaia.

Heidegger's answer to the 'question concerning technology' was an attempt to construct a new theory of being which represented an ecologically attuned version of modern existentialism. But his philosophy was essentially post-religious, and it is not unreasonable to note that his philosophy did not enable him to develop critical resistance to the extreme form of science-informed technocracy which characterised the Nazi era in Germany during which he continued to serve in senior capacities in a German university. Dietrich Bonhoeffer was a German philosopher and theologian of equivalent stature to Heidegger who did succeed in developing a philosophy and an ethics which, while validating modern philosophical claims about the distinctive nature of human ethical responsibility, nonetheless provided Bonhoeffer with philosophical and spiritual insights which underwrote his costly resistance to Nazi technocratic totalitarianism. Central to these resources were the belief that all life is divinely hallowed by having been created by God and that natural life has 'an element of independence and self-development' which is an expression of its innate freedom to evolve independently of direct divine sovereignty.[78] In this way, Bonhoeffer set himself apart from romantic nature religion in which human freedom is paramount, and in which nature plays a subordinate role in facilitating that freedom through human experiences of the natural sublime. Top-down attempts to control nature will therefore fail, since they represent a failure to respect the divine freedom of creatures.

Humans are also creatures, and they share in this freedom of all creatures. Their distinctiveness is situated for Bonhoeffer in the belief that humans are, more than other creatures, reflections of the 'divine image' and hence they are freer than other creatures to go against their divinely given nature, the essence of which for Bonhoeffer is that humans are rational individuals who nonetheless realise the truth of their existence through sociality, through living for others and not only for themselves. Bonhoeffer's ontology is therefore intrinsically relational: creatures live to cooperate with other creatures in realising the divinely given freedom of their nature. Humans also realise their freedom by living in sociality with other humans, and it is in their social relations with one another that they are able to discipline and exercise aright the higher powers with which their linguistic, rational, and latterly technological, capabilities have furnished them.[79]

For Bonhoeffer, the Incarnation of Jesus Christ was central to the restoration of the rightly directed moral capabilities of human beings in their relations with other persons *and* with all creatures. While as a Christian I want to affirm Bonhoeffer's claim, nonetheless I recognise that the influence of industrial culture, scientism, and technologies on most modern humans – and especially most Christians – has displaced the role of religion in orienting most Western peoples' ethics and practices, so the revival of religion per se is unlikely to achieve greater ecological consciousness.

Rappaport proposed a novel solution to this problem, which is to take up the classical Greek idea of *logos* as the ontological order represented by the conscious purposiveness of all life, and which, from the Gospel of John onwards, Christians identify with Jesus Christ, and situate it anew in 'post-modern' forms of religion and spirituality.[80] In particular, Rappaport proposed that contemporary cultures develop new/old beliefs and rituals in relation to new/old 'ultimate sacred postulates' which foster renewed care for Gaia and all creatures.[81] Rappaport's proposal is reminiscent of Bonhoeffer's statements in his prison letters about a revival of a post-institutional form of religion which resonated with many Western Christians in the post–Second World War era. It is also reminiscent of Paul Tillich's post-war theorisation of religion as 'ultimate concern.'

Rappaport's core thesis is that religious rituals were the ways in which prehistoric humans evolved mechanisms of social and ethical regulation to restrain excessive individualism and anti-social behaviours. Key to this process of regulation was the 'sanctification' of certain beliefs, behaviours, beings, and places. It is well attested in the literature on religion and ecology that among the places most often conserved for posterity, and from pollution, by ritual regulation are water sources and woodlands, as for example the cedar groves protected by the monks of the monastery of Saint Catherine of Sinai, and the veneration of water sources, and places where rivers meet, in cultures as far distant, and different, as Balinese Hinduism and Scottish Christianity. The oldest cathedral in Scotland is at Dunkeld, in a grove of trees adjacent to a right-angled meeting of two rivers: Dunkeld was the place of anointing for Scots kings in the Middle Ages, and the place to which Catholic monks brought the bones of Saint Columba from Iona to protect them from Viking invasions. Water is essential to life and humans themselves are composed of 60 per cent water. Woodlands are critical to the health of water catchments and hence drinking water. Religious sanctification of springs, wells, and places where rivers meet, together with woodlands, provides a social mechanism for regulating activities which might pollute or harm human water supplies. As Rappaport argues, 'sanctity is a functional replacement for genetic determination of patterns of behaviour, a determination which becomes increasingly specific as language emerges.'[82]

That individuals in pursuit of their own particular interests may harm the water sources of others, or use too much water, so limiting access to others, was the central social issue in the Bali water temple case study discussed above. No social system can successfully defend itself from extreme individualism – which is a product of the unique linguistic capacities of humans – without some means of collective regulation. As we have seen, globalists believe that the most effective means to such regulation is a top-down totalitarian and technologically enabled panopticon in which all human actions are surveilled and regulated by government statute and decree. But *all* totalitarian schemes to control humans, and ecosystems, from the top down have failed because, by destroying the evolved freedom of beings, they destroy the extremely complex evolved and symbiotic processes through which Life has continuously sustained the conditions for Life on Gaia. The environmental wreckage and toxification of much of the land of Russia after 70 years of communist collectivism – from 1917 to 1989 – is among the most powerful recent examples of this.

Rappaport's proposal of a post-religious approach to sanctity which has an analogous regulatory role to religious traditions is closely related to his appropriation of the Heraclitan idea of *logos* as the given order of the world which is generated by 'the rational relation of things to one another.'[83] English moral theologian Geoffrey Kirk summarised Heraclitus on the *logos* as follows: 'Only one activity can be described as genuinely wise, that is the understanding of the way in which everything in the world is part of an ordered whole: everything is guided along a determinate path so as to produce a complex and unified result.'[84] As I argue above, the concept of *logos* as the *way* of the world, the Tao, or wisdom which is imbued in the given structure of the world, finds correlates in all pre-modern religious traditions. In the Vedas as we have seen the Sanskrit word *Rta* corresponds closely to *log* in Islamic Iran and the Persian word *Asha* has the same meaning;[85] in ancient China the word *Tao* has the same meaning; in Christianity the word *logos* is itself taken up from the Heraclitan tradition as the unitive principle of all Being, and as Incarnate in a unique way in the person of Jesus Christ. But by what means can post-religious humans mobilise the ancient conception of logos or cosmic order in such a way as to fit their ways of life again, including the production of the things they need for nutrition, for dwelling, and for cultural enrichment, to the Tao or the *logos* so that the Earth flourishes alongside them?

The current globalist and technocratic proposal of an environmental 'ultimate sacred postulate,' in Rappaport's terminology, is the top-down regulation of activities associated with carbon dioxide emissions – which is all activities since carbon is the dominant material element of all life – and the dubious carbon accounts involved in the concept of 'net zero.' But, as we have seen, the premises for this approach to the ecological crisis are not well founded in the real causes of desertification, habitat destruction, local climate changes, and the declines in global species numbers which have rendered ecosystems less able to respond resiliently to environmental change. Carbon emissions is a globalist fetish – or 'factish' in Bruno Latour's insightful wording – and they are not the chief causative factor in global and local environmental degradation. Against the globalist fetish of carbon emissions, the most critical ultimate sacred postulates that moderns should adopt and for which they might construct rituals of care in every human community and habitat are water catchments, soil conservation, and species diversity.

Water-catchment governance is the oldest attested form of environmental regulation in human history, and arguably provides the best foundation for human political organisation as well as environmental regulation since water catchments represent a geographical scale of political organisation, as in the Bali case study, in which people may be known to one another, and they are therefore more resistant to capture by big governments and large corporations in partnerships which operate at the expense of citizens and their environments. The industrial capture of water sources for dams, and for commercial uses, was the originating spark of modern environmental protests in many domains, from England in the late eighteenth century to New Zealand in the 1960s, and India in the 1980s. What humans take from a water catchment, and what humans leave there as waste, are the core issues concerning the regulation of water as the essential prerequisite for life. Water catchment education is a practice that has been taken up in some of the States of the United States. Water educators go around schools, colleges, businesses, and households and teach people about what they should and should not put down the drains in their homes, and in the ground in their yards or into drains on the street. In Bali, where I am writing, the biggest threat to water catchments are plastic waste and artificial fertilisers. Both could potentially be addressed through the water-temple governance system but it would need adaptation and updating to include these issues in the traditions of water-catchment governance.

Education about water, including the neglected Gaian regulation of the climate via algae and phytoplankton in the ocean, and via cloud formation, should be foundational to all environmental and scientific curricula since without water there would be no human environment, no human sciences, and no biological life.[86] In addition to education, and regulation, a revived focus on water-catchment governance as the core geospatial and political unit for human collective arrangements would be advisable, though it runs directly counter to the global governance regime commended by the UN, the WEF, the Rockefellers, and other globalist agents. A revival of governance in each water catchment could only be achieved by returning to the communities that live on them the *ownership* and collective management of water tables from the globalist corporations who, in the main, now claim to own them under public-private partnerships with governments. In tandem with the relocalisation of water management, there should be revivals of traditional water-blessing ceremonies by the leaders of local religious traditions. In this way, the traditional liturgical function of religious ritual, as Rappaport theorises it, of upholding the world by shaping the ethos of the community towards honouring the givenness of *logos*, would be revived.[87] It is this combination of local water catchment governance, farming practices, and religious culture which is the source of the island of Bali's unique fusion of water temples, local community governance, and food production including as it does not only rice but an extensive variety of fruits, vegetables, and animal products.

The second 'ultimate sacred postulate' that should be foundational to local community governance, to educational curricula, and to the collective regulation of land use, is soil conservation. Of the three I propose here, soil conservation has been the most neglected by modern industrial societies, whereas there are extensive regulations in traditional religious texts, such as the Indic Vedas and the Hebrew Tanakh, concerning care of the soil. As we have seen, most of the Gaian chemistry that regulates the atmosphere of the planet at a balance of gases ideal for Life takes place below the surface of the soil and the oceans. As we have also seen, historically animals, not humans, played the critical Gaian role of maintaining soils, including in areas from which they are largely absent today and which are at risk of desertification. The interconnected practices that have led to soil erosion over much of the cultivatable land of the earth in the industrial era are industrial agricultural tillage and agrochemicals; forest conversion to monocultures; and farmer-initiated declines in numbers of wild animals. The conventional approach to this problem is to let farmers and plantation managers continue to wreck soils while excluding small subsistence farmers and indigenous peoples who ironically are far less prone to wrecking soils since their lives depend on its continuing fertility, so that these can be turned into 'protected areas' in which wild animals may reside unmolested by farmers. But the resultant environmental exclusion of millions of people, and their animals, leads to further environmental problems as people are either forced into polluted urban slums and informal settlements, often along riverbanks, or take to the roads and seas as refugees. At the same time, the protected-areas approach completely neglects the central role of industrial agricultural practices in wrecking habitats and hence driving down numbers of wild species, including large mammals, birds, insects, and the microbiota that once symbiotically maintained the huge atmospheric regulation engine of soils beneath the surface.

The neglect of soil conservation and soil microbiota by globalist agencies which promote 'sustainable development,' while they at the same time push on with their agenda to take over and industrialise food production, is a powerful indicator that globalist agencies are mainly dissembling in their claims to be concerned with the health of the planet and its peoples. But there are nonetheless significant numbers of people in the Global North and South who are now aware that they cannot trust globalists and industrial farmers and food corporations to grow and produce foods which are healthy for humans and in ways that do not wreck the Earth. These movements coalesce around global alliances such as La Via Campesina which is the largest social movement on Earth presently and campaigns for the rights of small farmers and indigenous peoples to grow their own food on their own lands. The other prominent movement is that of organic farming which grew out of the pioneering work of Rudolf Steiner in Europe in the nineteenth century and whose anthroposophical and astrological ideas promoted connections between food-growing, the health of the land, rivers, soils, the moon, the sun, and the stars. Steiner attempted to revive spiritual traditions, albeit in neopagan mode, that the churches in Europe in the course of industrialisation had dropped, which is to say, the deep interconnections between Christian liturgy and the agricultural, lunar, and solar years.[88]

Along with a new science-informed focus of politically generated environmental law on soil conservation as the key litmus test of land-use activities – both agricultural and urban – instead of the problematic focus on 'carbon emissions,' I suggest that churches and temples also have a responsibility to revive the now mostly forgotten ceremonies of blessing of the soils, food, and farming.

The third and final 'ultimate sacred postulate' I would propose for a post-religious retrieval of the traditional regulatory function of the sacred in ecological terms is species diversity. This third ultimate sacred postulate is perhaps the most far-reaching of all three since, if used to regulate and revise industrial practices, it would impact every kind of human activity including growing food and building houses, recreational activities such as sports, tourism and visits to 'wilderness,' and the arts and sciences. As I argued above, the disappearance of huge numbers of species on land and in the oceans is the real reason for micro-level and regional climatological changes across the Earth which are impacting on human as well as nonhuman habitats, and potentially contributing to global climate change. The main causes of species declines and extinctions are not 'people in general' but industrial agriculture and industrial fishing. Agriculture in principle could be regulated at the level of each water catchment, and as we have seen, this is not just a romantic idea. Bali still regulates agriculture at that level and the attempt of global agricultural engineers, agrochemical, seed, and fertiliser companies to change that ultimately failed. I am under no illusion that it will be easy to end the huge public–private partnerships which currently see public monies in almost every nation poured into destructive farming practices, in partnership with monopolistic seed and agribusiness corporations and globalist agencies such as the World Bank and the Food and Agriculture Organisation. But I have presented sufficient data in this book that this needs to happen urgently to prevent ecological breakdown since this partnership more than any other is responsible for the vast declines in species numbers, and hence reduced resilience of ecosystems, on land at an unprecedented scale in human history.

Industrial fishing is equally problematic, and even more difficult to fix since the oceans beyond territorial coastal zones are in legal terms 'free seas' and hence the enforcement of environmental regulations against sociopathic actors such as Chinese, Dutch, Russian, and Spanish trawlers, Italian tuna fishers, and Japanese whalers, is beyond the resources of most nations. Reducing species decimation in the marine environment is then arguably only likely to be achievable by globalist agencies although, as yet, none of these agencies have a plan for such a goal. But please note that I named the countries from which the trawlers emanate in identifying them. This is because ships must be registered on land. They cannot ply the oceans without land-based, and hence politically generated, papers and flags. Here then, there is a crucial role for national governments. But few, if any, national governments have been prepared to impose on their fishing industry regulations that will stop the destructive deep-sea trawling and line-fishing activities that are emptying the oceans of life with the aid of modern technologies.[89] I do not have a ready answer to this problem except popular pressure on governments to enforce environmental regulations on all the ships for which they provide flags, or permit safe harbours or passage in their national waters. That this remains extremely difficult to achieve is indicated by the extent to which the global fishing industry is not only ecologically destructive but also replete with large-scale employment and safety abuses including slavery. Beyond regulating fishing boats, the other urgently needed approach is a huge extension of marine protected areas to allow the seas to revive. But until today, current marine protected areas are poorly policed, and most still permit industrial fishing to take place within them albeit with stricter rules than elsewhere. A strategy for vast extensions of marine protected areas is proposed by Charles Clover and outlined in detail in his *Rewilding the Sea*, and his proposals to me as a non-maritime expert seem entirely viable provided nations commit to them.[90]

As with the first two ultimate sacred postulates – water catchments and soil conservation – religious traditions have important cultural roles to play in reviving citizen concern with the quality and quantity of life on land and in the oceans. As discussed above, the oldest aesthetic remnants of prehistoric *Homos*, cave paintings, reveal a profound awe and respect – almost to the point of divinisation – of the large mammals with which humans shared the Earth in ancient times and which lived in numbers moderns can hardly imagine. Deep respect for large mammals is also present in religious traditions, not least the Indian Vedas in which the cow, elephants, and primates such as monkeys are considered sacred. But this profound traditional religious respect for large mammals declined in the Christian tradition, particularly in the second millennium of the Christian era. As we have seen, this decline occurred in tandem with the rise in the Christian West of modern science and industrial technologies, the uses of which have increasingly tended to treat all animals as beast-machines, incarcerating huge numbers of animals in what are in effect animal prisons where they live out their miserable lives unable to fulfil most of the functions that they evolved to fulfil, including not least their Gaian roles as maintainers of soils and hence of the terrestrial cycle of life. Again, the globalist response to this problem, as in the problematic campaign for a meatless future advanced by Monbiot, Thunberg, the World Economic Forum, and others, is completely misguided. This is because, as the case studies above demonstrate, only a huge increase in the numbers of carefully managed domesticated animals, and reintroduced wild animals, on lands that are at risk of desertification, can prevent the continuing spread of deserts which has nothing to do with carbon dioxide emissions. Indeed, NASA data shows that in the last 20 years, the latter are helping to green lands at risk of desertification by pushing more moisture into the atmosphere, so causing more rainfall, as well as by increasing atmospheric availability of carbon dioxide for the process of photosynthesis.[91]

Religious traditions all have narratives of engagements with other animals as beings capable of befriending mystics and others similar to those of the desert tradition of Christianity. In some traditions, most notably the Tanakh, animals also act as messengers of God. The ecological crisis is one humans cannot resolve on their own, or even with 'smart' machines and artificial intelligence. If humans were to attempt alone to take control of planetary systems, to the point of geoengineering the climate as well as terraforming the land and modifying the genes of plants and animals, this would represent an intensification of the machine cosmology which is the source of the crisis between humans and nonhumans in industrial civilisation. Religious traditions urgently need to revive the old stories of peace between the saints and the beasts, and to make more prominent in their liturgies and teachings the symbiotic relations between humans and other animals which for most of human history have been central to their mutual flourishing. And so, regarding my third ultimate sacred postulate, religious traditions already have the resources to embrace an 'ontotheological turn' towards the revival of the co-agency of all species, and all Beings, including Mother Earth or Gaia, in recomposing Earthly habitats favourable to all kinds, and not only humans. The revival of numbers of animals, and microbiota, on the land and in the oceans should therefore be seen as the ecological goal above all others of the present era as they, and not Promethean technicians and climate modelers, are the ones Gaia urgently needs to replenish and revive her innate living powers of planetary healing (Figure 8.4).

FIGURE 8.4 Church apse, Rome. Author photo.

Notes

1 Geoffrey Parker, *Global Crisis: War, Climate Change and Catastrophe in the Seventeenth Century*. New Haven, CT, Yale University Press, 2013.
2 Gilbert Burnham, Riyadh Lafta, et al., 'Mortality after the 2003 Invasion of Iraq: A Cross-sectional Cluster Sample Survey.' *The Lancet* 6736 (2006) 69491–9.
3 'List of Wars: 1990–2002'; 'List of Wars 2003–2022' both published by Wikipedia online at https://en.wikipedia.org accessed June 21, 2022; on the faux 'war on terror' see Michael S. Northcott, *An Angel Directs the Storm: Apocalyptic Religion and American Empire*. London, I B Tauris, 2004.
4 See further Michael S. Northcott, 'The World Trade Organisation, Fair Trade and the Body Politics of Saint Paul.' In John Atherton ed. *Through the Eye of a Needle: Theology, Ethics and Economy*. London, Epworth Press, 2007, 169–188.
5 Nicolas Stern, et al., *The Economics of Climate Change*. Cambridge, Cambridge University Press, 2006, 56
6 James E. Lovelock, *A Rough Ride to the Future*. London, Penguin, 2014, 62.
7 Shilong Piao, Xuhui Wang, Taekin Park, et al., 'Characteristics, Drivers and Feedbacks of Global Greening.' *Nature Reviews: Earth and Environment* 1 (2020) 14–27.
8 Lovelock, *Rough Ride*, 74–5.
9 Helen Stockton and Ron Campbell, *Time to Reconsider UK Energy and Fuel Poverty Policies?* York, Joseph Rowntree Foundation, 2011 and published online at https://beatcold.org.uk/wp-content/uploads/2011/10/NEA-JRF-Viewpoint.pdf accessed June 22, 2022.
10 Björn Lomborg, *False Alarm: How Climate Change Panic Costs Us Trillions, Hurts the Poor, and Fails to Fix the Planet*. E-Edition, New York, NY, Basic Books, 2020, 12–27.
11 Lomborg, *False Alarm*, 48–52.
12 Lomborg, *False Alarm*, 55–68.
13 Joel Wainwright and Geoff Mann, *Climate Leviathan: A Political Theory of Our Planetary Future*. London, Verso, 2020.
14 Isaiah McCall, 'Blackrock: The Secret Company That Owns the World.' *Medium* October 26, 2021, and published online at https://medium.com/yardcouch-com/blackrock-the-secret-company-that-owns-the-world-b96111277e0f accessed June 22, 2022; Jan Fichtner, Eelke M. Heemskerk, and Javier Garcia-Bernardo 'Hidden Power of the Big Three? Passive Index Funds, Re-concentration of Corporate Ownership, and New Financial Risk.' *Business and Politics* 19 (2017) 298–326.
15 Daithí Mac Síthigh and Mathias Siems, 'The Chinese Social Credit System: A Model for Other Countries?' *Modern Law Review* 82 (2019) 1034–71.
16 Z. Zhongming, Z., L Linong, Y. Xiaona, et al., 'Can E-commerce Help Save the Planet?' Published online by the United Nations Environment Program (2022) at http://119.78.100.173/C666/handle/2XK7JSWQ/344770 accessed June 22, 2022.
17 Florian Skelton of Goethe-University, Frankfurt in a position paper 'Carbon Leviathan: A Planetary Sovereign Governing the Anthropocene.' Published online at https://withinenvironments2022.weebly.com/uploads/1/4/1/0/141089095/carbon_leviathan.pdf accessed June 22, 2022.
18 George Monbiot, *Regenesis: Feeding the World Without Devouring the Planet*. E-Edition, London, Penguin, 2022, 27–52.
19 George M. Church and Edward Regis, *Regenesis: How Synthetic Biology Will Reinvent Nature and Ourselves*. New York, NY, Basic Books, 2014.
20 Monbiot, *Regenesis*, 84–97.
21 For fuller discussion and references see Northcott, *Place, Ecology and the Sacred*, 12–46.
22 Michael S. Northcott, 'Reformed Protestantism and the Origins of Modern Environmentalism' *Philosophia Reformata*, 83 (2018), 19–33.
23 Willard A. Hanna, *A Brief History of Bali: Piracy, Slavery, Opium and Guns: The Story of an Island Paradise*. Kindle Edition, Singapore, Tuttle Publishing, 2016, 1007–17.

24 For a meticulous account see Geoffrey B. Robinson, *The Killing Season: A History of the Indonesian Massacres, 1965–66*. Princeton NJ, Princeton University Press 2018.
25 See the contrast between pre- and post- Green Revolution life in Stanley A. Freed and Ruth S. Freed, *Green Revolution: Agricultural and Social Change in a North Indian Village: Anthropological Papers of the American Museum of Natural History Monograph 85*. Washington, DC, American Museum of Natural History (2002), 1–312.
26 James C. Scott, *Weapons of the Weak: Everyday Forms of Peasant Resistance*. New Haven, CT, Yale University Press, 1985.
27 Vandana Shiva, *The Violence of the Green Revolution: Third World Agriculture, Ecology and Politics*. London, Zed Books, 1991.
28 Clifford Geertz, *Agricultural Involution: The Processes of Ecological Change in Indonesia*. Berkeley, CA, University of California Press, 1963.
29 Stephen Lansing, *Priests and Programmers: Technologies of Power in the Engineered Landscape of Bali*. Princeton, NJ, Princeton University Press, 1991, 144–5.
30 Lansing, *Priests and Programmers*, 147–8.
31 Lansing, *Priests and Programmers*, 149.
32 Lansing, *Priests and Programmers*, 117.
33 Roy Rappaport, *Ritual and Religion in the Making of Humanity*. Cambridge, Cambridge University Press, 1999, 478.
34 Erica Bornstein, *The Spirit of Development: Protestant NGOs, Morality, and Economics in Zimbabwe*. New York, NY, Routledge, 2003.
35 Lansing, *Priests and Programmers*, xxiv–xxv.
36 Lansing, *Priests and Programmers*, 150–6.
37 Lansing, *Priests and Programmers*, 157.
38 Lansing, *Priests and Programmers*, 116–7.
39 M.B.K. Darkoh, 'Towards Sustainable Development and Environmental Conservation in African Drylands.' *Journal of East African Research & Development* 23 (1993) 1–23.
40 Allan Savory, with Jody Butterfield, *Holistic Management: A New Framework for Decision Making*. E-Edition, Washington, DC, Island Press, 1999, 23–4.
41 Savory with Butterfield, *Holistic Management*, 33–8.
42 S.R. Verón, J.M. Paruelo, and M. Oesterheld, 'Assessing desertification.' *Journal of Arid Environments* 66 (2006) 751–63.
43 Savory, *Holistic Management*, 34.
44 Savory, *Holistic Management*, 40.
45 Savory, *Holistic Management*, 40.
46 Savory, *Holistic Management*, 48.
47 Savory, *Holistic Management*, 52.
48 Savory, *Holistic Management*, 57.
49 Savory, *Holistic Management*, 58.
50 Savory, *Holistic Management*, 57.
51 Savory, *Holistic Management*, 61.
52 Savory, *Holistic Management*, 63.
53 Savory, *Holistic Land Management*, 71.
54 James E. Lovelock, *Gaia: The Practical Science of Planetary Medicine*. London, Oxford University Press, 2000, 15.
55 Lovelock, *Gaia*, 29.
56 Lovelock, *Gaia*, 56.
57 Lovelock, *Gaia*, 68.
58 *Building a Common Vision for Sustainable Food and Agriculture: Principles and* Approaches. Food and Agriculture Organisation, Rome, Italy, 2014: For a critical overview of WEF and UN plans for controlling global food production see Tyler Durden, 'UN, World Economic Forum Behind Global "War on Farmers:: Experts.' *ZeroHedge* July 30 2022 https://www.zerohedge.com/geopolitical/un-world-economic-forum-behind-global-war-farmers-experts.

59 Ranil Wickremesinghe, 'Sri Lanka PM: This is how I will make my country rich by 2025.' World Economic Forum (August 29, 2018) https://www.weforum.org/agenda/2018/08/this-is-how-we-will-make-sri-lanka-rich-by-2025/; 'Explained: Why Sri Lanka's economy collapsed, and what's next for the island nation?' *The Indian Express*, July 10 2022 https://indianexpress.com/article/explained/explained-sri-lanka-economy-collapse-8020532/

60 Nick Ottens, 'The Netherlands farm crisis, explained.' *Atlantic Sentinel*, (June 12 2022) https://atlanticsentinel.com/2022/06/the-netherlands-farm-crisis-explained/

61 Nichola Daunton, 'Farmers drive tractors through Dublin city centre to protest Ireland's climate plans.' *Euronews.green* November 22 2021 https://www.euronews.com/green/2021/11/22/farmers-drive-tractors-through-dublin-city-centre-to-protest-ireland-s-climate-plans.

62 Anna Krajinska, *Magic Green Fuels: Why Synthetic Fuels Will Not Solve Europe's Pollution Problems*. Brussels, European Federation for Transport and Environment, 2021 and online at https://www.transportenvironment.org/wp-content/uploads/2021/11/2021_12_TE_e-fuels_cars_pollution.pdf accessed June 22, 2022.

63 For a detailed narrative, including some of the unanticipated scientific results of the 'wilding' of the Knepp Estate, see further Isabella Tree, *Wilding: How a Farm Returned to Nature*. London, Pan MacMillan, 2018.

64 Frans Vera, 'De sterfte in de Oostvaardersplassen in een internationaal kader,' *Natuur Vakbladbos Landschap* 2005, 20–22 at https://edepot.wur.nl/114495 accessed January 6, 2022.

65 The short documentary is at https://twitter.com/GretaThunberg/status/1396058911325790208?s=20 accessed January 6, 2022.

66 Stephen Corry, 'Scapegoats and Holy Cows: Climate Activism and Livestock.' Published online at https://www.theelephant.info/long-reads/2021/12/10/scape-goats-and-holy-cows-climate-activism-and-livestock/ accessed June 22, 2022.

67 Vanessa F. Jaiteh, Steve J. Lindfield, Sangeeta Mangubhai, et al., 'Higher Abundance of Marine Predators and Changes in Fishers' Behavior Following Spatial Protection within the World's Biggest Shark Fishery.' *Frontiers in Marine Science* 3 (2016), 1–15.

68 Marcello Rossi, 'Russian Scientists Are Trying to Revive Woolly Mammoths to Halt Siberia's Warming Crisis.' *Future Human: Medium* (April 2, 2021) and published online at https://futurehuman.medium.com/russian-scientists-are-trying-to-revive-wooly-mammoths-to-halt-siberias-warming-crisis-e1bd4fd3f6bb accessed June 22, 2022.

69 A. Zimov, N.S. Zimov, A.N. Tikhonov, et al., 'Mammoth Steppe: A High-Productivity Phenomenon.' *Quaternary Science Reviews* 57 (2012) 26–45.

70 'Pleistocene Park: Restore High Productive Grazing Ecosystems in the Arctic and Mitigate Climate Change.' https://pleistocenepark.ru/

71 Christian Beer, Nikita Zimov, Johan Olofsson et al., 'Protection of Permafrost Soils from Thawing by Increasing Herbivore Density.' *Nature: Scientific Reports* 10 (2020) 4170.

72 Rossi, 'Russian Scientists Are Trying to Revive Woolly Mammoths to Halt Siberia's Warming Crisis.'

73 Shaista Falak, Lo Ma Chiun, and Alvin Yeo Wee, 'Sustainable Rural Tourism: An Indigenous Community Perspective on Positioning Rural Tourism.' *Tourism*, 64 (2016) 311–27.

74 Hans Jonas, *The Imperative of Responsibility: In Search of an Ethics for the Technological Age*. Trans. David Herr and Hans Jonas, Chicago, IL, University of Chicago Press, 1985.

75 Roy Rappaport, *Ritual and Religion in the Making of Humanity*. Cambridge, Cambridge University Press, 1999, 4-29, 49-50.

76 Charles Taylor, *A Secular Age*. Cambridge, MA, Belknap Press, 2007.

77 Charles Taylor, *Sources of the Self: The Making of the Modern Identity*. Cambridge, Cambridge University Press, 1989.

78 Dietrich Bonhoeffer, *Ethics*, 121–3 as cited in Peter M. Scott, 'Christ, Nature, Sociality: Dietrich Bonhoeffer for an Ecological Age.' *Scottish Journal of Theology*, 53 (2000) 413–30.
79 For a fuller and well-referenced discussion see Scott, 'Christ, Nature, Sociality' to which I am indebted for an excellent summative account of Bonhoeffer's ecological theology.
80 Rappaport, *Ritual and Religion*, 344-52.
81 Rappaport, *Ritual and Religion*, 287-90.
82 Rappaport, *Ritual and Religion*, 418.
83 H. Kleinknecht, 'The logos in the Greek and Hellenistic world,' In Georg Kittel ed. *Theological Dictionary of the New Testament,* vol. 4. Grand Rapids, MI, Eerdmans, 1967 cited in Rappaport, *Ritual and Religion*, 371.
84 Geoffrey S. Kirk. *Heraclitus: The Cosmic Fragments.* Cambridge, Cambridge University Press, 1954 cited in Rappaport, *Ritual and Religion*, 372.
85 Rappaport, *Ritual and Religion*, 359–60.
86 On the critical role of marine algae in regulating the climate see Robert J. Charlson, James E. Lovelock, Meinrat O. Andreae, et al., 'Oceanic Phytoplankton, Atmospheric Sulphur, Cloud Albedo and Climate.' *Nature* 326 (1987) 655–61.
87 Rappaport, *Ritual and Religion*, 485.
88 For a curated collection of Rudolf Steiner's writings on ecology and spirituality see Rudolf Steiner, *Spiritual Ecology: Reading the Book of Nature and Reconnecting with the World.* Compiled and edited by Matthrew Barton, London, Rudolf Steiner Press, 2008.
89 For a fuller discussion see Charles Clover, *The End of the Line: How Overfishing Is Changing the World and What We Eat.* Berkeley, CA, University of California Press, 2006.
90 Charles Clover, *Rewilding the Sea: How to Save Our Oceans.* London, Ebury, 2022.
91 Zhenzhong Zeng, Zaichun Zhu, Xu Lian et al. 'Responses of land evapotranspiration to Earth's greening in CMIP5 Earth System Models.' *Environmental Research Letters* 11, 104006.

INDEX

Page numbers in *italics* indicate figures and references following "n" refer endnotes.

For Product Safety Concerns and Information please contact our EU representative GPSR@taylorandfrancis.com
Taylor & Francis Verlag GmbH, Kaufingerstraße 24, 80331 München, Germany

www.ingramcontent.com/pod-product-compliance
Ingram Content Group UK Ltd.
Pitfield, Milton Keynes, MK11 3LW, UK
UKHW022022180926
475701UK00005B/983
9780367627744